Inorganic Chemistry Concepts
Volume 7

Hans Rickert

Electrochemistry of Solids

An Introduction

With 95 Figures and 23 Tables

Springer-Verlag
Berlin Heidelberg New York 1982

Dr. Hans Rickert

Professor, Institute for Physical Chemistry I
University of Dortmund
P.O. Box 500 500
D-4600 Dortmund 50
Federal Republic of Germany

ISBN-13: 978-3-642-68314-5   e-ISBN-13: 978-3-642-68312-1
DOI: 10.1007/978-3-642-68312-1

Library of Congress Cataloging in Publication Data.

Rickert, Hans, 1928 —. Electrochemistry of solids.
(Inorganic chemistry concepts; v. 7)
Bibliography: p. Includes index.
1. Solid state chemistry. 2. Electrochemistry.
I. Title. II. Series.
QD478. R55   541'.042   81-23275
ISBN-13: 978-3-642-68314-5   AACR2

© Springer-Verlag Berlin, Heidelberg 1982
Softcover reprint of the hardcover 1st edition 1982

Bookbinding: Konrad Triltsch, Würzburg
2152/3020—543210

Preface

This book is the completely revised and extended version of the German edition "Einführung in die Elektrochemie fester Stoffe" which appeared in 1973. Since then, the subject of the electrochemistry of solids has developed further and a large number of new solid electrolytes have been discovered. With the help of solid electrolytes, i.e. solid ionic conductors, galvanic cells are constantly being built for thermodynamic or kinetic investigations and for technical applications. Though the book takes these new developments into consideration, its main aim is to provide an introduction to the electrochemistry of solids, emphasizing the principles of the subject but not attempting to present a complete account of the existing literature. The latter can be found in handbooks and specialists' reports of conferences in this field; these are referred to in the text. This book is written for scientists and graduate students who require an approach that will familiarize them with this field. It is assumed that the reader will be acquainted with the fundamentals of physical chemistry. The various chapters have been written so that most of them can be read independently of each other. Parts which may be omitted during a first reading are printed in small type.

Of vital importance for the publication of this English edition have been the comments, suggestions and the help of colleagues and co-workers. I would particularly like to express my thanks to Dr. Holzäpfel, Dr. Lohmar, Professor Mitchell, Dr. Schlechtriemen and Dr. Wiemhöfer for the valuable assistance they have given me during the translation and in working through many sections of the manuscript.

Dortmund, March 1982   Hans Rickert

Contents

VIII   Contents

List of Symbols

a_B activity of component B

$A \equiv U - TS$ Helmholtz energy

A neutral particle A on interstitial lattice site (Schottky new), area

$|A|$ neutral A-vacancy (Schottky new)

$A\circ$ neutral particle A on interstitial lattice site (Schottky old)

$A\square$ neutral A-vacany (Schottky old)

A_A particle A on A-site (Kröger-Vink)

A_B particle A on B-site (Kröger-Vink)

A_i particle A on interstitial lattice site (Kröger-Vink)

AB AB lattice molecule (Kröger-Vink/Schottky old, new)

$B \bullet (A)$ neutral particle B on A-site (Schottky old)

$B|A|$ neutral particle B on A-site (Schottky new)

B_m mechanical mobility

$c_B, [B]$ concentration of component B

D (Fick's) diffusion coefficient

D_K component diffusion coefficient

D_{Tr} tracer diffusion coefficient

$D(\varepsilon)$ density of states

e quasi-free electron (Schottky new, Kröger-Vink) elementary charge

$|e|$ electron defect (Schottky new)

E electromotive force (emf)

$\vec{E}$ electrical field

E_F Fermi potential, Fermi level or Fermi energy

E_C lower edge of conduction band

E_V upper edge of valence band

f correlation factor

$f(\varepsilon)$ Fermi-Dirac distribution function

F Faraday's constant

$\vec{F}$ force

$F_{1/2}$ Fermi-Dirac-integral "one half"

$\tilde{g}_i$ concentration independent contribution to the partial molar Gibbs energy of one defect

$\tilde{g}_M$ concentration independent contribution to the partial molar Gibbs energy of one lattice molecule

$G \equiv H - TS$ Gibbs energy

h Planck's constant; electron defect (Kröger-Vink)

$H \equiv U + pV$ enthalpy

i electrical current density

I electrical current

$\underline{I}$ electrical current in complex representation

j_B particle current density of component B

J_B particle current of component B

$k; \vec{k}$ Boltzmann's constant; wave vector

k_r rate constant

l length

L_{ik} phenomenological coefficient

m mass

m^* effective mass

M molecular mass

N_A Avogadro's number

$n = [e]$ concentration of quasi-free electrons

n_B number of moles of component B

N_B number of particles, particle density of component B

N_C effective density of states of the electrons

N_V effective density of states of the electron defects

$\vec{p}$ momentum

$p = [h]$ concentration of electron defects

Q	charge	$\overline{Z}_i$	partial molar quantity Z of component i
Q_i^*	heat of transport of particle type i	γ	activity coefficient
$\vec{r}_i$	i-th jump vector	Γ	surface concentration
R	gas constant; resistance	δ	deviation from ideal stoichiometry
$\vec{R}$	displacement vector		
S	entropy	ε	partial internal energy of the electrons; thermoelectric power
t	time		
t_B	transference number of component B	η_B	electrochemical potential of component B referred to one mole
T	(absolute) temperature		
u	electrical mobility	$\tilde{\eta}_B$	electrochemical potential of component B referred to one particle
U	internal energy		
$\underline{U}$	alternating voltage in complex representation		
v	reaction rate	μ_B	chemical potential of component B referred to one mole
$\vec{v}$	particle velocity		
V	Volta potential; volume	$\tilde{\mu}_B$	chemical potential of component B referred to one particle
V_A	A-vacancy (Kröger-Vink)		
V_i	interstitial lattice vacancy (Kröger-Vink)	$\mu_B^0, \tilde{\mu}_B^0$	chemical potential of component B in the standard state
V_m	molar volume		
w	probability density		
W	thermodynamic probability; work function	ν	stoichiometric coefficient; jump frequency
x_B	mole fraction of component B	ρ	charge density; resistivity
x, y, z	cartesian coordinates	σ	(electrical) conductivity
$\Delta X, X$	thickness of reaction or tarnishing layer	τ	transition time
X_i	generalized force	φ	electrical potential, Galvani potential
$\underline{Y}$	(complex) admittance	$\ominus$	quasi-free electron (Schottky old)
z_B	charge number of ion B (positive for cations, negative for anions)	$\oplus$	electron defect (Schottky old)
$\underline{Z}$	(complex) impedance	$\sim$	waved rule over a symbol means: referred to one particle
Z	number of states in a given energy interval		

1 Introduction

The development of the electrochemistry of solids was made possible because of the existence of solid electrolytes. These are solid ionic conductors, solid — generally crystalline — compounds in which the electric current is carried by charged atoms, i.e. by ions. The passage of current is thus accompanied by mass transport. The ionic conductivity in solid electrolytes is sometimes as high as that in concentrated liquid electrolytes or molten salts. These solid electrolytes are often called superionic conductors. Some of these solid electrolytes with very high ionic conductivity have become very important for scientific as well as technological applications. Concerning the scientific applications one must emphasize the use of solid electrolytes in galvanic cells. These are important for thermodynamic as well as kinetic investigations. The emf of galvanic cells gives information about Gibbs reaction energies, reaction enthalpies and entropies, and also about activities, chemical potentials and partial pressures. From the kinetic point of view it is important that the current through suitably built galvanic cells with solid electrolytes can be a measure of reaction rates. The combination of reaction rate measurements via electric current with determinations of thermodynamic quantities via the emf often permits the analysis of kinetic processes. These processes may be diffusion in solids and liquids, diffusion-controlled solid-state reactions and phase boundary reactions at solid/solid phase boundaries or at solid/gas phase boundaries, e.g. evaporation processes. The technical applications include batteries with solid electrolytes, high-temperature fuel cells, sensors for measuring partial pressures or activities, display units and recently the growing field of chemotronic components. Science and technology involving solid-state electrolytes are sometimes called solid-state ionics, in analogy to the field of solid-state electronics.

From the microscopic point of view, ionic conduction in solids is caused by the existence of defects in the volume of the crystalline material. These defects are vacancies or interstitials. The concentrations of defects obey laws of mass action similar to those holding for dissociation equilibria in liquid electrolytes. This is an important analogy to the electrochemistry of liquid electrolytes. Besides this, a variety of theoretical approaches and experimental techniques used in the electrochemistry of solids are in many respects comparable to those used in the electrochemistry of liquid electrolytes. This will become clearer in the different chapters of this book.

As a limiting case of disorder in solids all ions of one partial lattice may be quasi-disordered and are thus mobile. This so-called structural disorder with a quasi-molten state of one partial lattice explains why the ionic conductivity of certain solid electrolytes can reach high values which are normally to be found in concentrated liquid electrolytes or molten salts.

The history of the electrochemistry of solids originates in the discovery of the first solid ionic conductors. At the end of the last and at the beginning of this century, Warburg and Tegetmeier [1.1], Haber and Tolloczko [1.2] and Bruni and Scarpa [1.3] have carried out transference measurements. They could demonstrate the validity of Faraday's law for solid conductors, i.e. the amount of mass transported by the current in the solid was given by Faraday's law. In about 1920, Tubandt and co-workers [1.4] carried out systematic transference measurements. They showed, for example, that silver halides are cationic conductors. Since then, it is a well-known fact that α-AgI, being stable at temperatures above 149 °C, is a solid ionic conductor with very high conductivity similar to that of molten salts. Thus, AgI was one of the first known so-called "superionic conductors". The microscopic theory of ionic conductivity in solids has its origin in a paper by Joffé [1.5] who introduced the concept of interstitial ions and vacancies. This was the starting point for the theory of defects in solids. Frenkel [1.6] formulated the first laws of mass action which hold for interstitials and vacancies. In about 1930, Schottky and Wagner [1.7] reported an extended thermodynamic treatment of disorder in solids. They also took into account the possibility of equilibrium between anionic and cationic vacancies, the so-called Schottky disorder found in alkali halides. In 1933, Wagner [1.8] used the concept of mobile defects in solids to explain solid-state reactions. In this way, he treated the diffusion-controlled tarnishing reaction of metals.

Since 1930 attempts were made to explain the very high ionic conductivity of AgI. In 1934, Strock [1.9] investigated the structure of α-AgI. He postulated that the silver ions in AgI are distributed over a large number of nearly equivalent sites whereas the anions form a regular well-ordered structure. In the meantime, this picture has been somewhat modified but one can still speak of structural disorder which may be regarded as an extreme case of Frenkel disorder. It has been argued that one partial lattice of AgI or of other high-conductivity solid electrolytes is quasi-molten and that the conducting ions show a behaviour which is in several respects similar to that of a liquid. Up to now, a great number of solid electrolytes has been found, including some with very high ionic conductivity. In 1961, Reuter and Hardel [1.10] discovered the highly ionic conducting compound Ag_3SI which is stable above 235 °C. This was the first ternary compound with structural disorder for silver ions and a high ionic conductivity. After this, a great number of ternary compounds was investigated. In 1966, Bradley and Greene and in 1967 Owens and Argue reported on $RbAg_4I_5$ and other compounds with the general formula $M\,Ag_4I_5$ ($M = K$, Rb, NH_4) [1.11]. All these are solid electrolytes of high conductivity, which are stable at room temperature and below. Takahashi [1.12] found similar copper ion conductors.

Doped ZrO_2, an oxygen ion conductor, already known since the end of the last century, has found extensive use as a solid electrolyte since the work of Kiukkola and Wagner [1.13] in 1957 who carried out emf measurements to determine Gibbs' reaction energies. Another important electrolyte is β-Al_2O_3 which was first used as a solid ion conductor by Weber and Kummer [1.14] in 1967. β-Al_2O_3 with the formula $NaAl_{11}O_{17}$ is a sodium ion conductor and has found an important application as a solid electrolyte in the sodium-sulfur battery. Further conductors of sodium and other alkali metal ion conductors have been developed; one promising compound is the so-called Nasicon (Na superionic conductor) with the formula $Na_{1+x}Zr_2P_{3-x}Si_xO_{12}$ ($x \approx 2$) discovered by Goodenough, Hong and Kafalas [1.15] in 1976.

Thermodynamic investigations using galvanic cells with solid electrolytes have up to now been carried out on a very large number of systems. Furthermore, kinetic studies using galvanic cells with solid electrolytes have been performed by various authors. Kobayashi and Wagner [1.16] have investigated the reduction of Ag_2S by H_2. The author of this book has also reported kinetic investigations on solid electrolytes; these include, for example, electrochemical measurements of oxygen diffusion in metals [1.17], studies of a tarnishing reaction: the formation of solid nickel sulfide on nickel [1.18], electrochemical studies of phase boundary reactions: the transfer of silver, silver ions and electrons across the interface solid silver/solid silver sulfide [1.19], electrochemical investigations of evaporation processes: the evaporation of iodine from copper iodide, of sulfur from silver sulfide and selenium from silver selenide [1.20], investigations of the thermodynamics of gases using an electrochemical Knudsen-cell [1.21] and electrochemical measurements of the chemical diffusion coefficient: determination of the chemical diffusion coefficient of FeO and silver sulfide [1.22]. Further investigations have been carried out by Rapp [1.23], Raleigh [1.24], Fueki [1.25] and others.

Besides the already mentioned use of β-Al_2O_3 as a solid electrolyte in the sodium-sulfur battery, different batteries using other solid ion conductors have been built in the meantime. High-temperature fuel cells and electrolyzers using doped ZrO_2 as a solid electrolyte are further developed (see for example [1.26]). Weissbart and Ruka [1.27] proposed the application of doped ZrO_2 for oxygen partial pressure measurements in 1961. These sensors have found extensive technical application since. Chemotronic building units with solid electrolytes have been proposed by Takahashi and Yamamoto [1.28]. Further applications of solid electrolytes are in the field of display units.

A more detailed account of the development of solid-state electrochemistry is given in several review articles and monographs [1.29]. This topic will also be treated in this book.

1.1 References

[1.1]   Warburg, E.: Wiedemann Ann. Phys. *21*, 622 (1884); Warburg, E., Tegetmeier, F.:
        Wiedemann Ann. Phys. *32*, 455 (1888);
        Tegetmeier, F.: Wiedemann Ann. Physik *41*, 18 (1890)
[1.2]   Haber, F., Tolloczko, A.: Z. anorg. Chem. *41*, 407 (1904)
[1.3]   Bruni, G., Scarpa, G.: Rend. reale accad. naz. Lincei *22*, 438 (1913)
[1.4]   Tubandt, C.: Z. anorg. allgem. Chem. *115*, 105 (1921)
        Tubandt, C., Reinhold, H.: Z. Electrochem. *29*, 313 (1923)
[1.5]   Joffé, A.: Ann. Phys. (Leipzig) *72*, 461 (1923)
[1.6]   Frenkel, J.: Z. Phys. *35*, 652 (1926)

[1.7]   Wagner, C., Schottky, W.: Z. phys. Chem. *B11*, 163 (1930)
        Schottky, W.: Z. phys. Chem. *B29*, 335 (1935)
[1.8]   Wagner, C.: Z. phys. Chem. *B21*, 25 (1933)
[1.9]   Strock, L. W.: Z. phys. Chem. *B25*, 441 (1934)
        Strock, L. W.: Z. phys. Chem. *B31*, 132 (1935)
[1.10] Reuter, B., Hardel, K.: Naturwissenschaften *48*, 161 (1961)
        Reuter, B., Hardel, K.: Z. anorg. allg. Chem. *340*, 158 (1965)
[1.11] Bradley, J. N., Greene, P. D.: Trans. Faraday Soc. *62*, 2069 (1966)
        Bradley, J. N., Greene, P. D.: Trans. Faraday Soc. *63*, 424 (1967)
        Owens, B. B., Argue, G. R.: Science *157*, 308 (1967)
[1.12] Takahashi, T., Yamamoto, O., Ikeda, S.: J. Electrochem. Soc. *120*, 1431 (1973)
[1.13] Kiukkola, K., Wagner, C.: J. Electrochem. Soc. *104*, 308 (1957)
        Kiukkola, K., Wagner, C.: J. Electrochem. Soc. *104*, 379 (1957)
[1.14] Weber, N., Kummer, J. T.: Proc. Ann. Power Sources Conf. *21*, 37 (1967)
[1.15] Goodenough, J. B., Hong, H. Y.-P., Kafalas, J. A.: Mat. Res. Bull. *11*, 203 (1976)
[1.16] Kobayashi, H., Wagner, C.: J. Chem. Phys. *26*, 1609 (1957)
[1.17] Rickert, H., Steiner, R.: Z. phys. Chem. N.F. *49*, 127 (1966)
[1.18] Mrowec, S., Rickert, H.: Z. phys. Chem. N.F. *36*, 329 (1963)
[1.19] Rickert, H., O'Brian, C. D.: Z. phys. Chem. N.F., *31*, 71 (1962)
        Rickert, H., Wagner, C.: Z. phys. Chem. N.F. *31*, 32 (1962)
        Contreras, L., Rickert, H.: Ber. Bunsenges. Phys. Chem. *82*, 292 (1978)
[1.20] Mrowec, S., Rickert, H.: Z. Electrochem. *66*, 14 (1962)
[1.21] Ratchford, R.-J., Rickert, H.: Z. Electrochem. *66*, 497 (1962)
        Birks, N., Rickert, H.: Ber. Bunsenges. phys. Chem. *67*, 97 (1963)
        Detry, D. et al.: Z. phys. Chem. N.F. *55*, 314 (1967)
[1.22] Rickert, H., Weppner, W.: Z. Naturforsch. *29a*, 1849 (1974)
        Hartmann, B., Rickert, H., Schendler, W.: Electrochim. Acta *21*, 319 (1976)
[1.23] Rapp, R. A., Shores, D. A.: Techn. Met. Res. *2*, 123 (1970)
[1.24] Raleigh, D. O.: Progr. Solid State Chem. *3*, 83 (1967)
[1.25] Fueki, K., Mizusaki, J., Mukaibo, T.: Bull. Chem. Soc. Jap. *46*, 1663 (1973)
[1.26] Binder, H. et al.: Electrochim. Acta *8*, 781 (1963)
        Fischer, W. et al.: Chem.-Ing.-Tech. *43*, 1227 (1971)
        Fischer, W. et al.: Chem.-Ing.-Tech. *44*, 726 (1972)
        University of Oklahoma, Energy Alternatives: A Comprehensive Analysis, U.S.
        Government Printed Office, Washington, D.C. (1975)
[1.27] Weissbart, J., Ruka, R.: Rev. Sci. Instrum. *32*, 593 (1961)
[1.28] Takahashi, T., Yamamoto, O.: J. Appl. Electrochem. *3*, 125 (1973)
[1.29] Hladik, J.: Physics of Electrolytes. New York: Academic Press 1972
        van Gool, W.: Fast Ion Transport in Solids — Solid-State Batteries and Devices.
        Amsterdam: North-Holland 1973
        Kleitz, M., Dupuy, J.: Electrode Processes in Solid-State Ionics. Dordrecht: Reidel
        1976
        Alcock, C. B.: EMF-Measurements in High-Temperature Systems, Inst. Mining
        Metallurgy, London 1968
        Fischer, W. A., Jahnke, D.: Metallurgische Elektrochemie. Berlin, Heidelberg, New
        York: Springer 1975
        Mahan, G. D., Roth, W. L.: Superionic Conductors. New York, London: Plenum
        Press 1976
        Geller, S.: Solid Electrolytes. Berlin, Heidelberg, New York: Springer 1977
        Hagenmuller, P., van Gool, W.: Solid Electrolytes. New York, San Francisco,
        London: Academic Press 1978
        Vashishta, P., Mundy, J. N., Shenoy, G. K. (eds.): Fast Ion Transport in Solids,
        Electrodes and Electrolytes. New York, Amsterdam, Oxford: North-Holland 1979
        Subbarao, E. C.: Solid Electrolytes and Their Applications. New York, London:
        Plenum Press 1980
        Rapp, R. A., Shores, D. A.: Techn. Met. Res., Rapp, R. A. (ed.), Vol. 2, p. 123.
        Interscience, New York 1970

van Gool, W.: Annu. Rev. Mater. Sci. 1974, 311

Raleigh, D. O.: Progr. Solid State Chem. *3*, 83 (1967)

Worrell, W. L.: Am. Ceram. Soc. Bull. *53*, 425 (1974)

Heyne, L.: Electrochim. Acta *15*, 1251 (1970)

Steele, B. C. H.: J. Solid State Chem. *10*, 117 (1972)

Owens, B. B.: Adv. Electrochem. Eng. *8*, 1 (1971)

Holzäpfel, G., Rickert, H.: Festkörperprobleme XV (Advances in Solid State Physics). Qeisser, H. J. (ed.), p. 317. Braunschweig: Pergamon/Vieweg 1975

Rickert, H.: Angew. Chem. Int. Ed. Engl. *17*, 37 (1978)

2 Disorder in Solids

In the electrochemistry of solids the study of crystalline metal-non metal compounds plays a particularly important role. These compounds may be ionic, such as AgCl or NaCl, or may exhibit more covalent character as in many metal sulfides and oxides. The following considerations are, however, also valid for other ordered crystalline compounds such as ordered alloys.

Under real conditions, all ordered compounds exhibit deviations from a regular crystal structure which are of great importance, particularly for the understanding of thermodynamic and kinetic properties of crystalline compounds. This chapter is therefore concerned with irregularities in the structure of crystals, i.e. with lattice defects.

2.1 General Considerations Regarding Lattice Defects and Their Significance

Crystals are distinguished from gases and liquids by their so-called long-range order, i.e. by the uniform periodic arrangement of particles over distances, which are long in comparison with atomic dimensions. The smallest repeating unit, the building block from which the three-dimensional crystal is constructed, is called the unit cell. An ideal crystal is one in which the unit cell is repeated an infinite number of time in three dimensions. Crystal defects are deviations from this infinite periodicity. The existence of lattice defects has been confirmed by measuring ionic conductivities in solids, particularly at high temperatures, by comparing the density obtained by X-ray diffraction with that obtained by pycnometric methods, and by many results from solid-state physics. This chapter does therefore not present a detailed account of the arguments indicating their existence but merely describes them and gives a thorough thermodynamic treatment of point defects.

The most important lattice defects include: Zero-dimensional or point defects; irregular occupancy of lattice sites which will be discussed in Sect. 2.2. A distinction is made between vacancies (particles missing in a region of the lattice), interstitial particles (excess particles in a region of the lattice) or substitution particles (a lattice site is occupied by a foreign particle).

One-dimensional defects: edge or screw dislocations.
Two-dimensional defects: surfaces or grain boundaries.
Three-dimensional defects: e.g. cavities.

Further lattice defects are, among others, stacking faults and distortions, i.e. deviations from the regular lattice parameters. Because of the particular importance of lattice defects for many basic properties of crystals, they have been treated in considerable detail in the literature; a selection of references is given at the end of this chapter [2.1]. Dislocations, for example, are important both in determining the mechanical properties of crystals and the kinetics of crystal growth. Surfaces play an important role in the kinetics of heterogeneous systems, particularly in heterogeneous catalysis.

However, only zero-dimensional defects or point defects are properties of the thermodynamic equilibrium; we shall pay particular attention to these defects in the following discussion. Apart from point defects involving ions or atoms, leading to so-called material defects, we must also discuss electron disorder. While particle disorder, i.e. excess (interstitial particles), deficient (vacancies) or foreign particles (substitution particles), leads to deviations from the ideal crystal structure, electron disorder is due to the fact that energy levels occupied at low temperatures become empty at higher temperatures, while other levels, formerly empty, become occupied. Particularly important is the transition of electrons from the valence band, fully occupied at low temperatures, to the conduction band which, in semiconductors, is normally unoccupied under these conditions; this leads to the formation of quasi-free electrons in the conduction band and electron defects in the valence band. Though in the case of electrons we emphasize the role of energy levels, it must be noticed that the generation of material defects also involves energy changes. The zero-dimensional disorder of particles and the disorder of electrons are particularly important from the point of view of physical chemistry for

a) the thermodynamic behaviour of crystalline compounds: for example, the equilibrium partial pressures or the chemical potentials of the components in the compound depend on the concentrations of the defects,
b) transport processes in crystalline compounds, i.e. diffusion of or conduction by ions and electrons in the crystal.

Many properties of crystals, for example optical and magnetic properties, may also depend on the existence of lattice defects. Disorder equilibria determine the concentrations of the various defects present. However, first of all, an exact account of defects will be given.

2.2 Point Defects

As an example, Fig. 2.2.1 shows schematically a two-dimensional cut through a silver chloride crystal containing lattice defects. The various lattice defects are emphasized in Fig. 2.2.1 by the use of bold type.

The charges of defects and of regular lattice particles are only important with respect to the neutral unperturbed lattice. In the following discussion the charges of all point defects will therefore be stated relative to the neutral unperturbed lattice; positive excess charges will be symbolized by a point index, negative excess charges by a prime index and neutral point defects by a cross-like index;

in the case of neutral defects we shall, however, often use no symbol at all. The regular lattice particles of AgCl, i.e. Ag^+ and Cl^-, have no additional charge if referred to the neutral lattice, i.e. they are neutral with respect to the unperturbed lattice. Therefore, they have not been given a charge index. As can be seen in Fig. 2.2.1 some silver particles are missing from their lattice sites, i.e. the "crystal" contains so-called vacancies. These are negatively charged with respect to the unperturbed lattice, because a positive charge is missing, if a silver ion is removed from the crystal. At other points in the lattice there are excess silver ions, which may be considered to occupy interstitial sites. These are now positively charged with respect to the unperturbed lattice. A third possible type of defects in the silver ion sublattice would involve replacement of silver particles by other ions.

Interstitial particles may be those which are normally present in the crystal, yet usually occupying regular lattice sites, or foreign particles which may be brought from outside into a pure crystalline compound. Substitution particles are always foreign particles in metal/non-metal compounds, while this is not true for intermetallic compounds. Apart from simple defect sites there also exist complex defects consisting of several simple ones occupying neighbouring sites.

The terms "vacancy" and "interstitial particle" are often misunderstood. The presence of a vacancy does not imply that one lattice site is vacant while all surrounding lattice particles remain in their original positions. It may be assumed that the neighbouring particles are shifted from their regular positions so that in the neighbourhood of a vacancy a certain distortion of the lattice occurs, this makes a small contribution to the energetic effects accompanying the production of a vacancy. However, at a greater distance — several lattice sites away from the vacancy — the lattice is unperturbed. A more exact definition of the term vacancy is as follows: a vacancy occurs if in a volume element of the lattice a particle is missing with respect to the ideal lattice, independent of whether at this point or in its immediate vicinity other particles deviate from their normal positions or not.

The occurrence of an interstitial particle also generally leads to a large distortion of the surrounding region. It is even possible that a particle previously occupying a regular lattice site is shifted so far from its normal position that two particles, the so-called interstitial particle and the normal particle, now occupy equivalent positions in the neighbourhood of a regular lattice site. Thus,

Fig. 2.2.1. Section through an AgCl crystal containing lattice defects. V'_{Ag}: silver ion vacancy, negatively charged with respect to the unperturbed lattice. $Ag_i^{·}$: silver ion occupying interstitial site. The *arrows* indicate possible site exchange mechanisms (see also Sect. 6.2)

a more exact statement is that an interstitial particle is present if the lattice contains an excess particle, independent of whether this causes neighbouring particles to be shifted from their normal positions or not.

For simplicity, we shall generally assume below that the concentration of defects is so low that an undisturbed lattice region is usually present between two defects.

Electron disorder — the presence of quasi-free electrons and electron defects — will be discussed in detail in Chap. 4. Two different methods of describing crystals containing point defects exist at present.

a) Description in terms of structure elements. Point defects are defined relative to the empty space in which the lattice or interstitial positions are fixed with respect to an imaginary coordinate system.

b) Description in terms of building units or relative building units. The point defects are defined relative to the ideal crystal.

Atoms (or groups of atoms e.g. NO_3, SO_4) and vacancies, occupying certain positions in the lattice (either regular sites or interstitial sites), are called structure elements.

According to the notation proposed by Kröger and Vink [2.2] the sites occupied by the various particles are indicated by a lower index and the particles themselves are represented by their chemical symbols. If at a particular point no particle is present, i.e. there is a vacancy, this is given the symbol V. Vacancies as structure elements have no material significance, in particular not that of a negative particle, as is the case when a vacancy is considered as a building unit (as will be discussed below). In a crystal AB the A-sites are denoted by a lower index A, the B-sites by a lower index B and the interstitial sites by a lower index i. For example, in the AgCl lattice of Fig. 2.2.1, a silver ion vacancy is symbolized as V'_{Ag}. An interstitial silver ion is denoted by $Ag_i^{\cdot}$. A silver particle occupying a silver site is represented as Ag_{Ag}. For clarity, this notation has not been used in Fig. 2.2.1; the lower indexes of the particles occupying regular lattice sites are omitted. In Table 2.2.1 are listed symbols used in the Kröger system to describe structure elements in a crystal AB containing A- and B-particles, foreign particles C and vacancies occupying various sites. All structure elements are assumed to be neutral with respect to the unperturbed lattice.

Figure 2.2.2 is a schematic diagram of an AB lattice shown in terms of structure elements. No symbols are used to describe the interstitial vacancies, while bold type is used to emphasize the defects.

Table 2.2.1. The Kröger symbols for neutral structure elements in an AB lattice; the examples chosen are A- and B-particles, foreign particles C and vacancies at different sites

Site	Particle A	Particle B	Vacancy V	Foreign particle C
A-site	A_A	B_A	V_A	C_A
B-site	A_B	B_B	V_B	C_B
Interstitial site (i site)	A_i	B_i	V_i	C_i

Structure elements with the corresponding symbols proposed by Kröger and Vink have the advantage that they are easy to memorize; this notation is now also widely used in the literature. However, it is very important to note that the numbers of the various structure elements in a crystalline compound are not independent of one another. This is due to the fixed ratio of the number of A- and B-sites in an AB lattice, in the simplest case 1:1.

Thus, an AgCl lattice contains equal numbers of Ag- and Cl-sites. This ratio is not influenced by the surface area — especially for large crystals, which are generally considered in the disorder theory — since the surface area can always be neglected in comparison with the volume in such a case. It is therefore generally impossible to change the concentration of only one type of structure element in a crystal. Thus, to create a vacancy on an A-site one must either remove an A-particle or simultaneously add a B-particle to a new B-site, so that the ratio of A- to B-sites remains unchanged. The generation of an interstitial particle leads to the destruction of a vacancy in the interstitial lattice. Thus, if we mentally increase the size of a crystal or change the number of defects contained in it, we must either add or remove combinations of structure elements. Similarly, the use of structure elements to describe reactions of defects generally involves the use of combinations of structure elements.

It is not possible to calculate the change in the Gibbs energy for the addition of a structure element, and thus the chemical potential of a structure element. The latter is equal to the partial derivative of the Gibbs energy with respect to the particle number of the structure element considered keeping pressure, temperature and the numbers of the remaining structure elements constant. Kröger and Vink [2.2] have thus introduced virtual or quasi-potentials such that the sum of (generally) two virtual potentials yields one real chemical potential. However, since splitting the chemical potential into two virtual potentials is arbitrary and may therefore lead to misunderstandings, the introduction of virtual potentials remains unsatisfactory.

The second method of describing disorder in crystals involves building units and is due to Schottky [2.3]; defects are defined relative to the ideal crystal. Building units can — physically or mentally — be removed from or added to the crystal independent of other building units. They correspond to suitable combinations of structure elements. The number of building units necessary to

A_A B_B A_A B_B A_A B_B

   $C_i^{\cdot}$

B_B A_A B_B A_A $C_B^{\cdot\cdot}$ A_A

A_A B_B V_A' B_B A_A B_B

                                 $A_i^{\cdot}$

B_B A_A B_B A_A B_B A_A

   B_i'

A_A B_B C_A B_B B_A'' B_B

$A_B^{\cdot\cdot}$ A_A B_B A_A B_B A_A

Fig. 2.2.2. Section through an AB crystal, described in terms of structure elements using the Kröger symbols. Vacancies in the interstitial lattice are not shown. Disorder centres are shown by the use of bold type

Table 2.2.2. Notation of the lattice molecule and typical defects in an AB lattice in terms of building units using the old and new Schottky symbols compared to the corresponding structure elements according to Kröger and Vink

	Schottky old	Schottky new	Kröger/Vink
AB lattice molecule	AB	AB	AB
Neutral A-particle occupying interstitial site	A○	A	$A_i - V_i$
Neutral A-vacancy	A □	|A|	$V_A - A_A$
Neutral B-particle occupying A-site	B•(A)	B |A|	$B_A - A_A$
Neutral C-particle occupying A-site	C•(A)	C |A|	$C_A - A_A$
Quasi-free electron	⊖	e′	e′
Electron defect	⊕	|e|˙	h˙

Negative charge of a defect (with respect to the unperturbed lattice): upper prime index ′
Positive charge of a defect (with respect to the unperturbed lattice): upper point index ˙
Neutral defect site relative to the unperturbed lattice (often not used): upper cross index ˣ
The index showing the corresponding excess charge of electrons and electron defects is often not used so that one writes e and h instead of e′ and h˙.
Since the meaning of the symbols for electrons and electron defects is the same in all systems, only the symbols e and h will be used in this book.

describe the crystal is small; Schottky took lattice molecules, interstitial particles, vacancies and substitution particles as building units. Examples for the symbols used to describe them are given in Table 2.2.2. The equivalent combination of structure elements corresponding to the building unit is given in the last column. The lattice molecule, which in the simplest case is identical with the particles in the unperturbed unit cell, belongs to the building units; it can also correspond to a part of the unit cell. The lattice molecule, like a gas molecule, is represented by a chemical formula. When an infinite number of lattice molecules are placed next to one another in three dimensions, we obtain an ideal crystal which is free of defects. The introduction of defects or relative building units into this ideal crystal produces the real crystal. The introduction of a vacancy into a crystal involves the simultaneous removal of a particle (Table 2.2.2). Thus, the vacancy as a relative building unit has the significance of a negative particle, just as an electron defect has the significance of a negative electron.

Unfortunately, two different sets of symbols for the building units are used in the literature; these are the old Schottky notation, which is used in the older literature and in the first edition of Hauffe's book "Reaktionen in und an festen Stoffen" [2.4] and the new Schottky notation [2.5] which, for example, is used in the second edition of Hauffe's book [2.6]. According to the new Schottky notation [2.5], particles in excess of the number present in the unperturbed lattice are simply denoted by their chemical symbols, independent of whether or not such particles occur in the ideal lattice; e.g. an A-particle occupying an interstitial site is simply given the symbol A. Particles which are missing from the lattice are enclosed by vertical lines. A substitution particle is expressed by a combination of these two notations, i.e. the symbols for an additional and for a missing particle are combined: thus, a particle B occupying an A-site is denoted as B |A|. According to the old Schottky notation the interstitial symbol ○, the vacancy symbol □ and the substitution symbol ● are used so that, for example, an A-particle occupying an interstitial site is represented as A○, an A-vacancy as A□ and an X-particle occupying an A-site as X●(A).

The concentration of defects and their dependence on one another and on external parameters can be calculated as we shall show in the next section which deals with disorder equilibria and disorder reactions. It will be seen that in many respects defects may be considered as "chemical" particles so that one can justifiably speak of "imperfection chemistry". On practical grounds we shall first formulate disorder reactions using structure elements with the Kröger/Vink symbols. We shall then either rearrange the structure elements or add and subtract some to complete the equation in such a manner that we can combine structure elements to give building units: two examples, which we shall discuss in Sect. 2.5 in detail (Frenkel and Schottky equilibrium) will demonstrate this procedure where the correct combination of structure elements is emphasized by use of parentheses.

This will allow a rigorous thermodynamic treatment to follow since we can ascribe chemical potentials to building units, i.e. to suitable combinations of structure elements; it is then possible to calculate laws of mass action. The laws of mass action will be presented in the notation of both building units and structure elements. For the remaining part of the book we shall, when discussing disorder equilibria, leave out the intermediate steps or use only structure elements.

Since defects can either take up quasi-free electrons or electron defects from the lattice or give up these to it, i.e. in general electrons are also involved in disorder, the charge of the defects must also be characterized. This point is treated uniformly in the literature; a positive excess charge is denoted by a dot and a negative one by a prime, both as superscripts. According to this convention, quasi-free electrons should be symbolized by e' and electron defects by $h^{\cdot}$. However, since the charges of quasi-free electrons and electron defects with respect to the unperturbed lattices are always the same and unequivocal the superscripts are generally omitted in this case and electrons and electron defects are often denoted as e or h.

As an example of a disorder reaction we shall now consider the transition of a silver ion from a regular lattice site in the silver-chloride lattice to an interstitial site whereby a vacancy is formed:

$$Ag_{Ag} + V_i = V'_{Ag} + Ag_i^{\cdot}. \tag{2.2.1}$$

Rearrangement of Eq. (2.2.1) and combination of structure elements to give building units lead to the equation

$$0 = (V'_{Ag} - Ag_{Ag}) + (Ag_i^{\cdot} - V_i). \tag{2.2.2}$$

The building unit vacancy is expressed in the old Schottky notation (Table 2.2.2) as

$$(V'_{Ag} - Ag_{Ag}) \equiv Ag\square'$$

and according to the new Schottky notation

$$(V'_{Ag} - Ag_{Ag}) \equiv |Ag|'.$$

An interstitial particle as a building unit is expressed as

$$(Ag_i^{\cdot} - V_i)_{\text{Kröger}} \equiv (Ag\circ^{\cdot})_{\text{Schottky, old}}$$
$$\equiv (Ag^{\cdot})_{\text{Schottky, new}}.$$

Thus Eq. (2.2.1), which is written in terms of structure elements, can be represented in the old Schottky notation by

$$0 = Ag\square' + Ag\circ^{\cdot} \tag{2.2.3}$$

and using the new Schottky notation as

$$0 = |Ag|' + Ag^{\cdot}. \tag{2.2.4}$$

It is important to note that, using Schottky symbols for building units, a vacancy e.g. $|Ag|'$ has at the same time the significance of a negative particle. The treatment of electron defects as "building units" is generally accepted; it is unfortunate that the same is not true for the description of ionic and atomic defects. The fact that most authors treat ions and electrons in two different ways certainly causes some problems for the beginner. Unfortunately, however, in the present state of the literature, this cannot be avoided. Let us summarize once more:

Crystals and their defects may be described:

a) in terms of structure elements, e.g. A_A, V_A, A_i,

b) in terms of building units: molecule AB and relative building units (defects), e.g. $|A|$ (A-vacancy) and A (A on interstitial site).

The building units correspond to combinations of structure elements. Only building units, and thus suitable combinations of structure elements, can be introduced — mentally or physically — into a crystal independently of other building units so that chemical potentials can be ascribed to them. A vacancy as a building unit has the significance of a negative particle, just as an electron defect has the significance of a negative electron, while the vacancy as a structure element represented by the symbol V_A or V_B has no material significance. Since we shall use chemical potentials for calculating disorder equilibria or for deriving laws of mass action, we shall combine structure elements in an appropriate manner to give building units. This will be shown in Sect. 2.3. However, structure elements can also be of great importance, as will be shown for example in the kinetic treatment of disorder equilibria.

2.3 Disorder Equilibria

The theory of disorder equilibria in solid crystalline compounds is due to Frenkel (1926) [2.7] and Schottky and Wagner (1930) [2.8]. Considerations of disorder equilibria allow us to make statements regarding concentrations of defects, of electrons and electron defects. Such equilibria relate the concentrations of these quantities to one another and to external parameters, especially to the chemical potentials or equilibrium partial pressures of the components from which the crystal is built up. These relationships appear in the form of laws of mass action. From the thermodynamic point of view, solid crystalline compounds may be considered as ordered solutions in which the chemical potentials and the corresponding equilibrium partial pressures of the components can vary over wide ranges. These changes are accompanied by small deviations from the ideal stoichiometric composition, but usually by relatively marked changes in the concentrations of defects which thus often lead to drastic variations in the partial conductivities and other properties of the crystal. The number of chemical potentials which can be varied independently of one another is given by Gibbs' phase rule. For example, according to this rule a binary solid compound at a certain temperature and pressure has one further degree of freedom, i.e. one of the two chemical potentials of the two types of atoms is allowed to vary independently. If this potential is fixed, all disorder equilibria are determined. Disorder equilibria are nothing else than a special type of chemical equilibria. Defects may therefore be regarded as quasi-chemical particles, so that, as mentioned above, we can speak of imperfection chemistry. The derivation of disorder equilibria is carried out as for chemical equilibria. There is for example a strong similarity to the treatment of electrolytic dissociation in the electrochemistry of liquids. The same factors are involved in determining the conditions of a disorder equilibrium in a crystal as for any other thermodynamic equilibrium. The Gibbs energy G, which has a minimum value at equilibrium at fixed pressure and temperature, consists of two contributions, namely the enthalpy H and the entropy term $-T \cdot S$. Considering only the enthalpy the most stable state would be the one with the lowest possible energy. This is the case at absolute zero, the term $-T \cdot S$ then being zero; thus, at absolute zero a completely ordered crystal represents the most stable state, since the generation of lattice defects always requires a supply of energy. At higher temperatures the entropy term $-T \cdot S$ plays an equally important role, since the system (in our case the crystal) attempts to attain a state of high thermodynamic probability, i.e. a state with a large degree of disorder.

In the following these considerations will be discussed in detail on a quantitative basis. The equilibrium condition for a closed system — no exchange of particles with the surroundings is allowed — states that at a certain pressure p and temperature T the change in the Gibbs energy of the system, e.g. due for example to a chemical reaction, is equal to zero:

$$(\Delta G)_{p,T} = 0 \quad \text{(closed system)}. \tag{2.3.1}$$

A chemical reaction in which the species involved are denoted by A_1, A_2 etc. and the corresponding stoichiometric coefficients by ν_1, ν_2, etc. the stoichiometric coefficients of the products being positive and those of the reactants negative, may be written quite generally as follows:

$$0 = \nu_1 A_1 + \nu_2 A_2 + \dots . \tag{2.3.2}$$

This may be recast in the following form:

$$0 = \sum_i \nu_i A_i . \tag{2.3.3}$$

In order to apply the equilibrium condition (2.3.1) to such a chemical reaction, particles must be neither added nor removed from the system during the reaction, i.e. the masses and charges in Eq. (2.3.2) or (2.3.3) must balance; this is automatically the case if the reaction equation is correctly written. We are thus dealing with a process in which particles are exchanged within the boundaries of the system. It is of course possible to construct a system which consists not only of the crystal considered so far but also of the surrounding gas, or of another phase. This indeed becomes necessary if we do treat not only exchange processes within the crystal, so-called internal equilibria, but also processes involving exchange between the crystalline compound and its surroundings, i.e. equilibria with neighbouring phases.

The following discussion will not be concerned with chemical reactions in general, but with processes in the crystal or reactions between the crystal and the surroundings which can be written as disorder reactions; an example is the Frenkel disorder reaction discussed above. The change ΔG of the Gibbs energy corresponding to a per formula conversion can be expressed as follows:

$$\Delta G = \nu_1 \left(\frac{\partial G}{\partial n_1} \right)_{p,T,n_i \neq n_1} + \nu_2 \left(\frac{\partial G}{\partial n_2} \right)_{p,T,n_i \neq n_2} + \dots . \qquad (2.3.4)$$

The partial molar Gibbs energies $\partial G / \partial n_i$, i.e. the partial derivatives of G with respect to the mole numbers n_i of the various defects, are identical with the electrochemical potentials η_i of these defects if these are assumed to be charged. Otherwise, they are equal to the chemical potentials μ_i. Thus, using Eq. (2.3.4) we can write the equilibrium condition (2.3.1) as follows:

$$\Delta G = \nu_1 \eta_1 + \nu_2 \eta_2 + \dots = 0 \qquad (2.3.5)$$

or in a shorthand notation:

$$\sum_i \nu_i \eta_i = 0. \qquad (2.3.6)$$

Equation (2.3.6) is the basic equation applying to all disorder equilibria. It states that the sum of the electrochemical or — in the case of neutral defects — chemical potentials multiplied by their stoichiometric coefficients is equal to zero, for a properly formulated reaction equation, i.e. this must be written using building units as the reacting pacticles. Equation (2.3.6) becomes simpler in certain cases where chemical potentials may be used instead of electrochemical potentials, even if the defects are charged. On condition that the electrochemical potential η may be split into a chemical and an electrical term we can express η_i as follows:

$$\eta_i = \mu_i + z_i F \varphi, \qquad (2.3.7)$$

z_i is the charge number, μ_i the chemical potential of the various defects (building units), F Faraday's constant, and φ the electrical potential. Regarding disorder equilibria in volume elements with a constant electrical potential φ the latter is equal for all defects. In this case the terms $z_i F \varphi$ cancel since the charges in the reaction equation must be balanced. In the case of heterogeneous equilibria or equilibria between volume elements with different electrical potentials, the electrical terms also cancel if only particles or combinations of particles, which are electrically neutral, take part in the equilibria. In these cases the following equilibrium condition is valid

$$\sum_i \nu_i \mu_i = 0. \tag{2.3.8}$$

With respect to the restrictions discussed above Eq. (2.3.8) states the condition for various disorder equilibria, i.e. the sum of the chemical potentials of the reacting species multiplied by the stoichiometric coefficients is equal to zero. As before it is necessary that the reaction equation has been formulated correctly to fulfil the boundary conditions of Eq. (2.3.1); the reaction equation must also be written using building units or be rearranged to a form involving building units, so that the chemical potentials may be exactly defined. Thus, our problem of calculating disorder equilibria reduces to the calculation of chemical potentials of defects. If we can express the chemical potentials of the various defects as a function of their concentrations we can use Eq. (2.3.8) to establish laws of mass action corresponding to the reactions considered.

2.4 The Chemical Potentials of Defects

In this section we shall only be concerned with defects which are building units. Since structure elements can be combined to give building units, these are treated simultaneously. As Schottky [2.9] has stated, it can be shown that only a minimum number of building units is necessary to describe a crystal. These are first the lattice molecules which correspond to the smallest region of the lattice which can be expressed by a chemical formula, in the simplest case identical with the unit cell, so that the number of the lattice molecules N_M is equal to the number of unit cells or to a multiple thereof. The remaining building units, the relative building units, are the defects i, i.e. vacancies, interstitial particles and substitution building units. The real lattice may be represented as the sum of the lattice molecules and the defects which are present in the lattice.

We shall now attempt to obtain an expression for the Gibbs energy G of a real crystal containing the numbers N_M of the lattice molecules and N_i of the different defects. Taking the partial derivatives of G with respect to N_M or N_i gives us the corresponding chemical or electrochemical potentials. We assume that no interactions take place between the defects, i.e. that the concentrations of the defects are sufficiently small. Under these conditions — as will be derived below — the Gibbs energy of a real crystal can be composed of three contributions: a) A contribution, which is proportional to the number N_M of lattice molecules and which may be written as $N_M \tilde{g}_M$. b) A contribution of the different defects i, which are for each defect i proportional to the number N_i of these defects and may be written as the sum $\sum N_i \tilde{g}_i$. c) A contribution due to the configurational entropy resulting from the different possibilities of distributing the various defects over the possible lattice sites. W_i may be the number of possibilities for the distribution of N_i defects within the lattice. Then the contribution of each type of defect i is $-kT \ln W_i$ and that of all defects together $-kT \sum \ln W_i$.

k is Boltzmann's constant and T the absolute temperature. The quantities $\tilde{g}_M$ and $\tilde{g}_i$ contain the enthalpy change of the crystal due to the addition of various building units and also the entropy change without the configurational contribution. The entropy change is affected by a change in the vibrational modes of the crystal as building units are added to the lattice. The contribution due to the configurational entropy will lead to formulae similar to those of the Fermi-Dirac statistics. This is because of the fact that one lattice site can only be occupied at any time by one defect, just as one quantum state for an electron can only be occupied by one electron and one configuration gives only a contribution of one to W_i because the exchange of particles of the same kind within a certain arrangement does not yield a new distribution and thus affords no contribution to the number W_i. Thus, according to Schottky and Wagner [2.10], the following relation holds for the Gibbs energy of a lattice containing N_M lattice moecules and a total number of $\sum N_i$ of various defects i:

$$G_{N_M + \sum N_i} = N_M \tilde{g}_M + \sum_i N_i \tilde{g}_i - \sum_i kT \ln W_i. \tag{2.4.1}$$

Equation (2.4.1) may be derived in the following way: we start from the definition of the Gibbs energy given by

$$G \equiv A + pV, \tag{2.4.2}$$

i.e. the Gibbs energy equals the sum of the Helmholtz energy A and the product of pressure and volume, $p \cdot V$.

As shown by statistical thermodynamics the following relationship generally applies to closed systems

$$A = -kT \ln Z, \tag{2.4.3}$$

k is Boltzmann's constant, T the absolute temperature and Z the partition function.

The partition function is defined as

$$Z = \sum_l d_l \, e^{-E_l/kT}. \tag{2.4.4}$$

The partition function is obtained by summing all Boltzmann factors with respect to the energy levels E_l existing for the system considered whereby each exponential term of the sum is weighted by the degeneracy factor d_l representing the number of quantum states of energy E_l. The system dealt with is in our case a crystal, the Gibbs energy of which is to be calculated.

The Helmholtz energy of an ideal unperturbed crystal containing N_M lattice molecules of type M but no defects will be denoted by A_{N_M}. According to Eq. (2.4.4) the relationship between A_{N_M} and the corresponding partition function is as follows.

$$A_{N_M} = -kT \ln \sum_l (d_l \, e^{-E_l/kT})_{N_M}. \tag{2.4.5}$$

The terms in the sum on the right-hand side of Eq. (2.4.5) are also labeled by the index "N_M" emphasizing the fact that we consider the ideal crystal with N_M lattice molecules and the energy levels corresponding to it.

Dividing A_{N_M} by the number of lattice molecules N_M yields the Helmholtz energy $\tilde{a}_M$ with respect to one lattice molecule

$$\tilde{a}_M = \frac{A_{N_M}}{N_M}. \tag{2.4.6}$$

Rearrangement of Eq. (2.4.6) leads to

$$A_{N_M} = N_M \tilde{a}_M. \tag{2.4.7}$$

Equation (2.4.7) states that each lattice molecule M contributes equally to the Helmholtz energy of the ideal crystal. Taking into account the term $p \cdot V$ the Gibbs energy of the unperturbed crystal can be expressed with the aid of Eq. (2.4.2) in the following way:

$$G_{N_M} \equiv A_{N_M} + pV_{N_M} = N_M \left(\tilde{a}_M + p \frac{V_{N_M}}{N_M} \right). \tag{2.4.8}$$

The term in brackets may be denoted as $\tilde{g}_M$,

$$\tilde{g}_M = \tilde{a}_M + p \frac{V_{N_M}}{N_M}. \tag{2.4.9}$$

Then the Gibbs energy of the crystal G_{N_M} may be written as

$$G_{N_M} = N_M \tilde{g}_M. \tag{2.4.10}$$

Let us now consider a crystal containing N_M lattice molecules and in addition one single defect of type i located on a certain site. For simplicity, the possibility of distributing this defect among the various sites available for it will not be taken into account at the moment. The Helmholtz energy of the "perturbed" crystal will be denoted by $A_{N_M+1_i}$. A relation similar to Eq. (2.4.5) now applies:

$$A_{N_M+1_i} = -kT \ln \sum_1 (d_1' e^{-E_1'/kT})_{N_M+1_i}. \tag{2.4.11}$$

All energy levels E_1' together with the degeneracy factors d_1' corresponding to them are referred to the crystal containing N_M lattice molecules and one defect of type i whereby the prime index used expresses the fact that the addition of one defect to an ideal crystal causes the energy levels and the degeneracy factors to change compared to those of the ideal crystal. The index "$N_M + 1_i$" characterizing the terms of the partition function in Eq. (2.4.11) also emphasizes this point. The difference between the Helmholtz energy of the perturbed crystal and that of the ideal one is solely due to the presence of the defect, i.e. we may introduce the contribution $\tilde{a}_i$ of this single defects to the Helmholtz energy according to the relation

$$A_{N_M+1_i} - A_{N_M} = \tilde{a}_i. \tag{2.4.12}$$

With the aid of Eqs. (2.4.5) and (2.4.11) the following expression holds for the quantity $\tilde{a}_i$

$$\tilde{a}_i = -kT \left[\ln \sum_1 (d_1' e^{-E_1'/kT})_{N_M+1_i} - \ln \sum_1 (d_1 e^{-E_1/kT})_{N_M} \right]. \tag{2.4.13}$$

Rewriting Eq. (2.4.12) leads to

$$A_{N_M+1_i} = N_M \tilde{a}_M + \tilde{a}_i \tag{2.4.14}$$

whereby we have used Eq. (2.4.7) in order to eliminate the term A_{N_M}.

Similar expressions hold for the Gibbs energy of the "perturbed" crystal as will be seen in the following. Starting from the definition of $G_{N_M+1_i}$ according to which we can write

$$G_{N_M+1_i} \equiv A_{N_M+1_i} + pV_{N_M+1_i} \tag{2.4.15}$$

substitution of $A_{N_M+1_i}$ using Eq. (2.4.14) leads to

$$G_{N_M+1_i} = N_M \tilde{a}_M + \tilde{a}_i + pV_{N_M+1_i}. \tag{2.4.16}$$

The volume of the crystal composed of N_M lattice molecules and one defect may be split into two contributions, one due to the lattice molecules and denoted by V_{N_M}, the other due to the defect and denoted by $\tilde{v}_i$. Hence Eq. (2.4.16) may be rearranged to give

$$G_{N_M+1_i} = N_M \left(\tilde{a}_M + p \frac{V_{N_M}}{N_M} \right) + \tilde{a}_i + p\tilde{v}_i \tag{2.4.17}$$

or

$$G_{N_M+1_i} = N_M \tilde{g}_M + \tilde{g}_i. \tag{2.4.18}$$

In Eq. (2.4.18) we have introduced the Gibbs energy of the single defect i.e. $\tilde{g}_i$, which is equal to $\tilde{a}_i + p\tilde{v}_i$ and represents the contribution of this defect to the Gibbs energy of a crystal containing N_M lattice molecules and one defect of type i on a certain lattice site. This means that the configurational entropy has up to now not been taken into account.

Let us now consider the general case in which N_i defects of various types i are present in the crystal as well as N_M lattice molecules, and the defects may have the possibility of distributing themselves among the sites available. The sum of all defects shall fulfil the condition

$$\sum_i N_i \ll N_M, \tag{2.4.19}$$

i.e. we assume the number of all defects to be small compared to the number of lattice molecules. From the microscopic point of view this is equivalent to saying that the defects are separated from each other by a relatively large number of lattice molecules: hence, interaction between defects can be neglected. Except from the configurational contribution to the entropy the amount by which the Helmholtz or the Gibbs energy respectively is varied due to the addition of a single defect is then always the same for the type considered irrespective of the number of other defects present in the crystal. Or to put it in another way, incorporating N_i defects of type i involves a change in the energy and vibrational entropy of the crystal which is N_i times the change caused by the addition of one single defect of this type. Or to put it in other words, the contribution of one defect to the partition function of the crystal results from effects localised at the defect. In order to account for the configurational contribution to the Gibbs energy we have to introduce the term $-\sum kT \ln W_i$ into the expression for the Gibbs energy of the real crystal denoted by $G_{N_M + \sum N_i}$. The above arguments lead to the following equation for $G_{N_M + \sum N_i}$ (see also Eq. (2.4.1))

$$G_{N_M + \sum N_i} = N_M \tilde{g}_M + \sum_i N_i \tilde{g}_i - \sum_i kT \ln W_i. \tag{2.4.20}$$

The configurational contribution $-\sum kT \ln W_i$ may be derived as follows.
Considering a crystal which contains N_M lattice molecules and N_i defects of one type only, these being distributed among the sites in a random way, the following relationship for the Helmholtz energy $A_{N_M + N_i}$ is valid:

$$A_{N_M + N_i} = -kT \ln (W_i Z_{N_M + N_i}). \tag{2.4.21}$$

The term $Z_{N_M + N_i}$ appearing in Eq. (2.4.21) is equal to the partition function of a crystal with N_M lattice molecules and N_i defects, where the configurational contribution has not been taken into account, i.e. $Z_{N_M + N_i}$ corresponds to the fictitious situation that only one single distribution of the N_i defects is allowed to occur. Let W_i be the number of ways of realizing distinct arrangements of the N_i defects among the available sites. Then all terms in the partition function $Z_{N_M + N_i}$ are W_i times degenerate and must therefore be multiplied by W_i and thus $Z_{N_M + N_i}$ as well, to give the partition function of the crystal.

$$A_{N_M + N_i} = -kT \ln Z_{N_M + N_i} - kT \ln W_i. \tag{2.4.22}$$

With the aid of arguments similar to those used above one obtains an analogous relation for the Helmholtz energy of a crystal containing different types of defects. This relation is as follows:

$$A_{N_M + \sum N_i} = -kT \ln \left[\left(\prod_i W_i \right) Z_{N_M + \sum N_i} \right] \tag{2.4.23}$$

$$= -kT \ln Z_{N_M + \sum N_i} - \sum kT \ln W_i. \tag{2.4.24}$$

Since the Gibbs energy equals the sum of the Helmholtz energy and the product of pressure and volume the term $-\sum kT \ln W_i$ appearing in Eq. (2.4.24) for the Helmholtz energy is to be found in the corresponding expression for the Gibbs energy $G_{N_M + \sum N_i}$ as well.

The contribution due to quasi-free electrons and electron defects has not been taken into account here, since we shall calculate their chemical potentials in Chap. 4. We must now calculate the configurational contribution $\sum\limits_{i} kT \ln W_i$ from the number of lattice molecules N_M and the numbers N_i of defects i. We assume that there is one site per lattice molecule available for an i-defect. A generalization of this restriction has basically no influence on the calculations and on the equations obtained from them. As before, the number N_i of defects i shall be small in comparison to the number of lattice molecules N_M, i.e. the mole fraction $x_i = N_i/N_M$ is much smaller than unity. The following equation gives the number of different arrangements W_i for the distribution of N_i defects among the N_M sites.

$$W_i = \frac{N_M(N_M - 1) \cdot (N_M - 2) \cdot \ldots \cdot (N_M - N_i + 1)}{N_i!} \qquad (2.4.25)$$

which, since $N_i \ll N_M$, can be replaced by the approximation

$$W_i \approx \frac{N_M^{N_i}}{N_i!}. \qquad (2.4.26)$$

Equation (2.4.25) results from the following considerations: there are N_M sites available for the first i-particle. For every choice made by the first particle the second particle has $(N_M - 1)$ remaining possibilities, the third particle $(N_M - 2)$ in addition to every one of the $N_M(N_M - 1)$ possibilities for the first and second particle. Thus the total number of possible arrangements is given by the product of the numbers N_M, $(N_M - 1)$, $(N_M - 2)$ etc. to $(N_M - N_i + 1)$. Here, however, we have overcounted the possibilities since we have included all those distributions resulting from the sole exchange of the particles between the sites occupied by them. Dividing by the number of these distributions, which is equal to the number of possible arrangements of N_i particles on N_i sites, i.e. by $N_i! = 1 \cdot 2 \cdot 3 \cdot \ldots \cdot N_i$ leads to the correct result given in Eq. (2.4.25). Using the Stirling formula

$$\ln N_i! \approx N_i(\ln N_i - 1) \qquad (2.4.27)$$

we obtain the following expression for the configurational contribution to the Gibbs energy

$$-\sum\limits_{i} kT \ln W_i = -\sum\limits_{i} kT \ln (N_M^{N_i}/N_i!)$$

$$\approx -\sum\limits_{i} kT N_i(\ln (N_M/N_i) + 1)$$

$$= \sum\limits_{i} kT N_i(\ln x_i - 1). \qquad (2.4.28)$$

Substituting this expression into Eq. (2.4.20) gives the desired equation for the Gibbs energy of the real crystal

$$G_{N_M+\Sigma N_i} = N_M \tilde{g}_M + \sum\limits_{i} N_i \tilde{g}_i + \sum\limits_{i} kT N_i(\ln x_i - 1). \qquad (2.4.29)$$

Equation (2.4.29) states the Gibbs energy of the real crystal as a function of the number of building units present in the lattice, i.e. of the number of lattice molecules and of defects present in the lattice. Partial differentiation with respect to the number of the lattice molecules or defects gives the chemical potentials of the lattice molecules or the electrochemical potentials of charged defects or the chemical potentials of uncharged ones, referred in each case to a single particle. Electrochemical or chemical potentials with respect to a single particle will be denoted by the sign "$\sim$". Thus, according to (2.4.29) the electrochemical potential of the defects i is

$$\tilde{\eta}_i \equiv \left(\frac{\partial G}{\partial N_i} \right)_{N_M, N_{j \neq i}, p, T} = \tilde{g}_i + kT \ln x_i. \qquad (2.4.30)$$

Similarly, the chemical potential of the lattice molecule is given by

$$\tilde{\mu}_M \equiv \left(\frac{\partial G}{\partial N_M}\right)_{N_i, p, T} = \tilde{g}_M - kT \sum_i x_i. \tag{2.4.31}$$

Since the mole fractions x_i are much smaller than unity, the chemical potential $\tilde{\mu}_M$ of the lattice molecules is approximately constant; in fact it slightly depends on the sum of the concentrations of all defects (see Schottky [2.11]). This is for example the reason why a decrease of the vapour pressure and an increase of the freezing point occur by dissolving foreign particles. Since the electrochemical potential $\tilde{\eta}_i$ can be written as $\tilde{\eta}_i = \tilde{\mu}_i + z_i e\varphi$ the chemical potential $\tilde{\mu}_i$ of the defects can be obtained from Eq. (2.4.30):

$$\tilde{\mu}_i = \tilde{g}_i - z_i e\varphi + kT \ln x_i. \tag{2.4.32}$$

Thus the chemical potential of the defects exhibits the same concentration dependence as that shown by particles dissolved in solvents or by ions in electrolytes. For uncharged particles, i.e. $z_i = 0$ electrochemical and chemical potentials are identical.

Introducing a standard state corresponding to the chemical potential μ_i^0, the chemical potential μ_i of the defects i (referred to one mole) can be written as

$$\mu_i = \mu_i^0 + RT \ln \frac{c_i}{c_i^0}, \tag{2.4.33}$$

c_i is the concentration of particles considered, c_i^0 their standard concentration. Since μ_i is referred to one mole, Boltzmann's constant k in Eq. (2.4.32) is replaced in Eq. (2.4.33) by the gas constant R ($= k \cdot N_A$) where N_A is Avogadro's number.

Equations (2.4.30—33) are of course only valid for low defect concentrations, i.e. for $x_i \ll 1$. At higher concentrations these must be replaced by activities, since interaction between defects must be taken into account. In this case the following relation of general validity must be used:

$$\mu_i = \mu_i^0 + RT \ln a_i. \tag{2.4.34}$$

The activity a_i of the solute is usually defined in such a way that the activity is equal to c_i/c_i^0 if c_i approaches zero. It can be seen that the activity thus depends on the choice of the standard concentration c_i^0 and thus on the standard state. According to the definition of the standard state the activity a_i must be equal to unity in this state. An important reason for the deviation of the activity a_i from c_i/c_i^0 i.e. for non-ideal behaviour, are the weak Coulomb interactions. For example due to these electrostatic forces more positive defects (oppositely charged) are to be found in the neighbourhood of a negatively charged defect. This effect is known from the theory of liquid electrolytes and can be treated quantitatively using Debye-Hückel theory. Theoretical relations of this theory can be applied to treat defects and have been discussed for example by Kortüm [2.12] and Kröger [2.13]. They are, however, valid only for low concentrations of defects, i.e. for small deviations from the limiting case of ideal dilute solutions. At higher defect concentrations it may be necessary to consider the formation of defect associates such as $V''_{Me} + h^{\cdot} = V'_{Me}$ or $V''_{Me} + 2h^{\cdot} = V_{Me}$. Here the index Me denotes a Me-site in the partial lattice of the metal Me. We refer the reader to more detailed accounts in the literature [2.14].

Having calculated the chemical potential of defects and of lattice molecules as a function of the defect concentrations, we can now formulate disorder equilibria and thus laws of mass action for defects. This will be done in the following sections.

2.5 Internal Disorder Equilibria

Disorder equilibria in the volume of a crystalline compound involve relationships, written as laws of mass action, between the concentrations of the various defects. They thus provide information about the mutual dependencies of the concentrations of the various defects, as equilibria between ions in liquid electrolytes or equilibria between gases in a homogeneous phase describe the mutual dependence of the concentrations of the various types of ions or gas molecules.

For a solid compound in a state of internal equilibrium, which is generally the case at higher temperatures, the disorder equilibria to be discussed below will all be attained together and will depend on one another. It is, however, often the case that only certain types of defects predominate, so that for a discussion of crystal properties influenced by the disorder it is generally only necessary to consider only a few types of defects and their equilibria in the particular crystalline compound concerned.

The Equilibrium Between Vacancies and Interstitial Particles (*Frenkel Equilibrium*)

We shall consider the process in which a particle A moves from an A lattice site to an interstitial site whereby for example a (with respect to the unperturbed lattice) single positively charged interstitial particle $A_i^{\cdot}$ is formed, a vacancy V_i in the interstitial lattice is destroyed and a negative vacancy V'_A on an A-site is left; i.e. we are investigating the exchange of particles between normal lattice sites and interstitial sites and we wish to determine the equilibrium condition involved (compare Fig. 2.2.2). The exchange process can be written in the form of a chemical reaction, a so-called disorder reaction

$$A_A + V_i = A_i^{\cdot} + V'_A. \tag{2.5.1}$$

Here the reaction is formulated in terms of structure elements, the equilibrium condition for a chemical reaction proceeding in a single phase is always given by the relation $\sum_i \nu_i \mu_i = 0$. As discussed in Sect. 4 of this chapter, in order to calculate the chemical potentials unequivocally we rearrange Eq. (2.5.1) and combine structure elements to building units as follows:

$$(A_i^{\cdot} - V_i) + (V'_A - A_A) = 0. \tag{2.5.2}$$

Using the old Schottky notation this equation reads:

$$A\circ^{\cdot} + A\square' = 0 \tag{2.5.3}$$

and in the new Schottky notation

$$A^{\cdot} + |A|' = 0. \tag{2.5.4}$$

Applying the equilibrium condition

$$\sum_i \nu_i \mu_i = 0 \tag{2.5.5}$$

to Eq. (2.5.3) or (2.5.4) we thus obtain in terms of the new Schottky notation

$$0 = \mu_{A^{\cdot}} + \mu_{|A|'} \tag{2.5.6}$$

or

$$\mu_{A^{\cdot}} = -\mu_{|A|'}. \tag{2.5.7}$$

The chemical potential of an interstitial particle as a building unit is thus equal to the negative chemical potential of a building unit vacancy. Using the expression $\mu_i = \mu_i^0 + RT \ln (c_i/c_i^0)$ Eqs. (2.5.6) and (2.5.7) can be written in the following form in which the concentrations c_i are symbolized by square brackets [i]:

$$[A^{\cdot}] \cdot [|A|'] = [A^{\cdot}]^0 \cdot [|A|']^0 \exp\left[-(\mu_{A^{\cdot}}^0 + \mu_{|A'|}^0)/RT\right] \tag{2.5.8}$$

$$= [A^{\cdot}]^0 [|A|']^0 \exp\left(-\Delta G^0/RT\right) \tag{2.5.9}$$

or

$$[A^{\cdot}] \cdot [|A|'] = \text{const. (p, T)}. \tag{2.5.10}$$

The concentrations of the interstitial particles or vacancies as building units are identical with their concentrations as structure elements. We can thus write Eq. (2.5.10) as follows:

$$[A_i^{\cdot}] \cdot [V_A'] = \text{const. (p, T)}. \tag{2.5.11}$$

Hence we can see that for all crystals and in particular for crystalline compounds there exists at equilibrium a constant product of the concentrations of interstitial particles and the corresponding vacancies, the so-called Frenkel equilibrium depending on temperature and pressure. The equilibrium constant is almost always obtained empirically. The Frenkel equilibrium in crystals corresponds basically to the equilibrium between H^+- and OH^--ions in water, whereby the H^+-ions may be considered as excess particles, i.e. interstitial particles and the OH^--ions as vacancies.

For higher concentrations of defects the equations must be written in terms of activities; the general equation then becomes

$$a_{A^{\cdot}} \cdot a_{|A'|} = \text{const. (p, T)}. \tag{2.5.12}$$

In a binary crystalline compound such an equilibrium exists for both types of atoms or ions. It is of course not only valid for ideal stoichiometry but also for deviations therefrom. In this case the various concentrations of defect sites depend closely on the deviations from ideal stoichiometry or on external parameters, in particular on the chemical potentials of the components in the environment, as we shall show in the next section. For the special case of ideal stoichiometry the disorder is called intrinsic disorder.

In principle several types of disorder equilibria occur together, for example apart from the Frenkel equilibria for the various types of ions, the so-called Schottky equilibria which will be discussed below. If interstitial particles and vacancies of the same type of particle are the predominant defects, i.e. if the concentrations of all other defects are much lower, the concentrations of the interstitial particles and vacancies are equal for negligibly small deviations from the ideal stoichiometry. Disorder of this type is referred to as Frenkel disorder; it was first described by Frenkel [2.15] for AgCl: the silver halides are in fact the classical examples for this type of disorder [2.16]. An important experimental method of determining type and degree of disorder in such compounds is due to Koch and Wagner [2.17]: the concentrations of vacancies or interstitial particles are varied in a quantitative manner by doping i.e. by addition of compounds consisting of anions or cations with other valencies. Low amounts of the doping compounds then determine the concentrations of the vacancies or interstitial particles. Addition of $CdCl_2$ to AgCl for example raises the concentrations of cation vacancies while addition of Ag_2S leads to an increase in the number of interstitial ions. Measurements of the conductivity, in this case the partial ionic conductivity, as a function of the vacancy or interstitial particle concentration permits the determination of the mobilities of the defects. The relationship between concentration, mobility and partial conductivity will be discussed in detail in Chap. 6. Using mobilities determined as described above and ionic conductivities obtained for virtually ideal stoichiometry the degree of disorder can now be calculated, other compounds may be treated in a similar manner.

The Equilibrium Between A-Vacancies and B-Vacancies in a Binary AB Crystal (Schottky Equilibrium)

In this case we consider the exchange of particles between A-vacancies, B-vacancies and the crystal. This means that either particles A and B are transferred to the surface of the crystal, which is thus enlarged, while A- and B-vacancies are generated, or vice versa. For example the A-vacancies may be singly negatively and the B-vacancies singly positively charged. When the particles are brought to the surface, two of the particles present there will be converted to volume particles, i.e. the exchange effectively leads to the generation of a new lattice molecule and of two vacancies. This can be written in terms of structure elements as follows

$$A_A + B_B = V'_A + V^{\cdot}_B + AB \qquad (2.5.13)$$

where AB denotes the lattice molecule. This equation can be rewritten in terms of building units:

$$0 = (V'_A - A_A) + (V^{\cdot}_B - B_B) + AB \qquad (2.5.14)$$

or using the old Schottky notation

$$0 = A\square' + B\square^{\cdot} + AB \qquad (2.5.15)$$

and in the new Schottky notation

$$0 = |A|' + |B|^{\cdot} + AB. \qquad (2.5.16)$$

The equilibrium condition $\sum_i \nu_i \mu_i = 0$, applied to Eqs. (2.5.15) and (2.5.16), is expressed in the new Schottky notation as follows

$$\mu_{|A|'} + \mu_{|B|^{\cdot}} = -\mu_{AB}, \tag{2.5.17}$$

i.e. the sum of the chemical potentials of A- and B-vacancies is equal to the negative chemical potential of the lattice molecule. According to Eq. (2.4.31) the latter is dependent on the sum of the concentrations of all defects and thus at low defect concentration virtually constant.

$$\mu_{AB} = \mu_{AB}^0. \tag{2.5.18}$$

Inserting the relation $\mu_i = \mu_i^0 + RT \ln (c_i/c_i^0)$ for the building unit vacancies appearing in Eq. (2.5.17) gives, with the aid of Eq. (2.5.18) and after rearrangement:

$$[|A|'] \, [|B|^{\cdot}] = [|A|']^0 \, [|B|^{\cdot}]^0 \exp \left(-\frac{\mu_{AB}^0 + \mu_{|A|'}^0 + \mu_{|B|^{\cdot}}^0}{RT} \right) \tag{2.5.19}$$

or

$$[|A|'] \, [|B|^{\cdot}] = \text{const. (p, T)} \tag{2.5.20a}$$

or represented in the notation with structure elements due to Kröger

$$[V_A'] \, [V_B^{\cdot}] = \text{const. (p, T)}. \tag{2.5.20b}$$

Equations (2.5.19) and (2.5.20) thus state that the product of the concentrations of A- and B-vacancies in an AB crystal at a certain temperature and pressure is constant as long as the concentrations of the vacancies are low. For higher concentrations of the defects, Eqs. (2.5.19) and (2.5.20) must be rewritten in terms of activities

$$a_{|A|'} a_{|B|^{\cdot}} = \text{const. (p, T)}. \tag{2.5.21}$$

For a crystal which, in the case of ideal stoichiometry, contains particles A and B in the ratio k:l and lattice molecules represented by the formula $A_k B_l$, the corresponding disorder equation is as follows if we assume all vacancies to be electrically neutral:

$$kA_A + lB_B = kV_A + lV_B + A_k B_l \tag{2.5.22}$$

or after rearrangement:

$$0 = k(V_A - A_A) + l(V_B - B_B) + A_k B_l \tag{2.5.23}$$

and using the new Schottky notation:

$$0 = k \, |A| + l \, |B| + A_k B_l. \tag{2.5.24}$$

The corresponding law of mass action, which may be derived in a similar manner (using the new Schottky notation) states

$$[|A|]^k \, [|B|]^l = \text{const. (p, T)} \tag{2.5.25a}$$

or formulated with Kröger's symbols

$$[V_A]^k \, [V_L]^l = \text{const. (p, T)} \tag{2.5.25b}$$

or in the general case

$$a_{|A|}^k a_{|B|}^l = \text{const.} \, (p, T).$$ (2.5.26)

It must be emphasized that in contrast to the Frenkel equilibrium the lattice molecule is involved in the Schottky equilibrium. However, since its activity is practically constant, the latter can be included in the mass action constant. If the deviation from ideal stoichiometry is negligible and A- and B-vacancies are the predominant defects one speaks of Schottky disorder; this is to be found in the alkali metal halides such as NaCl and KCl [2.16].

The disorder in alkali metal halides is a particularly good example of the fact that in simple cases, especially for pure ionic crystals, it is possible to make quantitative estimates of disorder energies. Investigations of this type have been carried out by Jost [2.18], Schottky [2.19], Mott and Littleton [2.20] and Rittner, Hutner and Dupré [2.21]. This topic has also been reviewed by Barr and Lidiard [2.22] and Süptitz and Teltow [2.23]. The calculations are mainly based on the classical Born model known from the electrochemistry of liquids for the determination of solvation energies of ions.

Other Defect Equilibria in the Solid State

The various Frenkel equilibria and the Schottky equilibrium between vacancies of the various particles are next to electron disorder, the most important disorder equilibria in the volume of crystalline compounds containing no impurities or doping materials. Other types of disorder are in principle possible and the laws of mass action for these may be derived similarly; intermediate steps will be left out in the following discussion.

a) Equilibrium between different, e.g. neutral, interstitial particles; this is called the anti-Schottky-equilibrium.

We shall consider the following disorder reaction, written in terms of structure elements

$$A_i + B_i = AB + V_i + V_i.$$ (2.5.27)

After rearrangement into building units this equation reads:

$$(A_i - V_i) + (B_i - V_i) = AB$$ (2.5.28a)

or using the new Schottky notation

$$A + B = AB.$$ (2.5.28b)

The law of mass action for this reaction in terms of building units (Schottky new) is stated as follows:

$$[A] \, [B] = \text{const.} \, (p, T)$$ (2.5.29a)

or using structure elements

$$[A_i] \, [B_i] = \text{const.} \, (p, T).$$ (2.5.29b)

Compounds with predominant disorder of this type are apparently unknown.

b) Equilibrium between particles A on B-sites and particles B on A-sites (anti-structural equilibrium).

We shall consider the disorder reaction

$$A_A + B_B = B_A + A_B \tag{2.5.30}$$

or in terms of building units

$$0 = (B_A - A_A) + (A_B - B_B) \tag{2.5.31a}$$

or represented in the new Schottky notation

$$0 = B \,|A| + A \,|B|. \tag{2.5.31b}$$

In the case of ideal behaviour the law of mass action for this reaction using the new Schottky notation is as follows:

$$[B \,|A|] \,[A \,|B|] = \text{const. (p, T)} \tag{2.5.32a}$$

or with the Kröger notation:

$$[B_A] \,[A_B] = \text{const. (p, T)}. \tag{2.5.32b}$$

(c Equilibrium between A-vacancies and A-particles on B-sites [2.24].

Using the Kröger notation the disorder reaction considered reads as follows:

$$2A_A + B_B = A_B + AB + 2V_A \tag{2.5.33}$$

or after rearrangement

$$0 = (A_B - B_B) + 2(V_A - A_A) + AB \tag{2.5.34a}$$

and using the new Schottky notation

$$0 = A \,|B| + 2 \,|A| + AB. \tag{2.5.34b}$$

Applying the law of mass action to Eq. (2.5.34b) we have

$$[A \,|B|] \,[|A|]^2 = \text{const. (p, T)} \tag{2.5.35a}$$

or using structure elements

$$[A_B] \,[V_A]^2 = \text{const. (p, T)}. \tag{2.5.35b}$$

In a similar way we obtain the law of mass action corresponding to the equilibrium between A-particles on B-sites and B-particles on interstitial sites:

$$[A \,|B|] \,[B]^2 = \text{const. (p, T)} \tag{2.5.36a}$$

or in terms of structure elements

$$[A_B] \,[B_i]^2 = \text{const. (p, T)} \tag{2.5.36b}$$

Here again compounds in which these types of disorder predominate are apparently unknown.

Equilibrium Between Electrons and Electron Defects in a Semiconductor

In a semiconductor at absolute zero the conduction band is empty while all valence bands are completely occupied; this will be discussed in more detail in Chap. 4. At higher temperatures electrons are transferred from the highest valence band into the conduction band lying above. An equilibrium is attained between electrons in the valence and those in the conduction band. Due to the free exchange of electrons between valence and conduction band, the electrons in these bands have the same electrochemical potential (Fermi energy) and, since the electrical potential is the same for all electrons occupying different bands (homogeneous equilibrium) also the same chemical potential. However, it is not usual in a thermodynamic treatment to speak of electrons in the valence band since this would be tedious for a quantitative account. These are instead characterized by missing electrons, electron defects, in the valence band like building unit vacancies. Electron defects have the significance of negative particles, in this case negative electrons.

The transition of an electron from the valence to the conduction band is expressed in terms of structure elements (though this is not usual) as follows:

$$\text{Electron (valence band)} = \text{electron (conduction band)} \qquad (2.5.37)$$
$$+ \text{ vacancy (valence band)}.$$

The vacancies in the conduction band are neglected here. Equation (2.5.37) can be rearranged to give the following:

$$0 = \text{electron (conduction band)}$$
$$+ \text{[vacancy (valence band)-electron (valence band)]} \qquad (2.5.38)$$

According to common usage an electron in the conduction band will be symbolized as e and the term in square brackets of Eq. (2.5.38), i.e. an unoccupied state in the valence band as h (hole). Thus Eqs. (2.5.38) and (2.5.37) can be written as follows:

$$e' + h^\cdot = 0. \qquad (2.5.39)$$

The prime and point indices used denote the negative and positive charge respectively. Since the charge of an electron or electron defect is unique, we shall in future use no indices at all for e and h. Equation (2.5.39) represents a disorder reaction in terms of building units.

Formulations of this type are generally accepted in the treatment of electron disorder, contrary to the situation for ionic disorder. The significance of chemical or electrochemical potentials for e (free electrons or electrons in the conduction band) and h (electron defects) is completely clear.

If equilibrium has been attained, the following condition applies for reaction (2.5.39)

$$\mu_e + \mu_h = 0 \qquad (2.5.40)$$

or

$$\mu_e = -\mu_h, \qquad (2.5.41)$$

i.e. the chemical potential of the electrons is equal to the negative potential of the electron defects.

For low concentrations (Boltzmann approximation) the chemical potential of the electrons may be expressed as follows

$$\mu_e = \mu_e^0 + RT \ln ([e]/[e]^0) \tag{2.5.42}$$

while that of the electron defects is given by

$$\mu_h = \mu_h^0 + RT \ln ([h]/[h]^0), \tag{2.5.43}$$

as will be shown in Chap. 4. Thus using Eq. (2.5.40) or (2.5.41) we obtain the expression

$$[e] \cdot [h] = \text{const. (p, T)}, \tag{2.5.44}$$

i.e. the product of the concentrations of electrons and of electron defects is constant for a certain semiconductor at a given temperature and pressure.

Electrons in the conduction band may not only originate from the valence band but also from so-called donors i.e. from atomic defects in the crystal which can give off electrons. Similarly electron defects may be generated by acceptors, i.e. by atomic defects in the crystal which can take up electrons. If the free electrons originate only from the valence band, which leads to equal concentrations of electrons and electron defects, we speak of intrinsic disorder.

Equations (2.5.42) and (2.5.44) are valid for low concentrations, smaller than about 10^{18} electrons or electron defects per cm^3 at room temperature, which depend on the so-called effective masses. For higher concentrations the latter must be replaced by activities so that Eq. (2.5.44) becomes

$$a_e \cdot a_h = \text{const. (p, T)} \tag{2.5.45}$$

Ionisation Equilibria

Atomic defects may be either neutral, positively charged or negatively charged with respect to the unperturbed lattice. Transitions from one charge state to another may involve electron exchange with the conduction or valence band. Using structure elements due to Kröger such a transition is written as follows in the case of an interstitial particle

$$A_i = A_i^{\cdot} + e \tag{2.5.46a}$$

or using building elements in the new Schottky notation

$$A = A^{\cdot} + e. \tag{2.5.46b}$$

In this notation the equilibrium condition reads

$$\frac{[A]}{[A^{\cdot}] [e]} = \text{const. (p, T)} \tag{2.5.47a}$$

while in terms of the Kröger notation it is

$$\frac{[A_i]}{[A_i^{\cdot}] [e]} = \text{const. (p, T)}. \tag{2.5.47b}$$

Similarly the following equations describe the exchange with the valence band; using the Kröger notation

$$B_i = B'_i + h \tag{2.5.48a}$$

and with the new Schottky notation

$$B = B' + h. \tag{2.5.48b}$$

The equilibrium condition is then written as follows:

$$\frac{[B]}{[B'][h]} = \text{const. (p, T)} \tag{2.5.49a}$$

or

$$\frac{[B_i]}{[B'_i][h]} = \text{const. (p, T).} \tag{2.5.49b}$$

The reaction described by Eq. (2.5.46) is an example of a donor reaction, since an electron from a donor state is transferred to the conduction band; the reaction expressed by Eq. (2.5.48) is an example of an acceptor reaction.

Equilibria Between Simple and Complex Vacancies

If defects are present in the immediate vicinity of one another, this is accompanied by powerful energetic effects, so that it is reasonable to speak of new complex defects which consist of two or more simple defects. For example A- and B-vacancies may come close to each other; if we assume that the A-vacancies are singly negatively and the B-vacancies singly positively charged we can write (using the Kröger notation)

$$V'_A + V^._B = \{V_A V_B\} \tag{2.5.50a}$$

or using the new Schottky notation

$$|A|' + |B|^. = \{|A| \, |B|\}. \tag{2.5.50b}$$

The corresponding law of mass action in the Schottky or Kröger notation respectively is as follows:

$$\frac{[|A|'] [|B|^.]}{[\{|A| \, |B|\}]} = \text{const. (p, T)} \tag{2.5.51a}$$

$$\frac{[V'_A] [V^._B]}{[\{V_A V_B\}]} = \text{const. (p, T).} \tag{2.5.51b}$$

In this case the complex defect is neutral, since the charges of the simple defects are equal and opposite. However, all conceivable complex defects may be treated in a similar manner.

2.6 Kinetic Derivation of Disorder Equilibria

The thermodynamic treatment of disorder equilibria given in Sect. 2.5 emphasizes the great importance of electrochemical and chemical potentials of defects or lattice molecules. It is, however, also possible to derive laws of mass action on the basis of kinetic arguments by considering a dynamic equilibrium for which the condition applies that the rates of the forward and backward reactions are equal. Structure elements are particularly important for an approach of this kind. To clarify this point we shall discuss the Frenkel equilibrium for AgCl. In terms of structure elements this disorder reaction can be written as follows:

$$Ag_{Ag} + V_i \underset{k_2}{\overset{k_1}{\rightleftarrows}} V'_{Ag} + Ag_i^{\cdot}, \tag{2.6.1}$$

k_1 is the rate constant for the transition of silver ions from regular lattice sites to interstitial sites. For the rate of the forward reaction, i.e. v(for) the following relation holds:

$$v(for) = k_1[Ag_{Ag}][V_i] \tag{2.6.2}$$

since this rate is proportional to the concentration of silver ions on silver sites and proportional to the concentration of vacancies present in the interstitial lattice. The rate of the backward reaction i.e. v(back) with the rate constant k_2 can be written as

$$v(back) = k_2[V'_{Ag}] \cdot [Ag_i^{\cdot}]. \tag{2.6.3}$$

The rate of the backward reaction is similarly proportional to the concentration of silver ion vacancies and to that of the silver ions present in the interstitial lattice. Since we always assume that the number of defects is small in comparison with the number of lattice molecules, the concentration of silver ions on silver sites is virtually constant

$$[Ag_{Ag}] = const._1 (p, T). \tag{2.6.4}$$

The same applies to the vacancies in the interstitial lattice:

$$[V_i] = const._2 (p, T). \tag{2.6.5}$$

At equilibrium the rates of the forward and backward reactions are equal

$$v(for) = v(back). \tag{2.6.6}$$

Substituting Eqs. (2.6.4) and (2.6.5) into Eq. (2.6.2) for the forward reaction leads to the relation: v(for) = const. (p, T); because of the condition stated in Eq. (2.6.6) the same applies to the backward reaction i.e. both rates are independent of the concentrations of silver ion vacancies and of silver ions on interstitial sites. It therefore follows from Eq. (2.6.3) that

$$[Ag_i^{\cdot}][V'_{Ag}] = const. (p, T). \tag{2.6.7}$$

Thus the kinetic treatment leads to the same result obtained in Sect. 2.5 on the basis of thermodynamic arguments. However, the kinetic derivation of disorder equilibria is not always so simple and straightforward as in the example discussed above. Thus in general the thermodynamic treatment of equilibria and in particular of disorder equilibria is preferable. This is particularly true for heterogeneous equilibria e.g. equilibria between the volume of a crystal and a surrounding gas phase: in such cases one cannot easily formulate kinetic equilibria. One must assume that a dynamic equilibrium is attained between the gas phase and the surface and also between the surface and the volume of the crystal. Thus the complex equilibrium of the complete system can only be discussed if the various simple equilibria have first been described. This complication does not arise if we apply a thermodynamic treatment; in this case we immediately obtain clear results. The thermodynamic approach, however, has the disadvantage of yielding no detailed informations about single kinetic steps. As most remarkable result concerning the kinetic derivation of disorder equilibria it must be emphasized that structure elements have to be used for a quantitative account, and thus the introduction of structure elements has found its justification.

2.7 Disorder Equilibria Involving Neighbouring Phases

In this section we shall consider the dependencies of the concentrations of various disorder sites on the composition of the environment. This will generally be a gas phase in which one or several partial pressures of the components of the compound are fixed or varied. We shall for example treat the dependence of the concentrations of interstitial ions, vacancies, electrons or electron defects in an oxide or sulfide on the oxygen or sulfur partial pressure in the surrounding atmosphere when equilibrium between this atmosphere and the solid compound has been attained; at higher temperatures this generally occurs within a short period of time. We shall consider a metal/nonmetal compound whose formula is $MeX_{1+\delta}$, where Me denotes the metal, X the nonmetal e.g. oxygen or sulfur and δ the deviation from ideal stoichiometry. The component X has a certain partial pressure and thus a certain chemical potential in the surrounding gas phase. Equilibrium with respect to the exchange of the component X between the gas phase and the solid compound is attained if the change in the Gibbs energy ΔG for a transition of X from the gas phase into the solid compound at a certain temperature and pressure is equal to zero. The transition of X-particles from the gas phase into the solid compound may be regarded as a chemical reaction:

$$X(g) = X(s) \tag{2.7.1}$$

where g refers to the gas phase and s to the solid one. Application of the equilibrium condition $(\Delta G)_{T,p} = \sum_i \nu_i \mu_i = 0$ leads to the following relation

$$\mu_X(g) = \mu_X(s). \tag{2.7.2}$$

This statement is valid for all heterogeneous equilibria. Thus we are only faced with the problem of expressing the concentrations of the various defects as functions of the chemical potential of the component X in the solid.

If the chemical potential of one of the components, e.g. X, is fixed, the binary solid compound $MeX_{1+\delta}$ is, at a certain pressure p and temperature T, completely defined thermodynamically since according to the Gibbs phase rule there are no further degrees of freedom. It would also have been possible to define the state of the compound by fixing the chemical potential of the metal $\mu_{Me}(s)$ which is related to $\mu_X(s)$ by the Gibbs-Duhem equation. In the case of ternary compounds, e.g. spinels such as $NiCr_2O_4$, a further thermodynamic degree of freedom is present; it is then necessary to state the values of two chemical potentials in addition to temperature and pressure in order to define the compound uniquely.

While in general the chemical potentials in the compound or the equilibrium partial pressures of the components may vary over a wide range, the stoichiometric composition of a crystalline compound normally varies thereby but little. It is, however, reasonable and correct to consider a solid crystalline compound from the thermodynamic point of view as a solution. The large changes in the equilibrium partial pressures of the components are accompanied by relatively large changes in the concentrations of the defects, whereby small changes in the stoichiometric composition occur. Hence a solid compound may be characterized uniquely by the precise stoichiometric composition or by the statement of the deviation from the ideal stoichiometric composition as well. Whether one prefers to state the chemical potentials or equilibrium partial pressures of one or more components or the deviation from ideal stoichiometry depends on the circumstances.

Molecules or atoms migrating from the gas phase into the solid compound can react in various ways: the particles of type X may occupy a site in the interstitial lattice, they may occupy an X-site and thus destroy a vacancy in the X-sublattice or they may form together with a metal particle a new lattice molecule under formation of a metal vacancy or consumption of metal interstitial particles. In order to simplify the basic considerations involved we shall first neglect electron disorder and consider only neutral defects. According to the four possibilities formulated above, we can write the following four reactions or so-called incorporation equations:

a) X-atoms occupy interstitial sites in the compound MeX. In terms of structure elements the following equation applies:

$$X(g) + V_i = X_i. \tag{2.7.3}$$

b) X-atoms occupy regular X-sites thereby destroying vacancies:

$$X(g) + V_X = X_X. \tag{2.7.4}$$

c) X-atoms form together with a metal particle a new MeX lattice molecule under formation of a metal vacancy:

$$X(g) + Me_{Me} = MeX + V_{Me}, \tag{2.7.5a}$$

d) or under consumption of a metal interstitial particle

$$X(g) + Me_i = MeX + V_i. \tag{2.7.5b}$$

After rearranging these equations in order to formulate them in terms of building units we obtain the following expressions for the equilibrium conditions written with the corresponding chemical potentials:

$$\mu_X(g) = \mu_{(X_i-V_i)}(s),\tag{2.7.6}$$

$$\mu_X(g) = -\mu_{(V_X-X_X)}(s),\tag{2.7.7}$$

$$\mu_X(g) = \mu_{MeX} + \mu_{(V_{Me}-Me_{Me})},\tag{2.7.8a}$$

$$\mu_X(g) = \mu_{MeX} - \mu_{(Me_i-V_i)}.\tag{2.7.8b}$$

The situation can be summarized as follows

$$\mu_X(g) = \mu_X(s) = \mu_{(X_i-V_i)}(s)\tag{2.7.9a}$$

$$= -\mu_{(V_X-X_X)}(s)\tag{2.7.9b}$$

$$= \mu_{MeX} + \mu_{(V_{Me}-Me_{Me})}\tag{2.7.9c}$$

$$= \mu_{MeX} - \mu_{(Me_i-V_i)}.\tag{2.7.9d}$$

With respect to the situation considered these equations are of course only valid for neutral defects. If the defects are charged and if electrons or electron defects take part in the disorder reaction, the equilibrium conditions change accordingly. During our discussion of internal equilibria we had already established the equality of the chemical potential of interstitial particles with the negative potential of vacancies and with that of a corresponding combination of lattice molecules and vacancies or interstitials of the other type of particles. Since all these defects are in equilibrium with one another, it makes no difference which type of transition of particles from the gas phase into the solid is considered in order to calculate the chemical potentials. Equations (2.7.6—8) yield the corresponding laws of mass action which state how the concentrations of the disorder sites dealt with depend on the activity a_X of the component X in the gas phase:

$$[X_i] = K_1 a_X,\tag{2.7.10}$$

$$[V_X] = K_2/a_X,\tag{2.7.11}$$

$$[V_{Me}] = K_3 a_X,\tag{2.7.12}$$

$$[Me_i] = K_4/a_X.\tag{2.7.13}$$

Equations (2.7.10—13) are of course only valid for neutral defects.

Examples of equilibria involving charged defects, electrons and electron defects will be discussed below. We shall also discuss changes in stoichiometry resulting from changes in the concentrations of defects. Since in general one or two disorder sites predominate with respect to their concentrations only these defects are important for the discussion of the properties of the crystal which are affected by crystal disorder.

In order to demonstrate this let us consider the compound $MeX_{1+\delta}$ the predominant defect sites being single positively charged interstitial metal ions $Me_i^{\cdot}$, single negatively charged metal ion vacancies V'_{Me}, free electrons e and electron defects h. This means that at the stoichiometric point Frenkel disorder is present in the Me-sublattice and that free electrons and electron defects occur. We shall assume here that the concentrations of interstitial particles and vacancies are much larger than those of electrons and electron defects.

We wish to determine changes in the concentrations of the various defects if the partial pressure of component X, present in the gas phase in the form of X_2-molecules, varies. The transition of component X from the gas phase into the solid may be expressed by the equation

$$\frac{1}{2} X_2(g) + Me_{Me} \rightleftarrows MeX + V'_{Me} + h, \qquad (2.7.14)$$

i.e. an X-atom may be incorporated into the compound, whereby a new lattice molecule MeX, a metal ion vacancy and an electron defect are generated. Another possible incorporation equation involves the annihilation of metal ions on interstitial lattice sites and of free electrons accompanied by the generation of new lattice molecules MeX:

$$\frac{1}{2} X_2(g) + Me_i^{\cdot} + e \rightleftarrows MeX + V_i. \qquad (2.7.15)$$

The concentrations of the various defects must obey the condition of electroneutrality:

$$[e] + [V'_{Me}] = [h] + [Me_i^{\cdot}]. \qquad (2.7.16)$$

Since in the vicinity of the stoichiometric point the concentrations of metal ions on interstitial sites and of metal ion vacancies are much larger than the concentrations of electrons and electron defects, the relative changes of the

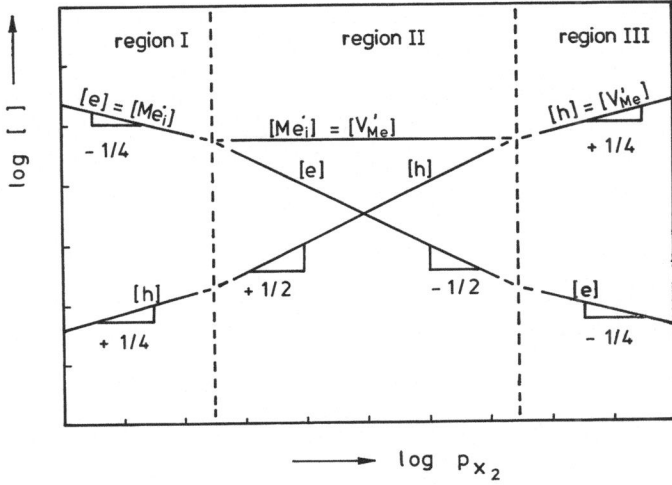

Fig. 2.7.1. Schematic diagram showing the equilibrium concentrations of defects in a compound $MeX_{1+\delta}$ with predominant Frenkel disorder as a function of the partial pressure of X_2 above the compound

concentrations of electrons and electron defects in this range (characterized in Fig. 2.7.1 as range II) are much larger than those of the interstitial ions and vacancies. Hence the concentrations of metal ion vacancies and metal ions in the interstitial lattice may be considered as being virtually constant:

$$[V'_{Me}] \cong \text{const.}, \tag{2.7.17}$$

$$[Me_i^{\cdot}] \cong \text{const.}.$$

Applying the law of mass action to Eq. (2.7.14) thereby noting the validity of Eq. (2.7.17) we obtain the relation

$$[h] \sim p_{X_2(g)}^{1/2}, \tag{2.7.18}$$

i.e. the concentration of electron defects is proportional to the square root of the partial pressure of the X_2-molecules. In a similar manner we may proceed to obtain an expression for the concentration of the electrons in this range starting from the incorporation Eq. (2.7.15):

$$[e] \sim p_{X_2}^{-1/2}. \tag{2.7.19}$$

The situation is different for large deviations from ideal stoichiometry. If the partial pressure of X_2 is very large the concentrations of electrons and metal ions on interstitial sites may be neglected. The electroneutrality condition (2.7.16) then reduces to:

$$[V'_{Me}] \cong [h]. \tag{2.7.20}$$

Applying the law of mass action to Eq. (2.7.14) and noting the relation above one obtains

$$[h] \sim p_{X_2}^{1/4}, \tag{2.7.21}$$

i.e. the concentration of electron defects, now equal to the concentration of metal ion vacancies, is proportional to the fourth root of the partial pressure of the X_2-molecules. It may similarly be shown that for very low X_2-partial pressures the concentration of free electrons which is identical with the concentration of metal ions on interstitial sites, is given by

$$[e] \sim p_{X_2}^{-1/4}. \tag{2.7.22}$$

These results are shown schematically in Fig. 2.7.1. Diagrams of this type are very useful; further examples can be found in the book of Kröger [2.25].

2.8 Disorder Equilibria Involving Surfaces

Many problems of interest to the physical chemist are connected with the adsorption of atoms, molecules or ions on the surface of solids. Two such topics are heterogeneous catalysis and the kinetics of evaporation and condensation. Complete thermodynamic equilibrium means that the particles adsorbed are in equilibrium with the surrounding gas phase and with the interior of the solid, i.e. with the various disorder sites. As an instructive example we shall discuss an ionisorption equilibrium involving the adsorption of O^--ions on the surface of a semiconducting metal oxide e.g. ZnO. We wish to determine the dependence of the concentration of adsorbed O^--ions on the oxygen partial pressure and on the electron concentration in the volume of the semiconductor. We consider the following reaction

$$\frac{1}{2} O_2(g) + e(se) = O^-(\sigma).$$
(2.8.1)

Particles in the gas phase have been denoted by (g), particles in the volume of the semiconductor by (se) and those on the surface by (σ). The equilibrium condition for this reaction is

$$\frac{1}{2} \mu_{O_2}(g) + \eta_e(se) = \eta_{O^-}(\sigma)$$
(2.8.2)

In Eq. (2.8.2) we have introduced electrochemical potentials for the electrons and for the O^--ions, because these particles are charged and occupy different positions in space, so that one must take into account that eventually different electrical potentials are to be attributed to them. Since arguments similar to those holding for defects in the volume of a crystalline compound may be applied the following equation for the chemical potential of adsorbed O^--ions is valid

$$\mu_{O^-}(\sigma) = \mu_{O^-}^0(\sigma) + RT \ln \Gamma,$$
(2.8.3)

Γ denotes the ratio of the number of adsorbed ions per unit area to the number of unit cells contained within that area. On condition that Γ is much smaller than unity, Eq. (2.8.3) is valid. If this condition is not fulfilled Γ must be replaced by the activity of the particles on the surface. Incorporating the expressions for the chemical or electrochemical potentials of the species involved into Eq. (2.8.2) we obtain

$$\frac{1}{2} \mu_{O_2}^0(g) + \frac{1}{2} RT \ln \frac{p_{O_2}}{p_{O_2}^0} + \mu_e^0(se) + RT \ln \frac{[e(se)]}{[e(se)]^0} - F\varphi(se)$$
$$= \mu_O^0-(\sigma) + RT \ln \Gamma - F\varphi(\sigma)$$
(2.8.4)

or after rearrangement

$$\Gamma = Kp_{O_2}^{1/2}[e(se)] \exp \{F[\varphi(\sigma) - \varphi(se)]/RT\}.$$
(2.8.5)

Quantities, which are constant, have been combined to give the factor K. Equation (2.8.5) states that the concentration of O^--ions adsorbed on a metal oxide surface is proportional to the square root of the oxygen partial pressure in the surrounding gas atmosphere, to the concentration of electrons in the volume of the semiconductor and to an exponential expression which contains the difference of the electrical potentials between the volume of the semiconductor and its surface. The concentration of the electrons in the volume of the semiconductor depends of course on the oxygen partial pressure in the surrounding atmosphere as well.

The exponential term may be calculated using the Poisson equation and the equations for the electrochemical potential of the electrons together with the corresponding boundary conditions, as has been shown in competent treatments of surface and semiconductor physics [2.26].

Equation (2.8.5) could have been obtained by considering first the equilibrium of the electrons between the volume of the semiconductor and the surface and then the equilibrium between electrons at the surface, the O^--ions in the adsorption layer and oxygen in the gas phase. It would then have become evident that the exponential expression is to be found in the relation for the concentration of the electrons at the surface.

Since it is often assumed in heterogeneous catalysis that charged particles are important as intermediates or that a take-up of electrons (acceptor reactions) or a release of electrons (donor reactions) by adsorbed particles are important rate-determining steps, the significance of adsorption equilibria becomes clear. For the detailed treatment of chemisorption and heterogeneous catalysis the reader is referred to the literature [2.26].

2.9 References

[2.1]   Bueren, H. G. van: Imperfections in Crystals. Amsterdam: North-Holland Publ. Comp. 1961
Dekker, A. J.: Solid-State Physics. London: Macmillan 1952
Eyring, H., Henderson, D., Jost, W. (ed.): Physical Chemistry, An Advanced Treatise, Vol. X, New York: Academic Press 1970
Gool, W. van: Principles of Defect Chemistry of Crystalline Solids. New York, London: Academic Press 1966
Gray, T. J.: The Defect Solid State, New York: Interscience Publ. Inc. 1957
Hauffe, K.: Reaktionen in und an festen Stoffen, 2nd edit. Berlin, Heidelberg, New York: Springer 1966
Hedvall, J. A.: Einführung in die Festkörperchemie, Braunschweig: Vieweg 1952
Joffé, A. F.: Physik der Halbleiter. Berlin: Akademie Verlag 1960
Kittel, C.: Introduction to Solid-State Physics, 5th edit. New York: John Wiley and Sons, Inc. 1976
Kröger, F. A.: The Chemistry of Imperfect Crystals, 2nd edit. Amsterdam: North-Holland Publ. Comp. 1974
Mott, N. F., Gurney, R. W.: Electronic Processes in Ionic Crystals. Dover: Publ. Inc. New York 1964
Reiss, H. (ed.): Progress in Solid State Chemistry. Oxford, London: Pergamon Press 1964
Spenke, E.: Electronic Semiconductors. New York: McGraw-Hill Book Company 1958
Stasiw, O.: Elektronen- und Ionenprozesse in Ionenkristallen. Berlin, Göttingen, Heidelberg: Springer 1959
Greenwood, N.: Ionic Crystals, Lattice Defects and Nonstoichiometry. London: Butterworth 1968
Schmalzried, H.: Solid State Reactions, 2nd edit. Weinheim: Verlag Chemie 1981
Kofstad, P.: Nonstoichiometry, Diffusion and Electrical Conductivity in Binary Metal Oxides. New York, London, Sydney, Toronto: Wiley 1972
Jarzbski, Z. M.: Oxide Semiconductors. Oxford, New York, Toronto, Sydney, Braunschweig: Pergamon 1973
Mrowec, S.: Defects and Diffusion in Solids. Amsterdam, Oxford, New York: Elsevier 1980
[2.2]   Kröger, F. A.: The Chemistry of Imperfect Crystals, 2nd edit., Chap. 7, 8, p. 207. Amsterdam: North-Holland Publ. Comp. 1974
Kröger, F. A., Stieltjes, F. H., Vink, H. J.: Philips Res. Repts. *14*, 557 (1959)
Kröger, F. A., Vink, H. J.: Solid-State Physics, Vol. 3, Seitz, F., Turnbull, D., (eds.), pp. 307—435. New York: Academic Press 1956
[2.3]   Schottky, W.: Halbleiterprobleme. Schottky, W., (ed.), Vol. IV, p. 235. Braunschweig: Vieweg 1958
[2.4]   Hauffe, K.: Reaktionen in und an festen Stoffen, 2nd edit. Berlin, Heidelberg, New York: Springer 1966
[2.5]   Schottky, W.: Halbleiterprobleme. Schottky, W., (ed.), Vol. IV, p. 235. Braunschweig: Vieweg 1958
[2.6]   Hauffe, K.: Reaktionen in und an festen Stoffen, 2nd edit. Berlin, Heidelberg, New York: Springer 1966
[2.7]   Frenkel, J.: Z. Phys. *35*, 652 (1926)

[2.8]   Wagner, C., Schottky, W.: Z. phys. Chem. *B 11*, 163 (1930), see also the following
        review articles and books respectively
        Gool, W. van: Principles of Defect Chemistry of Crystalline Solids. New York,
        London: Academic Press 1966
        Hauffe, K.: Reaktionen in und an festen Stoffen, 2nd edit., Berlin, Heidelberg,
        New York: Springer 1966
        Kröger, F. A.: The Chemistry of Imperfect Crystals, 2nd edit. Amsterdam: North-
        Holland Publ. Comp. 1974
        Kröger, F. A.: Physical Chemistry, An Advanced Treatise. Eyring, H., Henderson,
        D., Jost, W., (eds.), Vol. X, p. 229. New York, London: Academic Press 1970
        Schottky, W.: Halbleiterprobleme. Sauter, F., (ed.), Vol. I, p. 139. Braunschweig:
        Vieweg 1954
        Schottky, W.: Halbleiterprobleme. Schottky, W., (ed.), Vol. IV, p. 235. Braun-
        schweig: Vieweg 1958
[2.9]   Schottky, W.: Halbleiterprobleme. Sauter, F., (ed.), Vol. I, p. 139. Braunschweig:
        Vieweg 1954
        see also:
        Schottky, W.: Halbleiterprobleme. Schottky, W., (ed.), Vol. IV, p. 235. Braun-
        schweig: Vieweg 1958
[2.10] Schottky, W.: Z. phys. Chem. *B29*, 335 (1935)
        Wagner, C.: Z. phys. Chem. Bodenstein-Festband *1931*, 177
        Wagner, C.: Z. phys. Chem. *B22*, 181 (1933)
        Wagner, C., Schottky, W.: Z. phys. Chem. *B11*, 163 (1930)
[2.11] Schottky, W.: Halbleiterprobleme. Schottky, W., (ed.), Vol. IV, pp. 235, 285.
        Braunschweig: Vieweg 1958
[2.12] Kortüm, G.: Treatise on Electrochemistry. Amsterdam: Elsevier Publ. Comp. 1951
[2.13] Kröger, F. A.: Physical Chemistry, An Advanced Treatise. Eyring, H., Henderson,
        D., Jost, W., (eds.), Vol. X, p. 229. New York, London: Academic Press 1970
[2.14] Lidiard, A. B.: Rep. Conf. on Defects in Crystalline Solids, p. 283. London: The
        Physical Society 1955
        Lidiard, A. B.: Handb. Phys. *20*, 246, 298 (1957)
        Pick, H.: Springer Tracts in Modern Physics *38*, 1 (1965)
        Teltow, J.: Ann. Phys. (6) *5*, 63, 71 (1949)
        Teltow, J.: Halbleiterprobleme. Schottky, W., (ed.), Vol. III, p. 26. Braunschweig:
        Vieweg 1956
        Wagner, C., Hammen, H.: Z. phys. Chem. *B40*, 137 (1938)
[2.15] Frenkel, J.: Z. Physik *35*, 652 (1926)
[2.16] Barr, L. W., Lidiard, A. B.: Physical Chemistry, An Advanced Treatise, Eyring,
        H., Henderson, D., Jost, W., (eds.), Vol. X. New York, London: Academic Press
        1970
        Süptitz, P., Teltow, J.: Phys. Stat. Sol. *23*, 9 (1967)
[2.17] Koch, E., Wagner, C.: Z. phys. Chem. *B38*, 295 (1937)
[2.18] Jost, W.: J. phys. Chem. *1*, 466 (1933)
[2.19] Schottky, W.: Z. phys. Chem. *B29*, 335 (1935)
[2.20] Mott, N. F., Littleton, H. J.: Trans. Faraday Soc. *34*, 485 (1938)
[2.21] Rittner, E. S., Hutner, K. A., Du Pré, K. F.: J. Chem. Phys. *17*, 198, 204 (1949),
        18, 379 (1950)
[2.22] Barr, L. W., Lidiard, A. B.: Physical Chemistry, An Advanced Treatise, Vol. X,
        p. 152. New York: Academic Press 1970
[2.23] Süptitz, P., Teltow, J.: Phys. Stat. Sol. *23*, 9 (1967)
[2.24] Kröger, F. A.: Physical Chemistry, An Advanced Treatise, Eyring, H., Henderson,
        D., Jost, W., (eds.), Vol. X, p. 229. New York, London: Academic Press 1970
[2.25] Kröger, F. A.: The Chemistry of Imperfect Crystals. 2nd edit. Amsterdam: North-
        Holland Publ. Comp. 1974
        Kröger, F. A.: Physical Chemistry, An Advanced Treatise. Eyring, H., Henderson,
        D., Jost, W., (eds.), Vol. X, p. 229. New York, London: Academic Press 1970

[2.26] Doehlemann, E.: Z. Elektrochem. *44*, 180 (1938)

Engell, H. J.: Halbleiterprobleme. Sauter, F., (ed.), Vol. I, p. 249. Braunschweig: Vieweg 1954

Hauffe, K.: Reaktionen in und an festen Stoffen, 2nd edit., especially Chap. 4. Berlin, Heidelberg, New York: Springer 1966

Hauffe, K., Schottky, W.: Halbleiterprobleme. Schottky, W., (ed.), Vol. V, p. 203. Braunschweig: Vieweg 1960. (For further references see especially Sect. 3.1)

Schwab, G. M., (ed.): Handbuch der Katalyse, Vol. 1 (1941), Vol. 2 (1940), Vol. 3 (1941), Vol. 4 (1943), Vol. 5 (1957), Vol. 6 and 7 (1943). Berlin, Göttingen, Heidelberg: Springer

Wagner, C.: J. Chem. Phys. *18*, 69 (1950)

Wagner, C., Hauffe, K.: Z. Elektrochem. *44*, 172 (1938), *45*, 409 (1939)

Wolkenstein, T.: Elektronentheorie der Katalyse an Halbleitern. Berlin: VEB Deutscher Verlag der Wissenschaften 1964

3 Examples of Disorder in Solids

In this chapter we shall discuss a few typical examples of disorder in solids. We have not attempted to give a complete review, since a number of other authors have already given good account of these topics [3.1].

As already mentioned in the previous chapter, the classical examples for compounds with a Schottky-type disorder are alkali halides, whereas those of the Frenkel-type are silver halides such as silver chloride and silver bromide. An example of the Frenkel disorder involving anions, i.e. nearly equal concentrations of anions on interstitial sites and anion vacancies, is CaF_2 which is also an important solid electrolyte used in galvanic cells. Important examples of the occurrence of defect associates, especially of complex formation between anion and cation vacancies, are the so-called F'-centres [3.2] present in alkali halides below certain temperatures, and the defects in solid lead chloride which have been examined by Simkovich [3.3].

At higher temperatures one finds in some compounds many more lattice sites than ions present to occupy them. Here the ions are more or less statistically distributed among the available sites. In this case we refer to structural disorder, and one can speak of a quasi-molten partial lattice; this is found in silver iodide at temperatures above 149 °C, in compounds of the type Ag_4RbI_5 even at room temperature.

This type of disorder will be discussed in more detail at the end of this chapter. These compounds are often important solid electrolytes. Some of them will be treated in more detail in Chap. 7.

We shall now demonstrate by a few examples how the concentrations of defects in the solid considered depend on the partial pressure of one component in the gas phase when equilibrium conditions apply. The first examples to be discussed are doped zirconium dioxide and doped thorium dioxide, these being solid electrolytes, which have become particularly important because of their prevailing oxygen ion conduction.

As examples for compounds with predominant electronic conduction we shall consider zinc oxide, copper (I) oxide and copper (II) oxide. For all these compounds the activities of the defects can to a first approximation be replaced by concentrations, i.e. from the thermodynamic point of view the defects behave in an ideal manner. We shall treat deviations from the ideal behavior in the case of electrons in Chaps. 4 and 5.

As a qualitative rule it is found that an excess of electron defects, and therefore electron defect conduction, occurs particularly in oxides, sulfides and halides where the cations show a tendency to increase their valency. Examples are FeO, CuO, NiO, ZnO, Cu_2S, Cu_2O and CuI. In the other case the electronic conduction exhibited is mainly due to free electrons, for example in ZnO, CdO, Ag_2S, Ag_2Se. Readers should consult Hauffe's book [3.4] for references to original publications in which these compounds are discussed.

3.1 Disorder in Doped Zirconium Dioxide and Thorium Dioxide

In this section we shall discuss how the concentrations of the most important disorder centres in doped zirconium dioxide and doped thorium dioxide depend on the partial pressure of oxygen in the surrounding gas phase. Zirconium dioxide and thorium dioxide, doped with about ten per cent of the oxides CaO, MgO or Y_2O_3, have an ordered cation lattice; however, there exist oxygen ion vacancies in the anion lattice: these may be considered as double positively charged compared to the unperturbed lattice.

Doping with CaO, MgO or Y_2O_3 results in an increase of the ratio between cations and anions. In principle, either additional cations may be incorporated into interstitial sites or oxygen ion vacancies may occur. The comparison between the values for the lattice constant obtained by X-ray methods with the experimentally measured pycnometric density allowed the conclusion that in this particular case the disorder model involving oxygen ion vacancies is the correct one. [3.5].

Variations in the concentration of oxygen ion vacancies due to variations in the oxygen partial pressure may be neglected in comparison with the vacancy concentration caused by doping. In this case the concentration $[V_O^{\cdot\cdot}]$ of oxygen ion vacancies may therefore be considered as being approximately constant over a wide range of oxygen partial pressures.

$$[V_O^{\cdot\cdot}] \cong \text{const.} \tag{3.1.1}$$

Apart from oxygen ion vacancies, quasi-free electrons and electron defects are the next most important disorder centres. At the stoichiometric point the concentrations of quasi-free electrons and electron defects are equal. This point corresponds to a certain oxygen partial pressure, being dependent on temperature. At lower oxygen partial pressures quasi-free electrons are in excess, at higher oxygen partial pressures electron defects. The dependence of the concentration of electrons on the partial pressure may be obtained with the aid of the following equation which describes the incorporation of oxygen into the lattice:

$$\frac{1}{2} O_2(g) + V_O^{\cdot\cdot} + 2e = O_O. \tag{3.1.2}$$

The law of mass action for this reaction is given by:

$$p_{O_2}^{1/2}[V_O^{\cdot\cdot}][e]^2 = K_1, \tag{3.1.3}$$

p_{O_2} denotes the partial pressure of oxygen for which equilibrium with zirconium dioxide is attained. Since the concentration of oxygen ion vacancies is constant, the following relation applies:

$$[e] \sim p_{O_2}^{-1/4}, \tag{3.1.4}$$

i.e. the concentration of quasi-free electrons is proportional to the inverse fourth root of the partial pressure of oxygen with which zirconium dioxide is in equilibrium.

The concentration of electron defects may be obtained from the following equation

$$\frac{1}{2} O_2(g) + V_O^{\cdot\cdot} = O_O + 2h. \tag{3.1.5}$$

The corresponding law of mass action is given by

$$p_{O_2}^{1/2}[V_O^{\cdot\cdot}][h]^{-2} = K_2 \tag{3.1.6}$$

so that when the concentration of oxygen ion vacancies is constant (Eq. (3.1.1)) we obtain

$$[h] \sim p_{O_2}^{1/4}, \tag{3.1.7}$$

i.e. the concentration of electron defects is proportional to the fourth root of the oxygen partial pressure. Supposing the mobilities of the vacancies, the electrons and the electron defects to be constant, the following dependences of the partial conductivities in zirconium dioxide on the oxygen partial pressure result if we take into account that the partial conductivity σ_i of the particle type i is proportional to the corresponding concentration [i]:

$$\sigma_{O^{2-}} = \text{const.}, \tag{3.1.8}$$

i.e. the partial conductivity of the oxygen ions which is due to the partial conductivity of oxygen ion vacancies is constant.

$$\sigma_e \sim p_{O_2}^{-1/4}, \tag{3.1.9}$$

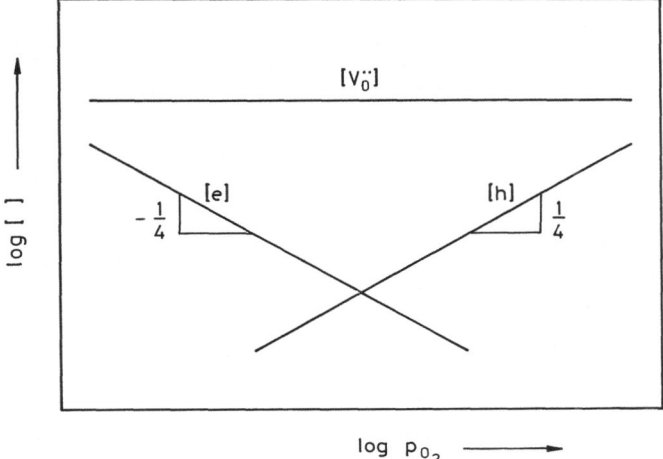

Fig. 3.1.1. Dependence of the concentrations of oxygen ion vacancies, electrons and electron defects on the oxygen partial pressure in doped zirconium and thorium dioxides

the partial conductivity of the electrons is proportional to the inverse fourth root of the oxygen partial pressure, and

$$\sigma_{\mathrm{h}} \sim p_{O_2}^{1/4} \tag{3.1.10}$$

the partial conductivity of the electron defects is proportional to the fourth root of the oxygen partial pressure. These results based on the considerations stated above are shown schematically in Fig. 3.1.1.

Because of the fact that oxygen ion vacancies are the predominant disorder centres over a wide range of oxygen partial pressures, doped zirconium dioxide and thorium dioxide exhibit predominant ion conduction for oxygen ions over a large range of p_{O_2}. These compounds have therefore become very important during the last few years as solid electrolytes in galvanic cells. Further details may be found in Chaps. 8, 9 and 11. The dependences shown in Fig. 3.1.1 have been experimentally confirmed within the partial pressure ranges over which it was possible to carry out measurements [3.6].

3.2 Disorder in ZnO, Cu₂O and CuO

The oxides ZnO, Cu$_2$O and CuO are also at higher temperatures predominant electron or electron defect conductors.

An important method of determining the type of disorder in a certain compound is the measurement of the partial conductivities of electrons and ions as a function of the partial pressure of one component of the compound, e.g. for oxides as a function of the oxygen partial pressure. Assuming the mobility of electrons or electron defects to be independent of their concentrations, then (as will be shown in Chap. 6), the partial conductivities of the various charge carriers are proportional to their concentrations: this means that the partial conductivity due to a certain type of defects is a measure of its concentration in the compound under study. It is thus possible to relate changes in conductivities to corresponding variations in concentrations. Such conductivity measurements have been carried out on ZnO, Cu$_2$O and CuO, and the results are

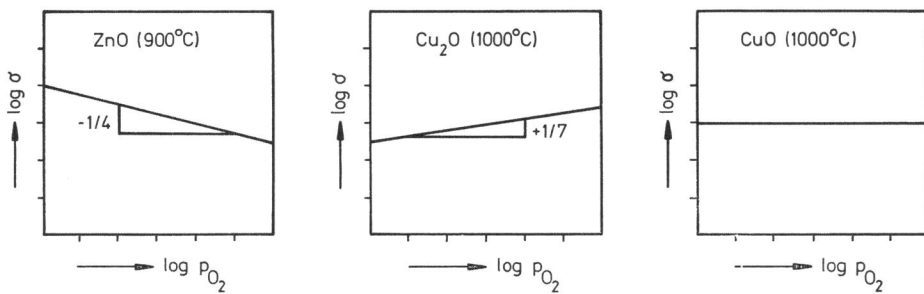

Fig. 3.2.1. Dependence of the conductivity of ZnO, Cu$_2$O and CuO on the oxygen partial pressure

shown schematically in Fig. 3.2.1. The results of the conductivity measurements shown in Fig. 3.2.1 can be explained using the following disorder models:

ZnO

We can assume that at 600 °C the most important disorder centres in this compound are single positively charged zinc ions on interstitial sites and free electrons. The following equation representing the incorporation of oxygen into ZnO may therefore be formulated.

$$\frac{1}{2} O_2 \text{ (g)} + Zn_i^{\cdot} + e = V_i + ZnO. \tag{3.2.1}$$

On the grounds of electrical neutrality it follows that in ZnO

$$[Zn_i^{\cdot}] \cong [e], \tag{3.2.2}$$

i.e. the concentration of zinc ions on interstitial sites is virtually equal to that of free electrons. The law of mass action corresponding to Eq. (3.2.1) is given by

$$p_{O_2}^{1/2}[e] [Zn_i^{\cdot}] = \text{const.} \tag{3.2.3}$$

which may be simplified using Eq. (3.2.2) to yield

$$p_{O_2}^{1/2}[e]^2 = \text{const.}, \tag{3.2.4}$$

i.e.

$$[e] \sim p_{O_2}^{-1/4}. \tag{3.2.5}$$

According to Eq. (3.2.5), the concentration of free electrons is proportional to the inverse fourth root of the oxygen partial pressure; this has been confirmed experimentally [3.7]. The conductivity of ZnO, which is mainly due to the partial conductivity of the free electrons, decreases with increasing oxygen partial pressures according to Eq. (3.2.5).

Cu$_2$O at 1000 °C

At 1000 °C the conductivity of Cu$_2$O increases with increasing oxygen partial pressure. The quantitative dependence may be explained by assuming that electron defects and copper ion vacancies with an excess charge of -1 are the predominant defects in Cu$_2$O. It is therefore possible to write the following equation for the incorporation of oxygen into Cu$_2$O:

$$\frac{1}{2} O_2(g) + 2 Cu_{Cu} = Cu_2O + 2 V'_{Cu} + 2h. \tag{3.2.6}$$

It follows from the condition of electrical neutrality that the concentration of copper ion vacancies equals that of the electron defects, i.e. one can write

$$[V'_{Cu}] \cong [h]. \tag{3.2.7}$$

Inserting Eq. (3.2.7) into the law of mass action holding for Eq. (3.2.6) the following relationship results

$$p_{O_2}^{1/2} = \text{const.} \cdot [h]^4 \qquad\qquad\qquad (3.2.8)$$

or

$$[h] \sim p_{O_2}^{1/8}, \qquad\qquad\qquad (3.2.9)$$

i.e. the concentration of electron defects is proportional to the eighth root of the oxygen partial pressure. If we assume that the mobility of electron defects is independent of their concentration the electronic conductivity (in this case equal to the partial conductivity of electron defects) must be proportional to $p_{O_2}^{1/8}$; measurements have shown that this is approximately valid [3.8]; according to these measurements the conductivity of Cu_2O is proportional to $p_{O_2}^{1/7}$.

Disorder in CuO at 1000 °C

The conductivity of CuO is found to be independent of the oxygen partial pressure (see Fig. 3.2.1) [3.9].

Such behaviour is to be expected if the concentrations of electrons and electron defects are much larger than those of the ionic defects. In this case the disorder of electrons is not influenced by small deviations from ideal stoichiometry.

3.3 Structural Disorder

The concentrations of defects in crystals increase with rising temperature. The limiting case is reached if the concentrations of vacancies and interstitials in such crystals become comparable; then a statistical distribution of particles among normal lattice sites and interstitial sites occurs. To describe this structural state it may no longer be reasonable to distinguish between regular and interstitial lattice sites. The total number of positions a single type of particle can occupy may be several times higher than the number of such particles in the crystal. There are crystals having equivalent lattice sites available for one type of ion several times as much as ions are present in the lattice. Then at sufficiently high temperatures the ions may be statistically distributed among these lattice sites. This state may be called structural disorder — a partial lattice of the crystal may be considered as quasi-molten. This state can be reached in a continuous process extending over a larger range of temperature. CaF_2 is an example for such a behaviour and will be discussed below. This state of structural disorder — quasi molten state — may also be reached in a discontinuous manner. AgI is an example for this case. At 149 °C AgI makes a transition from the β- into the α-phase, the iodide lattice thereby changing into a cubic body-centered structure while the silver ions are randomly distributed among the region between the iodide ions.

On the basis of X-ray structure analyses carried out by Strock [3.10] the silver ions in α-AgI were formerly supposed to occupy fixed positions thereby being statistically distributed as shown in Fig. 3.3.1. Although there are 42 positions available for two silver ions, the very size of these ions precludes independent occupation of all these positions. Discussion of the correlation factor for tracer diffusion in Ag_2S [3.11], which exhibits an ionic disorder analogous to that in AgI, as well as investigations by microwave measurements and by neutron scattering [3.12] have established that it is more reasonable to speak of regions between which the silver ions virtually perform random motions as sketched in Fig. 3.3.2. The idea that the silver ions in α-AgI are already in a quasi-molten state is supported by the values of thermodynamic quantities and diffusion coefficients, the latter being of the order of 10^{-5} cm^2 s^{-1} similar to those found for liquids. The entropy for the phase transition from β-AgI to α-AgI amounts to 14.5 JK^{-1} mol^{-1}, the entropy of fusion to 11.3 JK^{-1} mol^{-1}. Melting of silver iodide may therefore be assumed to take place in two stages. At the $\beta \rightarrow \alpha$ transition the silver partial lattice achieves structural disorder — thus leading to a quasi-molten state connected with this large entropy effect, whereas the iodide partial lattice does not become liquid below the actual melting point. Comparable crystals that do not show the special

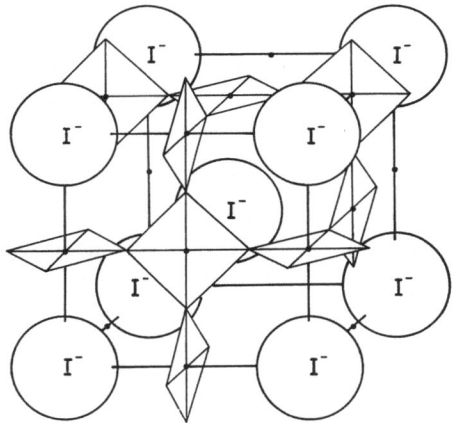

Fig. 3.3.1. Silver iodide lattice with the assumption of point sites for the silver ions (for two silver ions there are 42 point sites)

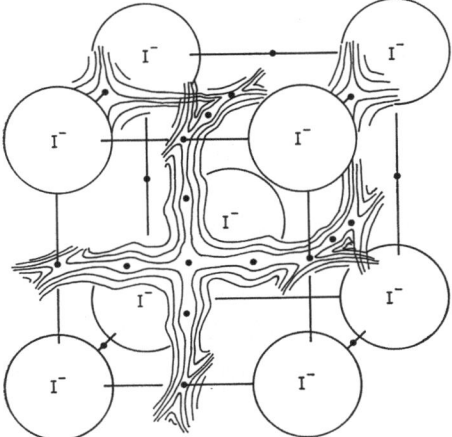

Fig. 3.3.2. Silver iodide lattice with the assumption of regions in which the silver ions can move

Table 3.3.1. Solid-state transition entropy $\Delta_t S$ and melting entropy $\Delta_m S$ of several compounds (from [3.13])

Compound	T_t/K	Solid transition $\Delta_t S/J\ K^{-1}\ mol^{-1}$	T_m/K	Melting $\Delta_m S/J\ K^{-1}\ mol^{-1}$
AgI	419	14.5	830	11.3
Ag_2S	452	9.3	1 115	12.6
CuBr	664	9.0	761	12.6
$SrBr_2$	918	13.3	930	11.3
$BaCl_2$	1 193	14.4	1 233	13.3
LuF_3	1 230	20.4	1 457	20.8
YF_3	1 350	24.0	1 428	19.6

feature of structural disorder before actual melting occurs have an entropy of fusion about as large as the sum of the transition entropy and the entropy of fusion in AgI, i.e. about twice as large as the residual entropy of fusion in AgI. An analogous situation is found to apply for other crystalline compounds that show structural disorder or a quasi-molten partial lattice after passing through a transition point. This was especially shown by O'Keefe and Hyde [3.13]. In Table 3.3.1 entropies of transition and of fusion of typical examples are listed.

However, it should be mentioned that in all cases where an ordered state of one partial lattice is transformed into a disordered state, structural changes also occur in the other partial lattice. It must remain an unsolved problem how the silver ions would behave if the iodide partial lattice of α-AgI remained unchanged at lower temperatures, i.e. if it were possible to supercool α-AgI. It is conceivable that in this case the silver ions would continuously become more and more ordered as the temperature is lowered.

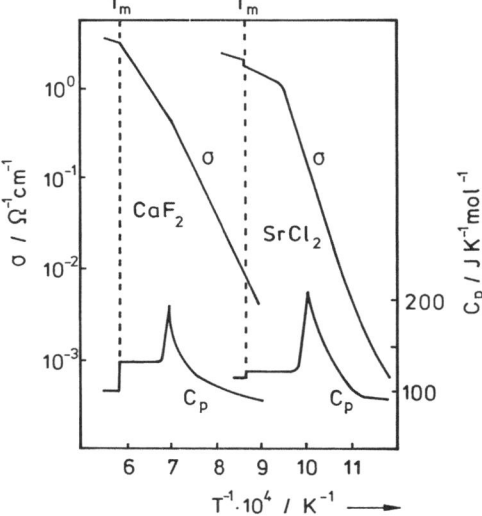

Fig. 3.3.3. Ionic conductivity, σ (*upper curves*), and heat capacity, C_p (*lower curves*), of $SrCl_2$ and CaF_2 as a function of reciprocal temperature. T_m = melting temperature

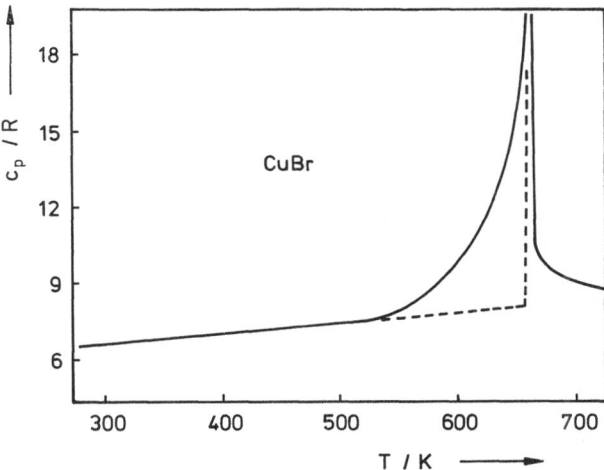

Fig. 3.3.4. Heat capacity of CuBr as a function of temperature

As mentioned above compounds of another group, for example CaF_2, undergo on heating a continuous transition from the ordered to the strongly disordered state. This continuous transition of a partial lattice from an ordered into a quasi-molten state without a change of the partial lattice of the other type of ions is also accompanied by a characteristic behaviour of certain thermodynamic quantities: in a temperature range, where high disorder starts to occur, an abnormally high specific heat is exhibited [3.13]. Hence, an increase in disorder requires the supply of additional energy. This is shown for CaF_2 and $SrCl_2$ in Fig. 3.3.3, which represents the results of measurements performed by Dworkin and Bredig [3.14]. Similar investigations have been carried out on AgBr, CuI and CuBr by Jost, Kubaschewski and Nölting [3.15]. All these compounds exhibit a considerable increase in their heat capacity with increasing temperature already in the low temperature range, which indicates that a continuous transition to structural disorder takes place, if the temperature is raised. As an example Fig. 3.3.4 shows the heat capacity of CuBr as a function of temperature. The values represented are based on measurements carried out by Nölting [3.15].

3.4 References

[3.1]   Eyring, H., Henderson, D., Jost, W.: Physical Chemistry, An Advanced Treatise, Vol. X, p. 229. New York, London: Academic Press 1970
Hauffe, K.: Reaktionen in und an festen Stoffen, 2nd edit. Berlin, Heidelberg, New York: Springer 1966
Kröger, F. A.: Chemistry of Imperfect Crystals, 2nd edit. Amsterdam: North-Holland Publ. Comp. 1974
Pick, H.: Ann. Phys. (5) *31*, 365 (1938)
Steele, B. C. H.: Solid-state chemistry, in: MTP International Review of Science (ed.) Roberts, L. E., Vol. X, p. 117. London, Butterworth/Baltimore, University Park Press 1972

[3.2]   Glaser, G.: Göttinger Nachr. *3*, 31 (1937)
Hilsch, R., Pohl, R. W.: Trans. Faraday Soc. *34*, 883 (1938)
Mott, N. F., Gurney, R. W.: Electronic Processes in Ionic Crystals. Oxford University Press 1940. 2nd edit. New York: Dover Publications 1964
[3.3]   Simkovich, G.: J. phys. Chem. Solids *24*, 213 (1963)
[3.4]   Hauffe, K.: Reaktionen in und an festen Stoffen, 2nd edit. Berlin, Heidelberg, New York: Springer 1966
[3.5]   Hund, F.: Z. Elektrochem. *55*, 363 (1951)
Wagner, C.: Naturwissenschaften *31*, 265 (1943)
[3.6]   Burke, L. D., Rickert, H., Steiner, R.: Z. phys. Chem. N.F. *74*, 146 (1971)
Patterson, J. W., Bogren, E. C., Rapp, R. A.: J. Electrochem. Soc. *114*, 752 (1967)
[3.7]   Baumbach, H. H. v., Wagner, C.: Z. phys. Chem. *B22*, 199 (1933)
[3.8]   Dünwald, H., Wagner, C.: Z. phys. Chem. *B22*, 212 (1933)
Gundermann, J., Wagner, C.: Z. phys. Chem. *B37*, 155 (1937)
[3.9]   Baumbach, H. H. v., Dünwald, H., Wagner, C.: Z. phys. Chem. *B22*, 226 (1933)
[3.10]   Strock, L. W.: Z. phys. Chem. *B25*, 441 (1934)
Strock, L. W.: Z. phys. Chem. *B31*, 132 (1935)
[3.11]   Rickert, H.: Z. phys. Chem. N.F. *24*, 418 (1960)
[3.12]   Jost, W., Funke, K.: Z. Naturforsch. *A25*, 983 (1970)
Funke, K., Jost, A.: Ber. Bunsenges. phys. Chem. *75*, 436 (1971)
Funke, K., Hackenberg, R.: Ber. Bunsenges. phys. Chem. *76*, 885 (1972)
Funke, K., Kalus, J., Lechner, R.: Solid-State Commun. *14*, 1021 (1974)
[3.13]   O'Keefe, M., Hyde, B. G.: Philos. Mag. *33*, 219 (1976)
[3.14]   Dworkin, A. S., Bredig, M. A.: J. phys. Chem. *67*, 697 (1963); J. phys. Chem. *72*, 1277 (1968)
[3.15]   Nölting, J.: Angew. Chem. *82*, 498 (1970)
Jost, W., Kubaschewski, P.: Z. phys. Chem. N.F. *60*, 69 (1968)
Nölting, J., Troe, J., Rein, D.: Nachr. Akad. Wiss. Göttingen, II math.-physik. Kl., p. 31, 1969

4 Thermodynamic Quantities of Quasi-Free Electrons and Electron Defects in Semiconductors

4.1 General Considerations

In the previous chapters we have discussed disorder equilibria which often involved quasi-free electrons and electron defects. We should therefore discuss the thermodynamic quantities of these particles, in particular the electrochemical potential $\bar{\eta}$ and the chemical potential $\bar{\mu}$, both considered with respect to a single particle. The electrochemical potential $\bar{\eta}_e$ of the electrons is referred to by semiconductor physicists as Fermi level or Fermi energy and generally given the symbol E_F. We shall be particularly interested in the relation between these quantities and the concentrations of the quasi-free electrons and electron defects; we shall also be interested in the relations between the terminology used in physical chemistry, particularly chemical thermodynamics, and that used in semiconductor physics, such as the upper edge or top of the valence band, lower edge or bottom of the conduction band etc. In the course of the treatment of phase boundaries or surfaces we shall also discuss the terms work function, Volta potential and Galvani or macropotential. A detailed treatment of the band model of crystalline solids is beyond the scope of this book, and we refer readers to the accounts given in the literature [4.1]. We note only that electrons in crystalline solids are located within certain energy ranges, the so-called energy bands. Electrons cannot have energy values which lie between the bands. The conduction band of semiconductors contains no electrons at absolute zero, but for metals it is partially occupied. The highest fully occupied band at absolute zero in semiconductors is the so-called valence band; at higher temperatures, thermal excitation causes some electrons to leave this band and enter the next higher conduction band. The electrical conductivity which is thus made possible is called intrinsic conductivity. Electrons in the conduction band are called quasi-free or simply free electrons; the unoccupied energy states in the valence band are called electron defects. Quasi-free electrons can also be thermally excited from so-called donor levels, which lie between the valence and conduction bands.

During our discussion of equilibria in Sects. 2.3 and 2.5 we found that the relation $\sum_i \nu_i \bar{\eta}_i = 0$ is valid for equilibrium between charged particles; ν_i are the stoichiometric coefficients and $\bar{\eta}_i$ the electrochemical potentials referred to one particle of species i. If we consider equilibria in volumes with a constant mean electrical potential (Galvani potential), this relation is identical with $\sum_i \nu_i \bar{\mu}_i = 0$ where $\bar{\mu}_i$ are the corresponding chemical potentials (referred to one particle), since the electrical terms cancel out. Equilibria can therefore be directly expressed when the electrochemical or chemical potentials of the reactants are known. We shall first consider the electrochemical potential of the electrons.

4.2 Electrochemical Potential $\tilde{\eta}_e$ of Electrons

The electrochemical potential $\tilde{\eta}_e$ of the electrons with respect to one electron which is identical with the Fermi energy E_F is defined according to the equation

$$\tilde{\eta}_e = E_F = \left(\frac{\partial G}{\partial N_e}\right)_{p,T,N_{j+e}} = \left(\frac{\partial A}{\partial N_e}\right)_{V,T,N_{j+e}}, \tag{4.2.1}$$

N_e is the number of electrons, N_{j+e} the number of particles other than electrons present in the system considered, in our case a crystal; G is the Gibbs energy and A the Helmholtz energy of the crystal. We wish to obtain the relation between the electrochemical potential of the electrons and their concentration; we are considering semiconductors for which the bandmodel with free electrons in the conduction band can be used.

As mentioned above, the free electrons are in the conduction band and in different energy states, which however lie very close together. In order to obtain the total number N_e of the electrons between two energies ε_1 and ε_2, we can use the following equation:

$$N_e = \int_{\varepsilon_1}^{\varepsilon_2} N_e(\varepsilon)\, d\varepsilon, \tag{4.2.2}$$

$N_e(\varepsilon)\, d\varepsilon$ is the number of electrons in the energy interval between ε and $\varepsilon + d\varepsilon$. This quantity is obtained by multiplying the number of states $D(\varepsilon)\, d\varepsilon$ in an energy interval between ε and $\varepsilon + d\varepsilon$ which are available to electrons by the probability $f(\varepsilon)$ with which an energy state of energy ε is occupied by an electron with a certain spin. $D(\varepsilon)$ is called the state density and the probability $f(\varepsilon)$ the Fermi distribution function, sometimes also called Fermi-Dirac-distribution function. Since each state can be occupied by two electrons with opposite spin, we can convert Eq. (4.2.2) into the following expression for N_e

$$N_e = 2 \int_{\varepsilon_1}^{\varepsilon_2} f(\varepsilon)\, D(\varepsilon)\, d\varepsilon. \tag{4.2.3}$$

It is now necessary to obtain expressions for the Fermi distribution function $f(\varepsilon)$ and for the state density $D(\varepsilon)$.

4.3 The Fermi Distribution Function $f(\epsilon)$

The Fermi distribution function $f(\varepsilon)$ gives the probability that an energy state which is available for electrons will be occupied by an electron with a defined spin. This probability depends on the difference between the electrochemical potential

$\bar{\eta}_e$ of the electrons and the value of the energy of the state considered. According to Schottky [4.2] we can calculate $\bar{\eta}_e$ as follows: the electrochemical potential $\bar{\eta}_e$ of the electrons or the Fermi energy E_F, which are defined according to Eq. (4.2.1), may be split into two terms corresponding to the division of the Helmholtz energy $A = U - TS$ into the internal energy U and the entropy term $-TS$.

$$\bar{\eta}_e = E_F = \left(\frac{\partial A}{\partial N_e}\right)_{V,T,N_{j\pm e}} = \left(\frac{\partial U}{\partial N_e}\right)_{V,T,N_{j\pm e}} - T\left(\frac{\partial S}{\partial N_e}\right)_{V,T,N_{j\pm e}}, \qquad (4.3.1)$$

$\left(\dfrac{\partial U}{\partial N_e}\right)_{V,T,N_{j\pm e}}$ is the partial derivative of the internal energy U with respect to the number of electrons N_e, where the subscript implies the constancy of volume, temperature and all other particles numbers N_j except N_e. $\left(\dfrac{\partial S}{\partial N_e}\right)_{T,V,N_{j\pm e}}$ is the corresponding derivative of the entropy S.

All electrons have the same electrochemical potential $\bar{\eta}_e = E_F$, which is independent of their internal energy. Let us consider a group of electrons "i", which all have virtually the same energy ε_i within an infinitesimal range $d\varepsilon$. The number of these electrons is N_i, the energy interval $d\varepsilon$ very small compared to kT. If we were to consider only one distinct energy state, this could be occupied according to the Pauli principle by not more than two electrons with opposite spin. However, for a finite energy interval the number of energy states for the electrons is very large. The partial derivative of the internal energy U of the crystal with respect to the number N_i of the electrons considered above gives the partial internal energy of these electrons often referred to as their energy, i.e. the following equation is valid:

$$\left(\frac{\partial U}{\partial N_i}\right)_{V,T,N_{j\pm i}} = \varepsilon_i. \qquad (4.3.2)$$

The partial entropy, i.e. the entropy change of the crystal caused by a variation of the number N_i of electrons of energy ε_i is described by the following equation, since a change in the number N_i can only change the entropy of the "i"-electrons:

$$\left(\frac{\partial S}{\partial N_i}\right)_{V,T,N_{j\pm i}} = \left(\frac{\partial S_i}{\partial N_i}\right)_{V,T,N_{j\pm i}} \qquad (4.3.3)$$

or, because of the relation $S_i = k \ln W_i$

$$\left(\frac{\partial S}{\partial N_i}\right)_{V,T,N_{j\pm i}} = k\left(\frac{\partial \ln W_i}{\partial N_i}\right)_{V,T,N_{j\pm i}}, \qquad (4.3.4)$$

W_i is the thermodynamic probability or degeneracy of the electrons at energy ε_i. It is equal to the number of possibilities of distributing the N_i electrons between the Z_i energy states which are available in the interval $d\varepsilon$ at ε_i; the electrons cannot be distinguished from one another. From Eqs. (4.3.1—4), we obtain

$$\bar{\eta}_e = E_F = \varepsilon_i - kT\left(\frac{\partial \ln W_i}{\partial N_i}\right)_{V,T,N_{j\pm i}}. \qquad (4.3.5)$$

Thus, we have expressed the electrochemical potential or the Fermi energy E_F of the electrons in terms of the energy ε_i of the "i"-electrons and their thermodynamic probability W_i. Eq. (4.3.5) can be rearranged to give

$$\left(\frac{\partial \ln W_i}{\partial N_i}\right)_{V,T,N_{j \neq i}} = \frac{\varepsilon_i - E_F}{kT}. \tag{4.3.6}$$

We can now use this relation to calculate the Fermi-Dirac distribution function.

Taking into account the Pauli principle, i.e. the statement that an energy state can only be singly occupied by an electron with a given spin orientation (this restriction leads to Fermi-Dirac statistics), we obtain the following relationship:

$$W_i(N_i) = \frac{Z_i(Z_i - 1) \cdot (Z_i - 2) \cdot \ldots \cdot [Z_i - (N_i - 1)]}{N_i!}, \tag{4.3.7}$$

Z_i is the number of energy states ε_i in a certain interval which can be occupied by electrons. Equation (4.3.7) is obtained from the following considerations: for the first electron Z_i states are available, for the second one $(Z_i - 1)$ each time after the first electron has occupied one of the Z_i states. These two numbers must be multiplied together to give $Z_i \cdot (Z_i - 1)$. This product must be extended for N_i electrons as far as $Z_i - (N_i - 1)$. Since we cannot distinguish between the electrons, we must divide by the number $N_i!$ of possible permutations.

Extension of Eq. (4.3.7) by the factor $(Z_i - N_i)!$ gives

$$W_i(N_i) = \frac{Z_i!}{(Z_i - N_i)! \, N_i!}. \tag{4.3.8}$$

Similarly, the thermodynamic probability for $(N_i + 1)$ particles in Z_i states is given by

$$W_i(N_i + 1) = \frac{Z_i!}{[Z_i - (N_i + 1)]! \, (N_i + 1)!}. \tag{4.3.9}$$

From equations (4.3.8) and (4.3.9) we obtain the ratio of the thermodynamic probabilities of $(N_i + 1)$ to N_i electrons occupying Z_i states

$$\frac{W_i(N_i + 1)}{W_i(N_i)} = \frac{Z_i - N_i}{N_i + 1}. \tag{4.3.10}$$

If we consider the smallest real possible variation $\partial N_i = 1$ of the particle number, which is small compared to the number N_i of the electrons, we can obtain the derivative $\dfrac{\partial \ln W_i}{\partial N_i}$:

$$\frac{\partial \ln W_i}{\partial N_i} = \frac{\ln W_i(N_i + 1) - \ln W_i(N_i)}{1} = \ln \frac{W_i(N_i + 1)}{W_i(N_i)} \tag{4.3.11}$$

or according to Eq. (4.3.10)

$$\frac{\partial \ln W_i}{\partial N_i} = \ln \frac{Z_i - N_i}{N_i + 1}. \tag{4.3.12}$$

Since the number of electrons N_i is assumed to be much larger than one, we can write Eq. (4.3.12) as:

$$\frac{\partial \ln W_i}{\partial N_i} = \ln \frac{Z_i - N_i}{N_i} \tag{4.3.13}$$

or, using Eq. (4.3.6)

$$\frac{Z_i}{N_i} - 1 = \exp \frac{\varepsilon_i - E_F}{kT}. \tag{4.3.14}$$

Equation (4.3.14) contains the relationship between the number of electrons N_i and the number of available energy states Z_i, without consideration of the different spin orientations. The ratio between both numbers is equal to the probability of occupying one state, i.e. to the Fermi-Dirac distribution function $f(\varepsilon_i)$

$$\frac{N_i}{Z_i} = f(\varepsilon_i) = \frac{1}{\exp\left[(\varepsilon_i - E_F)/kT\right] + 1}. \tag{4.3.15}$$

Since these considerations apply to all different energies ε_i, we have solved the first part of our problem. We can see from Eq. (4.3.14) that the Fermi energy E_F (identical with the electrochemical potential $\tilde{\eta}_e$ of the electrons) denotes a level which is constant for the crystal. It is therefore often called the Fermi level. For energies ε_i smaller than E_F, the probability that these energy states are occupied by electrons approaches unity, while for energies larger than E_F, the probability of occupation suddenly becomes very small, particularly at low temperatures.

The probability of an energy state not to be occupied is identical with the probability that an electron defect is present. We can therefore write the Fermi-Dirac distribution function f_h for electron defects (or holes) as

$$f_h = 1 - f_e \tag{4.3.16}$$

where $f_e \equiv f(\varepsilon_i)$ is the Fermi-Dirac distribution function for free electrons. From Eq. (4.3.15) and (4.3.16) we obtain

$$f_h = \frac{1}{\exp\left[-(\varepsilon_i - E_F)/kT\right] + 1}. \tag{4.3.17}$$

This is identical with the corresponding expression for the electrons except that the minus sign occurs in the exponential function.

4.4 The Density of States D(ε)

The density of states $D(\varepsilon)$ is the number of energy states per unit energy available to electrons at energy ε. Therefore, the product $D(\varepsilon)\,d\varepsilon$ is equal to the number of available energy states in the energy interval from ε to $\varepsilon + d\varepsilon$. To calculate these we start from the quantum mechanical model of a free point particle in a box which is taken to be a cube of volume V. The energy states of such a particle are given by

$$E_{kin} = \frac{1}{2m} p^2 = \frac{h^2}{8mV^{2/3}} n^2 = \frac{h^2}{8mV^{2/3}} (n_x^2 + n_y^2 + n_z^2), \qquad (4.4.1)$$

E_{kin} denotes the kinetic energy of the particle, m its mass, p its momentum, h Planck's constant and n_x, n_y and n_z integral positive numbers. As can be seen from Eq. (4.4.1) n^2 is equal to the sum of the squares of the quantum numbers n_x, n_y, n_z; i.e. $n^2 = n_x^2 + n_y^2 + n_z^2$. For the derivation of this relation we refer to standard text-books of physics or physical chemistry [4.1].

The energy states of quasi-free electrons in the conduction band of solids can be expressed in a similar way [4.1]

$$E_{kin} = \frac{p^2}{2m_e^*} = \frac{\hbar^2 k^2}{2m_e^*}$$

$$= \frac{h^2}{8m_e^* V^{2/3}} n^2 = \frac{h^2}{8m_e^* V^{2/3}} (n_x^2 + n_y^2 + n_z^2) = \varepsilon - E_C, \qquad (4.4.2)$$

E_C denotes the energy of the lower edge of the conduction band, E_{kin} the kinetic energy of the quasi-free electrons, $\vec{k} = \vec{p}\hbar^{-1}$ the wave vector, m_e^* the effective mass of the electrons and $\hbar = h/2\pi$. The deviation of the effective mass of the electrons from the mass of the free electron reflects the interaction between the quasi-free electrons and the lattice of the semiconductor compound. In our calculation we will use an effective mass which is independent of the direction in the crystal: this corresponds to the simple case of a spherically symmetric band model. We shall also assume that m_e^* is independent of the wave vector $\vec{k}$. Thus our considerations are restricted to energies and $\vec{k}$-values which are not too high.

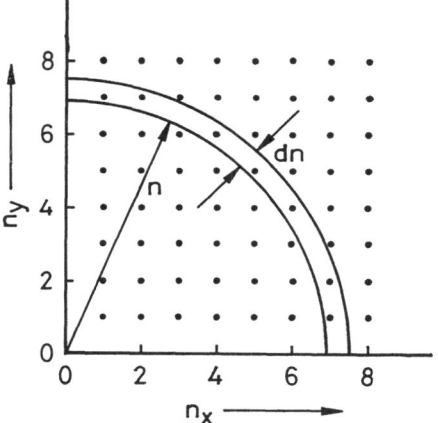

Fig. 4.4.1. Grid of points in the n_x, n_y — space (section through the n_x, n_y, n_z — space) representing allowed quantum states of the quasi-free electrons. Each state "occupies" the area of a single unit square. The area confined between the circles with radius n and n + dn respectively in the first quadrant corresponds to the number of quantum states of the electrons with energy between E and E + dE

Our calculation is therefore no longer universally valid but still applies for many semiconductors particularly if the conduction band contains only a relatively small number of electrons. To determine $D(\varepsilon)$ we first calculate the density of states $D(n)$ in a three-dimensional space, the coordinates of which are the quantum numbers n_x, n_y and n_z.

In this lattice the cube of unit volume represents one quantum state. All quantum states are represented by the cubes in the first octant of the coordinate system, since n_x, n_y and n_z are integral positive numbers. The cross section of this space i.e. the two dimensional n_x, n_y-space is shown in Fig. 4.4.1. The number of quantum states which lie between n and n + dn is equal to 1/8 of the volume of the spherical shell with radius n and thickness dn, namely

$$D(n)\,dn = \frac{4\pi}{8}\,n^2\,dn. \tag{4.4.3}$$

Expressing n in terms of energy with the aid of Eq. (4.4.2) we obtain

$$D(\varepsilon)\,d\varepsilon = 4\pi 2^{1/2}\,\frac{V}{h^3}\,(m_e^*)^{3/2}\,(\varepsilon - E_C)^{1/2}\,d\varepsilon \tag{4.4.4}$$

which is the desired expression for the number of states available to the electrons between ε and $\varepsilon + d\varepsilon$.

4.5 Relation Between the Concentration of Electrons or Electron Defects and Their Electrochemical Potential

Using the expressions for $f(\varepsilon)$ and $D(\varepsilon)$ in Eq. (4.3.15) and (4.4.4), we can now calculate from Eq. (4.2.3) the number of quasi-free electrons N_e in a crystal of volume V within a certain energy interval. First we determine the number of quasi-free electrons in the conduction band. The integration in Eq. (4.2.3) must therefore be carried out between the lower and upper edge of the conduction band, that is between E_C and E_C'. In carrying out this calculation we assume the effective mass $m_e^*(\varepsilon)$ to be constant, i.e. that ε shows a parabolic dependence on $|\vec{p}|$ (see Eq. (4.4.2)),

$$N_e = 2\int_{E_C}^{E_C'} f_e(\varepsilon)\,D_e(\varepsilon)\,d\varepsilon$$

$$= 2\int_{E_C}^{E_C'} \frac{1}{\exp\left(\dfrac{\varepsilon - E_F}{kT}\right) + 1}\,4\pi\,2^{1/2}\,\frac{V}{h^3}\,m_e^{*3/2}(\varepsilon - E_C)^{1/2}\,d\varepsilon. \tag{4.5.1}$$

Since we presume that electrons are only to be found in the vicinity of the lower edge E_C of the conduction band, we can integrate to $\varepsilon = \infty$ instead of taking the upper edge E_C' of the conduction band as the upper integration limit without changing the result of the calculation. By substituting the integration variable ε

in the following way

$$\frac{\varepsilon - E_C}{kT} = \xi, \quad d\varepsilon = kT \, d\xi \tag{4.5.2}$$

we obtain from Eq. (4.5.1)

$$N_e = \left(\frac{2\pi m_e^* kT}{h^2}\right)^{3/2} 4V\pi^{-1/2} \int_0^\infty \frac{\xi^{1/2}}{\exp\left(\xi + \dfrac{E_C - E_F}{kT}\right) + 1} \, d\xi \tag{4.5.3}$$

or, using the abbreviations $N_C = 2\left(\dfrac{2\pi m_e^* kT}{h^2}\right)^{3/2}$ and

$$F_{1/2}\left(\frac{E_F - E_C}{kT}\right) = \int_0^\infty \frac{\xi^{1/2}}{\exp\left(\xi + \dfrac{E_C - E_F}{kT}\right) + 1} \, d\xi:$$

$$N_e = V N_C \, 2\pi^{-1/2} F_{1/2}\left(\frac{E_F - E_C}{kT}\right). \tag{4.5.4}$$

N_C is called the effective density of states, sometimes also the degeneracy concentration of the electrons. The meaning of this quantity will become clear in the course of our discussion. $\left(\dfrac{h^2}{2\pi m_e^* kT}\right)^{1/2}$ is identical to the De Broglie wavelength λ. $F_{1/2}$ is the Fermi-Dirac integral one half; the "one half" refers to the fact that in the numerator of the integrand, the integration variable has the exponent $\frac{1}{2}$. The Fermi-Dirac integral $F_{1/2}$ cannot be solved explicitly. Tables containing numerical values of $F_{1/2}$ and also of derivatives of $F_{1/2}$ have been compiled by McDougall and Stoner [4.3]. N_C has a value of

$$N_C = 2\left(\frac{2\pi m_e^* kT}{h^2}\right)^{3/2} = 2.5 \cdot 10^{19} \left(\frac{m_e^*}{m_e}\right)^{3/2} \left(\frac{T}{300 \text{ K}}\right)^{3/2} \text{cm}^{-3}, \tag{4.5.5}$$

i.e. at room temperature and if the effective mass m_e^* of the electrons is equal to the actual mass of a free electron, N_C has the value $2.5 \cdot 10^{19}$ cm^{-3}.

In a similar way we obtain an expression for the number N_h of electron defects or holes in the valence band:

$$N_h = V N_V \, 2\pi^{-1/2} F_{1/2}\left(\frac{E_V - E_F}{kT}\right) \tag{4.5.6}$$

where E_V is the upper edge of the valence band and N_V the effective density of states of the electron defects in the valence band:

$$N_V = 2\left(\frac{2\pi m_h^* kT}{h^2}\right)^{3/2}, \tag{4.5.7}$$

m_h^* is the effective mass of the electron defects. Equation (4.5.6) expressing the number of electron defects differs from Eq. (4.5.4) for electrons: E_V instead of E_C appears in the argument of the Fermi-Dirac integral, and the signs of E_F and E_V are different.

Approximations of the Number of Quasi-free Electrons N_e and Electron Defects N_h. If the value of $(E_C - E_F)/kT$ is sufficently large, i.e. if E_F lies far enough below the value of E_C the number 1 in the denominator of the integrand can be neglected in comparison to the exponential function. The integral can now be solved explicitly and we obtain from Eq. (4.5.4)

$$\frac{N_e}{V} = n = N_C \exp\left(-\frac{E_C - E_F}{kT}\right), \tag{4.5.8}$$

n is the concentration or density of the quasi-free electrons. In this case we speak of the so-called Boltzmann or classical approximation. The expression for the concentration p of electron defects can be obtained for the case of the Boltzmann approximation from Eq. (4.5.6)

$$\frac{N_h}{V} = p = N_V \exp\left(\frac{E_V - E_F}{kT}\right). \tag{4.5.9}$$

The product of the concentrations of electrons and electron defects which is a mass action constant (see. Sect. 2.5), can be obtained in the case of the Boltzmann approximation from Eq. (4.5.8) and (4.5.9):

$$pn = N_C N_V \exp\left(-\frac{E_C - E_V}{kT}\right) = n_i^2, \tag{4.5.10}$$

n_i is called the intrinsic density. If the densities of electrons and electron defects are equal, we speak of intrinsic disorder. In this case, i.e. if $n = p = n_i$, the Fermi energy E_F for the Boltzmann approximation may be expressed in terms of E_C, E_V, N_C and N_V by setting the right hand side of Eq. (4.5.8) equal to that of Eq. (4.5.9):

$$E_F = \tilde{\eta}_e = \frac{E_C + E_V}{2} - \frac{1}{2} kT \ln \frac{N_C}{N_V}. \tag{4.5.11}$$

It follows from Eq. (4.5.11) that if the effective densities of states N_C of the electrons and N_V of the electron defects are equal, which is equivalent to the statement that they have equal effective masses, the Fermi energy E_F or the electrochemical potential $\tilde{\eta}_e$ of the electrons lies exactly half way between the upper edge of the valence band and the lower edge of the conduction band, i.e. just in the middle of the band gap.

4.6 Chemical Potential of Electrons and Standard States

We shall now proceed to establish the relationship to the quantities "chemical potential $\tilde{\mu}_e$ of the electrons" and "chemical potential $\tilde{\mu}_e^0$ in the standard state" used in chemical thermodynamics. Equation (4.5.8) is the basis for the following discussion. As discussed above it represents the Boltzmann approximation for the quasi-free electrons in the conduction band; this equation can be rewritten as

$$E_F - E_C = kT \ln \frac{n}{N_C} \tag{4.6.1}$$

or by introducing a standard concentration n^0

$$E_F - E_C = kT \ln \frac{n^0}{N_C} + kT \ln \frac{n}{n^0}. \tag{4.6.2}$$

The general expression for the chemical potential $\bar{\mu}$ (see also Sect. 2.4) is as follows:

$$\bar{\mu}_e = \bar{\mu}_e^0 + kT \ln a_e \tag{4.6.3}$$

where a_e is the activity of the electrons. For ideal behaviour of the electrons, which means that the Boltzmann approximation is valid, their activity a_e is equal to the ratio of their concentration and a chosen standard concentration:

$$\bar{\mu}_e = \bar{\mu}_e^0 + kT \ln \frac{n}{n^0}. \tag{4.6.4}$$

The chosen standard concentration defines the standard state and thus also the chemical potential $\bar{\mu}_e^0$ in the standard state. From Eqs. (4.6.2) and (4.6.4) we obtain

$$E_F - E_C = \bar{\mu}_e - \bar{\mu}_e^0 + kT \ln \frac{n^0}{N_C}. \tag{4.6.5}$$

This equation connects $(E_F - E_C)$ with the chemical potential of the electrons $\bar{\mu}_e$ and that in the standard state $\bar{\mu}_e^0$, the effective density of states N_C in the conduction band and the standard concentration n^0. If we choose the effective density of states N_C for the standard concentration n^0, the expression $kT \ln (n^0/N_C)$ vanishes and we obtain

$$E_F - E_C = \bar{\mu}_e - \bar{\mu}_e^0 (n^0 = N_C), \tag{4.6.6}$$

i.e. $(E_F - E_C)$ is equal to $\bar{\mu}_e - \bar{\mu}_e^0$.

Since E_F is identical with the electrochemical potential $\bar{\eta}_e = \bar{\mu}_e - e\varphi$, it follows from Eq. (4.6.6) that for $n^0 = N_C$:

$$E_C = \bar{\mu}_e^0 (n^0 = N_C) - e\varphi, \tag{4.6.7}$$

i.e. we have an expression for the lower edge of the conduction band E_C. This is equal to the chemical potential of the electrons in the standard state minus $e\varphi$, if we take $n^0 = N_C$ as the standard concentration. Thus E_C contains the negative electrical potential multiplied by e and varies in an opposite direction compared to the electrical potential. If we select another standard concentration n^0, we obtain a more general expression for the lower edge of the conduction band E_C from Eq. (4.6.5), where we again take into account that $E_F = \bar{\mu}_e - e\varphi$

$$E_C = \bar{\mu}_e^0 - e\varphi - kT \ln \frac{n^0}{N_C}. \tag{4.6.8}$$

Equations (4.6.5—8) which have been obtained for the case of the Boltzmann approximation, have a more general validity as long as the quantities E_C and N_C do not vary with the concentration of the electrons; i.e. these equations are still valid in the case of degeneracy, if only the effective mass of the electrons is constant so that Eq. (4.5.4) can be used. These relations are, however, no longer valid if further effects, e.g. Coulomb interactions, have to be taken into account, so that edges of the conduction band and the effective densities of states are changed.

As standard concentrations we could take 1/cm³ as well as N_C and the intrinsic concentration $n_i = (np)^{1/2}$. The corresponding standard states and chemical potentials in that standard states are discussed below (see also Ref. [4.4]):

a) $n^0 = N_C$; i.e. the standard concentration is equal to the effective density of states. In this case, according to Eq. (4.6.8) $E_C = \bar\mu_e^0 - e\varphi$. The chemical potential in the standard state $\bar\mu_e^0$ lies $e\varphi$ above the lower edge of the conduction band E_C. If φ is taken as zero, we have $\bar\mu_e^0 = E_C$. This is, however, only possible, if only one phase is considered, or, if several phases are considered, only in one of these phases.

b) $n^0 = 1/cm^3$, in this case, according to Eq. (4.6.8)

$$E_C = \bar\mu_e^0 - e\varphi - kT \ln (1/N_C \; cm^3),$$

i.e. $\bar\mu_e^0$ lies above the lower edge of the conduction band by an amount $e\varphi +$ $kT \ln (1/N_C \; cm^3)$. However, since $kT \ln (1/N_C \; cm^3)$ is negative and equal to about -1.5 eV, $\bar\mu_e^0$ lies about 1.5 eV below E_C when φ is taken as zero.

c) $n^0 = n_i = (N_C N_V)^{1/2} \exp\left(-\dfrac{E_G}{2kT}\right)$, where $E_G = E_C - E_V$. It then follows from equation (4.6.8) that for the conditions $N_C = N_V$ and $\varphi = 0$ $\bar\mu_e^0$ lies exactly between E_V and E_C, i.e. in the centre of the band gap. If φ is not taken as zero, $\bar\mu_e^0 - e\varphi$ lies exactly in the centre of the band gap for $N_C = N_V$.

4.7 The Activity Coefficient of Electrons and Electron Defects in the Case of Degeneracy

If the Boltzmann approximation is no longer applicable and the Fermi-Dirac integral must be used, we speak of degeneracy of the electrons, which then no longer behave in an ideal manner. In chemical thermodynamics the activity coefficient is introduced to take into account deviations from ideal behaviour; this coefficient is here given the symbol γ_e.

$$\bar\mu_e = \bar\mu_e^0 + kT \ln \gamma_e \frac{n}{n^0} \tag{4.7.1}$$

or

$$a_e = \gamma_e \frac{n}{n^0}. \tag{4.7.2}$$

Incorporation of the activity coefficient leads to the following expression for the Fermi energy or the electrochemical potential

$$E_F = \bar{\eta}_e = \bar{\mu}_e - e\varphi = \bar{\mu}_e^0 - e\varphi + kT \ln \gamma_e \frac{n}{n^0}. \tag{4.7.3}$$

We can insert into Eq. (4.7.3) the expression for the concentration n of the electrons, taking into account Fermi-Dirac statistics according to Eq. (4.5.4) and using Eq. (4.6.8) which shows how E_C, $\bar{\mu}_e^0$ and $e\varphi$ are related. The following equation then results

$$E_F - E_C = kT \ln \gamma_e + kT \ln \left[2\pi^{-1/2} F_{1/2} \left(\frac{E_F - E_C}{kT} \right) \right]. \tag{4.7.4}$$

Using the abbreviation

$$\frac{E_F - E_C}{kT} = \frac{\bar{\mu}_e - \bar{\mu}_e^0 + kT \ln \dfrac{n^0}{N_C}}{kT} = \xi \tag{4.7.5}$$

the activity coefficient of the electrons can be obtained from Eq. (4.7.4)

$$\gamma_e = \frac{\pi^{1/2} \exp \xi}{2F_{1/2}(\xi)}. \tag{4.7.6}$$

Similar considerations for the electron defects lead to the following expression

$$\gamma_h = \frac{\pi^{1/2} \exp \zeta}{2F_{1/2}(\zeta)} \tag{4.7.7}$$

with

$$\zeta = \frac{E_V - E_F}{kT}. \tag{4.7.8}$$

4.8 The Phase Boundary Solid/Vacuum. The Work Function W

Let φ^I and φ^{Vac} denote the electrical potentials in the solid (I) and in the vacuum just above the surface, yet outside the range of influence of the atoms; $\bar{\mu}_e^I$ and $\bar{\mu}_e^{Vac}$ are the corresponding chemical potentials of the electrons. When equilibrium has been attained at the phase boundary solid/vacuum, the electrochemical potentials $\bar{\eta}_e$ of the electrons are equal on both sides of the phase boundary:

$$\bar{\eta}_e^I = \bar{\eta}_e^{Vac} \tag{4.8.1}$$

and, since $\bar{\eta}_e = \bar{\mu}_e - e\varphi$

$$\bar{\mu}_e^I - e\varphi^I = \bar{\mu}_e^{Vac} - e\varphi^{Vac}. \tag{4.8.2}$$

From Eq. (4.8.2) we can obtain an expression for the chemical potential of the electrons above the surface of the solid:

$$\bar{\mu}_e^{Vac} = \bar{\mu}_e^I - e(\varphi^I - \varphi^{Vac}) \tag{4.8.3}$$

or, in terms of the electrochemical potential η_e^I of the electrons in the solid

$$\bar{\mu}_e^{Vac} = \bar{\eta}_e^I + e\varphi^{Vac}. \tag{4.8.4}$$

The chemical potential $\bar{\mu}_e^{Vac}$ of the electrons above the surface is referred to as the negative work function:

$$\bar{\mu}_e^{Vac} = -W. \tag{4.8.5}$$

The energy zero-point is so chosen that an electron at rest in the vacuum has the energy $-e\varphi^{Vac}$. In this case when the standard concentration in Eq. (4.6.4) is chosen as $n^0 = N_C$, the chemical potential of the electrons in the standard state $\bar{\mu}_e^0$ is equal to zero, and $\bar{\mu}_e^{Vac}$ or $-W$ are a direct measure for the concentration of electrons above the surface. For a dilute electron gas [4.5] the concentration of electrons in the vacuum directly above the surface is given by

$$n^{Vac} = N_{Vac} \exp\left(-\frac{W}{kT}\right). \tag{4.8.6}$$

N_{Vac} is the effective density of states of the electrons in the vacuum

$$N_{Vac} = 2\left(\frac{2\pi m_e kT}{h^2}\right)^{3/2}. \tag{4.8.7}$$

According to Eq. (4.8.6) one can determine the work function in the case of the Boltzmann approximation by measuring the concentration of the electrons above the surface in a system at thermodynamic equilibrium; it is also possible to use a measurement which involves this expression, for example that of the saturation current of the electron emission (see Schottky [4.5]). Let us summarize: the work function W, identical with the negative chemical potential of the electrons above the solid surface when the system is at thermodynamic equilibrium (Eq. (4.8.5)), is a measure of the equilibrium concentration of the electrons above a surface (Eq. (4.8.6)). It can be expressed as the difference of the chemical potentials of the electrons in the solid and the electrical potential difference $(\varphi^I - \varphi^{Vac})$ between the interior of the solid and the vacuum immediately above the surface, multiplied by the charge e, or as the sum of the electrochemical potential of the electrons in the solid and the electrical potential in the vacuum above the surface multiplied by e (Eq. (4.8.4)).

4.9 The Volta Potential or Contact Potential

The Volta potential $V_{I,II}$ is defined as follows:

$$\varphi^{Vac,II} - \varphi^{Vac,I} = V_{I,II}. \tag{4.9.1}$$

It is the difference between the electrical potentials in the vacuum of a point just above the surface II and a second point just above the surface I. At thermodynamic equilibrium, the electrochemical potential $\bar{\eta}_e$ of the electrons in the vacuum is constant, i.e. the following relation is valid for $\bar{\eta}_e$ above both surfaces

$$\bar{\mu}_e^{Vac,I} - e\varphi^{Vac,I} = \bar{\mu}_e^{Vac,II} - e\varphi^{Vac,II}. \tag{4.9.2}$$

Using Eq. (4.8.5), we obtain the following relation involving the Volta potential

$$e(\varphi^{Vac,II} - \varphi^{Vac,I}) = eV_{I,II} = \bar{\mu}_e^{Vac,II} - \bar{\mu}_e^{Vac,I} = W^I - W^{II}, \tag{4.9.3}$$

i.e. the Volta potential multiplied by the elementary charge e is equal to the difference between the chemical potentials of the electrons above the semiconductor surfaces at thermodynamic equilibrium and is therefore equal to the difference of the two work functions.

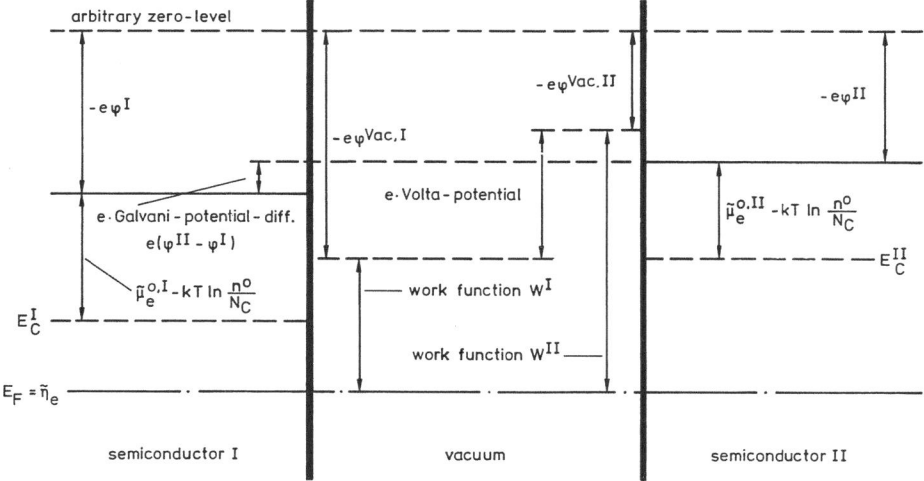

Fig. 4.9.1. The phase boundaries solid/vacuum of two solids, which are at electronic equilibrium

If we insert the expression for the chemical potential $\tilde{\mu}_e^{Vac}$ from Eq. (4.8.3) into Eq. (4.9.3), we obtain

$$e(\varphi^{Vac,II} - \varphi^{Vac,I}) = eV_{I,II} = \tilde{\mu}_e^{II} - e(\varphi^{II} - \varphi^{Vac,II})$$
$$- \tilde{\mu}_e^{I} + e(\varphi^{I} - \varphi^{Vac,I}) \qquad (4.9.4)$$

where $\tilde{\mu}_e$ refers to the interior of the semiconductor. Thus the Volta potential can be expressed in terms of differences in the electrical potentials between points above the surface and in the interior of the phases concerned, and the difference of the chemical potentials of the electrons in the phases; the latter is equal to the Galvani potential difference, as will be shown below. The Volta or contact potential does not generally indicate whether, if two substances are brought in contact, negative charge is transfered from compound I to II or vice versa. Figure 4.9.1 shows the situation which occurs when two semiconductors, joined by an external electronic contact (the electrons are in equilibrium), are placed in a vacuum.

4.10 The Phase Boundary Solid I/Solid II. The Galvani Voltage

When the electrons have reached equilibrium

$$\tilde{\eta}_e^{I} = \tilde{\eta}_e^{II} \qquad (4.10.1)$$

and, using the relation $\tilde{\eta}_e = \tilde{\mu}_e - e\varphi$

$$\tilde{\mu}_e^{I} - e\varphi^{I} = \tilde{\mu}_e^{II} - e\varphi^{II} \qquad (4.10.2)$$

or

$$e(\varphi^{II} - \varphi^{I}) = \tilde{\mu}_e^{II} - \tilde{\mu}_e^{I}. \qquad (4.10.3)$$

The difference of the inner electrical potentials $\varphi^{II} - \varphi^{I}$ is called the Galvani voltage. No methods for measuring this quantity are apparently known.

Potential differences at phase boundaries are due on the one side to rigid double layers, which in turn are due to dipoles present in the structure or to charge transfer, on the other side to diffuse double layers, which are also termed space charge layers.

4.11 References

[4.1] Sommerfeld, A.: Z. Phys. *47*, 1 (1928)

Bethe, H., Sommerfeld, A.: Elektronentheorie der Metalle. Berlin, Heidelberg, New York: Springer 1967

Smith, R. A.: Semiconductors, 2nd edit. Cambridge: University Press 1978

Slater, J. C.: The electronic structure of solids, Handb. d. Physik 19/1, Flügge, S. (ed.). Berlin, Göttingen, Heidelberg: Springer 1956

Spenke, E.: Electronic Semiconductors. New York: McGraw-Hill Book Company 1958

Dekker, A. J.: Solid State Physics. New York: Engelwood Cliffs Prentice-Hall Inc. 1957

Kittel, C.: Introduction to Solid State Physics, 5th ed. New York: John Wiley and Sons, Inc. 1976

Madelung, O.: Grundlagen der Halbleiterphysik. Berlin, Heidelberg, New York: Springer 1970

Ashcroft, N. W., Mermin, N. D.: Solid State Physics. New York: Holt, Rinehart and Winston 1976

[4.2] Schottky, W.: Halbleiterprobleme I. Sauter, F., (ed.), p. 139. Braunschweig: Vieweg 1954

[4.3] McDougall, Stoner, E. C.: Phil. Trans. *A 237*, 67 (1929)

[4.4] Harvey, W. W.: J. Phys. Chem. Solids *23*, 1545 (1962)

Harvey, W. W.: Phys. Rev. *123*, 1666 (1961)

Pearson, G. L., Bardeen, J.: Phys. Rev. *75*, 865 (1949)

Rosenberg, A. J.: J. Chem. Phys. *33*, 665 (1960)

[4.5] Schottky, W., Rothe, H.: Handbuch der Experimentalphysik Wien, W., Harms, F., (eds.), Vol. 13, Part 2. Leipzig: Akademische Verlagsges. 1928

5 An Example of Electronic Disorder.
Electrons and Electron Defects in α-Ag$_2$S

5.1 Disorder in α-Ag$_2$S and Coulometric Titration Curve

An instructive example to discuss electronic disorder in solid compounds is α-Ag$_2$S, the cubic high-temperature modification of silver sulfide being stable above 179 °C; this compound has been studied by Wagner [5.1], Miyatani [5.2], the author, Wedde and Sattler [5.3] and more recently by Bonnecaze, Lichanot and Gromb [5.9]. In 1931, Tubandt and Reinhold [5.4] found that the conductivity of α-silver sulfide being in equilibrium with metallic silver is about 20 times higher than that of α-Ag$_2$S which has been equilibrated with sulfur. This behaviour has been substantiated by more recent measurements, which show that the concentration of the free electrons varies also by a factor of about 20. However, for a rigorous thermodynamic treatment, which will be given below, it is essential to note that the thermodynamic activity of the electrons changes by a factor of about 100 concomitant to the concentration change stated above. This means that the thermodynamic behaviour of the electrons in α-Ag$_2$S cannot be understood on the basis of the Boltzmann approximation. It is, however, possible to explain their behaviour quantitatively if the electrons are assumed to be degenerate thus making the use of Fermi-Dirac statistics necessary. Proceeding in this way theory and experiment come together, i.e. quantitative agreement will be established.

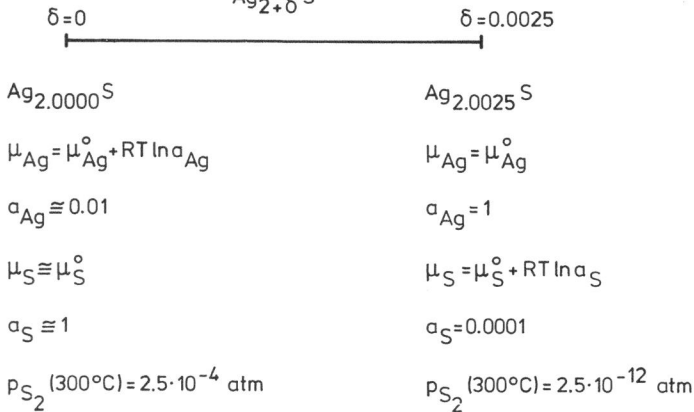

Fig. 5.1. Limits of the stoichiometric existence of Ag$_{2+\delta}$S together with the most important thermodynamic data

At 300 °C the stoichiometric width of existence of α-Ag₂S lies between the limit Ag$_{2.0000}$S for equilibrium with sulfur and the limit Ag$_{2.0025}$S for equilibrium with silver; this is represented schematically in Fig. 5.1. δ denotes the deviation from the ideal stoichiometric composition. As will be shown below, Ag₂S is approximately in equilibrium with sulfur if δ is equal to zero. The chemical potential of sulfur in Ag₂S corresponding to this situation is the same as that of liquid sulfur, i.e. $\mu_S(Ag_2S) = \mu_S^0(l)$. With proper choice of the standard state for the chemical potential of sulfur in Ag₂S, i.e. $\mu_S^0(Ag_2S) \equiv \mu_S^0(l)$, it follows that the thermodynamic activity of sulfur denoted by $a_S(Ag_2S)$ is equal to 1 whereby that of silver denoted by $a_{Ag}(Ag_2S)$ is approximately 0.01. The partial pressures of the different sulfur molecules have values like those in saturated sulfur vapour, e.g. the partial pressure of S₂-molecules at 300 °C is approximately 2.5×10^{-4} atm.

If α-Ag₂S is in equilibrium with silver, the deviation δ from ideal stoichiometry is 0.0025. In this case the chemical potential of silver in Ag₂S is equal to that of metallic silver, i.e. $\mu_{Ag}(Ag_2S) = \mu_{Ag}^0(s)$. Defining the standard state of the chemical potential of silver in Ag₂S according to the relation $\mu_{Ag}^0(Ag_2S) \equiv \mu_{Ag}^0(s)$ the activity of silver in Ag₂S is equal to 1.

According to X-ray studies carried out by Rahlfs [5.5] the sulfur ions constitute a body-centred cubic lattice with two S²⁻-ions in the conventional unit cell, while the corresponding four Ag⁺-ions are more or less statistically distributed over 42 lattice sites. Because of this large number of lattice sites available for the silver ions in Ag₂S one may conclude, as Wagner [5.1] has pointed out, that small deviations from the ideal stoichiometry do not cause any virtual change in the chemical potential of the silver ions. Therefore, the following is valid

$$\mu_{Ag^+} \cong const. \ (T, p). \tag{5.1}$$

For notational simplicity, we have written μ_{Ag^+} instead of $\mu_{Ag^+}(Ag_2S)$ implying the latter. If there is no explicit indication of the phase referred to, this abbreviated notation will be applied in the following equations.

The conclusion in Eq. (5.1) is in accordance with the fact that the partial conductivity of the silver ions is virtually independent of changes in stoichiometry [5.6]. The chemical potential of silver in Ag₂S can be split into that of silver ions and that of electrons:

$$\mu_{Ag} = \mu_{Ag^+} + \mu_e. \tag{5.2}$$

Since μ_{Ag^+} is virtually constant, changes in the chemical potential of silver must be due to changes in the chemical potential of the electrons; therefore, the following is true

$$d\mu_e = d\mu_{Ag}, \tag{5.3}$$

i.e. the thermodynamic behaviour of Ag₂S is solely determined by changes in the concentration of the free electrons. The variation of the chemical potential of silver resulting from changes in stoichiometry may be studied very precisely by means of an electrochemical method, the so-called electrochemical or coulometric titration. Wagner [5.1] used the following solid-state galvanic cell

$$Pt \mid Ag \mid AgI \mid Ag_2S \mid Pt. \tag{5.I}$$

Silver iodide mereley serves as a virtually pure ionic conductor for silver ions. It will be shown in Chap. 8 that the emf E of this cell is related to the difference between the chemical potential of silver in Ag$_2$S, $\mu_{Ag}(Ag_2S)$, and that of metallic silver, conveniently taken as the standard state, i.e. $\mu_{Ag}^0(Ag_2S) \equiv \mu_{Ag}^0(s)$.

The following relationship applies:

$$\mu_{Ag}(Ag_2S) - \mu_{Ag}^0(s) = -EF, \tag{5.4}$$

F is Faraday's constant. From Eqs. (5.3) and (5.4) we obtain the following using the abbreviated notation mentioned above:

$$d\mu_e = d\mu_{Ag} = -F\,dE, \tag{5.5}$$

i.e. a change in the measured emf of the galvanic cell (5.I) reflects a change in the chemical potential of the electrons in Ag$_2$S.

Using the galvanic cell (5.I), the stoichiometry of Ag$_2$S can be varied definitely by allowing a certain current to flow through the cell for a certain period of time. Applying the positive pole on the left-hand side of cell (5.I), silver ions will flow through silver iodide and electrons via the platinum lead on the right-hand side into Ag$_2$S so that the amount of silver in Ag$_2$S is increased according to Faraday's law stated in the following form:

$$\Delta n_{Ag} = \int_0^t \frac{I\,dt}{F}, \tag{5.6a}$$

Δn_{Ag} is the number of moles of silver incorporated in Ag$_2$S and I is the instantaneous electric current flowing through the cell at time t. For stationary current flow, i.e. I = const., Eq. (5.6a) may be simplified to give

$$\Delta n_{Ag} = \frac{It}{F}. \tag{5.6b}$$

The corresponding stoichiometry change of Ag$_{2+\delta}$S is given by

$$\Delta\delta = \frac{It}{n_S F}. \tag{5.7}$$

It is caused by this addition of silver, where n_S is the number of moles of sulfur in Ag$_{2+\delta}$S. As a result of conductivity measurements and thermodynamic calculations, the additional silver is completely dissociated into ions and electrons, i.e. addition of each silver atom causes either the production of a free electron or the annihilation of an electron defect. At the so-called stoichiometric point, where the deviation δ from ideal stoichiometry is zero, the concentration n of the electrons may be assumed to be equal to that of the electron defects p. If there are deviations from ideal stoichiometry, the expression for δ involving the concentrations of electrons and electron defects is given by

$$\delta = \frac{V_m}{N_A}\,(n - p), \tag{5.8}$$

N_A is Avogadro's number and V_m the molar volume of Ag$_2$S.

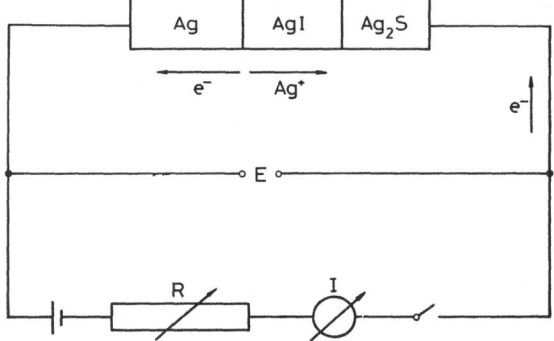

Fig. 5.2. Schematic diagram of the set-up used by C. Wagner [5.1] for the electrochemical titration of Ag₂S

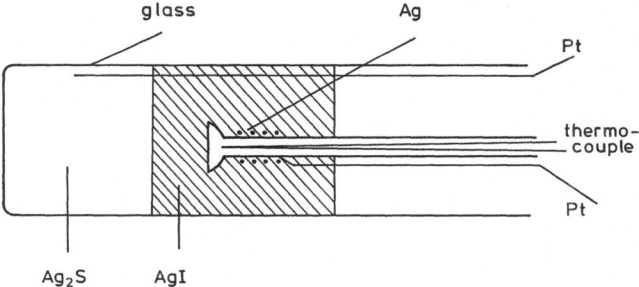

Fig. 5.3. Experimental set-up for the electrochemical titration of Ag₂S using a gas-tight sample [5.3]

A schematic representation of the set-up for the electrochemical titration of Ag₂S is shown in Fig. 5.2. Figure 5.3 shows an arrangement suitable for carrying out this electrochemical titration up to equilibrium with liquid sulfur. The Ag₂S sample is sealed so that the evaporation of sulfur out of Ag₂S, which causes stoichiometric changes, is made impossible even at high sulfur activities.

5.2 Quantitative Evaluation of Coulometric Titration Curves of α-Ag₂S

Figure 5.4 shows the emf of cell (5.I) as a function of the deviation from ideal stoichiometry at 160, 200 and 300 °C. These curves can be quantitatively interpreted using the above arguments together with the results obtained in Chap. 4. According to Eq. (4.5.4) the concentration of free electrons is given by

$$n = \left(\frac{2\pi m_e^* kT}{h^2}\right)^{3/2} \cdot \frac{4}{\pi^{1/2}} F_{1/2}\left(\frac{E_F - E_C}{kT}\right) \tag{5.9}$$

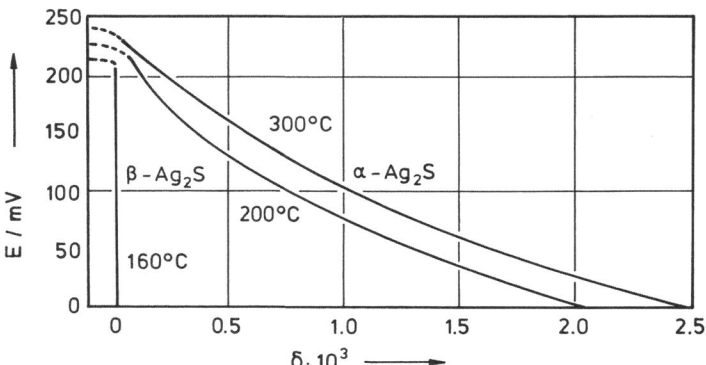

Fig. 5.4. Coulometric titration curves for Ag$_2$S at 160, 200 and 300 °C [5.3]

(m_e^* = effective mass of the electron, k = Boltzmann's constant, h = Planck's constant). The argument of the Fermi-Dirac integral "one-half" is related to the activity of the electrons according to the following equation

$$\frac{E_F - E_C}{kT} = \frac{\mu_e - \mu_e^0}{RT} = \ln a_e. \tag{5.10}$$

In this case the effective state density N_C of the electrons in the conduction band is chosen as the reference concentration n^0, thereby defining the standard state of the electrons. The standard state of the electrons where $\mu_e = \mu_e^0$ and the standard state of silver ($\mu_{Ag} = \mu_{Ag}^0$) therefore refer to two completely different states of Ag$_2$S. While $\mu_{Ag} = \mu_{Ag}^0$ holds for equilibrium of Ag$_2$S with pure silver, μ_e^0 is determined by choosing a standard concentration of the electrons, in this case N_C, which is not the value of the electron concentration in silver sulfide being equilibrated with silver. The chemical potential of the electrons in Ag$_2$S at equilibrium with silver will be denoted by μ_e^*:

$$\mu_e = \mu_e^* \quad \text{if} \quad \mu_{Ag} = \mu_{Ag}^0. \tag{5.11}$$

Integration of Eq. (5.3) leads to the relation

$$\mu_e - \mu_e^* = \mu_{Ag} - \mu_{Ag}^0. \tag{5.12}$$

With the aid of Eq. (5.4) the term on the right-hand side of Eq. (5.12) can be replaced to give

$$\mu_e = \mu_e^* - EF. \tag{5.13}$$

As was already stated above the emf of the galvanic cell (5.I) is a measure of the chemical potential of the electrons in Ag$_2$S. In order to carry out the following calculations it is useful to modify Eq. (5.13) by introducing the quantity μ_e^0:

$$\mu_e - \mu_e^0 = \mu_e^* - \mu_e^0 - EF. \tag{5.14}$$

Noting the validity of Eqs. (5.10) and (5.14) we may now introduce the emf of cell (5.I) according to Eq. (5.14) into the argument of the Fermi-Dirac integral

$F_{1/2}$, which occurs in Eq. (5.9):

$$n = \left(\frac{2\pi m_e^* kT}{h^2}\right)^{3/2} \frac{4}{\pi^{1/2}} F_{1/2}\left(\frac{\mu_e^* - \mu_e^0 - EF}{RT}\right). \tag{5.15}$$

Equation (5.15) represents a relationship between the concentration of free electrons in Ag$_2$S and the emf of cell (5.I), the latter being determined by the chemical potential of silver in Ag$_2$S. Up to now, Eq. (5.15) contains two undetermined quantities:

a) The effective mass of the quasi-free electrons m_e^*
b) The difference between the chemical potential of the electrons in Ag$_2$S at equilibrium with metallic silver μ_e^* and that belonging to the standard state μ_e^0.

Two findings are available to calculate these two quantities:

a) The concentration n* of the electrons in Ag$_2$S at equilibrium with silver. This concentration is virtually equal to the deviation δ* from ideal stoichiometry at that point. Using Eq. (5.8), we may neglect the relatively small concentration of the electron defects at this point compared to that of the electrons. Hence, the following approximation applies:

$$n(\mu_{Ag} = \mu_{Ag}^0) = n* = \frac{N_A}{V_m} \delta*. \tag{5.16}$$

From experimental data we obtain

$$n*(200\,°C) = 3.6 \times 10^{19} \text{ cm}^{-3},$$
$$n*(300\,°C) = 4.4 \times 10^{19} \text{ cm}^{-3}.$$

b) The slope of the graph of the function $E = f(n)$ at point n* which, in this case, is virtually proportional to the slope of the plot of E versus δ at point δ*.

With the aid of n* and the gradient of the titration curve at δ*, i.e. at equilibrium with silver, we obtain the ratio of the derivative of the Fermi-Dirac function with respect to its argument and the Fermi-Dirac function itself for the argument $(\mu_e^* - \mu_e^0)/RT$:

$$\frac{1}{n*}\left(\frac{dn}{d\left(\frac{EF}{RT}\right)}\right)_{\mu_{Ag} = \mu_{Ag}^0} = \frac{1}{\delta*}\left(\frac{d\delta}{d\left(\frac{EF}{RT}\right)}\right)_{\mu_{Ag} = \mu_{Ag}^0}$$

$$= -\frac{F_{1/2}'\left(\frac{\mu_e^* - \mu_e^0}{RT}\right)}{F_{1/2}\left(\frac{\mu_e^* - \mu_e^0}{RT}\right)}. \tag{5.17}$$

With the aid of the experimental data we can calculate the left-hand side of Eq. (5.17). By referring to the compilations of $F_{1/2}$- and $F_{1/2}'$-values found in the tables of Dougall and Stoner [5.7] it is now possible to determine the argument $(\mu_e^* - \mu_e^0)/RT$. Proceeding in this way the following values for $\mu_e^* - \mu_e^0$ result: 4.1 RT at 200 °C and 3.7 RT at 300 °C. According to Eq. (5.10), these values correspond to the difference between the Fermi energy E_F and the lower edge of the conduction band E_C, i.e. they give the position of the Fermi energy in the band scheme, e.g. at 300 °C E_F lies 0.16 eV above the lower edge of the conduction band if Ag_2S is in equilibrium with silver.

If the values of n* and $\mu_e^* - \mu_e^0$ are inserted into Eq. (5.15), i.e. the concentration of the electrons in the equilibrium of Ag_2S with silver and the chemical potential of the electrons at this point minus that in the standard state, we can calculate the effective mass m_e^* of the electrons which is equal to 0.23 times the actual mass of the electrons. This value is in good agreement with that obtained by Geserich et al. [5.8] from optical measurements and is virtually independent of deviations from ideal stoichiometry.

After the determination of the values of the two parameters, which occur in Eq. (5.15), namely the difference $\mu_e^* - \mu_e^0$ and the effective mass of the free electrons m_e^*, we can use Eq. (5.15) to evaluate the concentration n of the free electrons as a function of the difference $\mu_e - \mu_e^0$ which is equivalent to the difference $E_F - E_C$, i.e. the Fermi energy minus the lower edge of the conduction band. The results are given in Fig. 5.5 which contains further information in addition to the concentration of the free electrons that will be discussed in the following. We start with the discussion of the concentration of the electron defects as a function of the chemical potential of the electrons or of the emf. E. Since in the case of metal excess the concentration of electron defects is much smaller than that of the electrons, it is reasonable to assume that the Boltzmann approximation may be applied to describe the behaviour of the electron defects in the complete range of existence of Ag_2S. From the equilibrium between electrons and electron

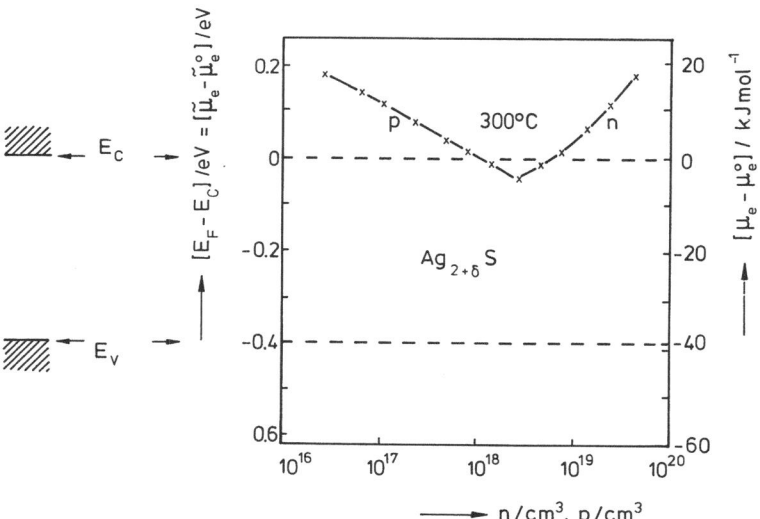

Fig. 5.5. Variation of the concentrations of the quasi-free electrons n and the electron defects p respectively as a function of the chemical potential of the electrons μ_e and the Fermi energy E_F respectively at 300 °C. E_C, lower edge of the conduction band; E_V, upper edge of the valence band

defects according to

$$e + h = 0 \qquad (5.18)$$

it follows for the chemical potentials μ_e of the electrons and μ_h of the electron defects

$$\mu_e = -\mu_h \qquad (5.19)$$

Using this equation and the Boltzmann approximation, we get the following for the concentration of the electron defects

$$p = p' \exp\left(-\frac{\mu_e - \mu_e'}{RT}\right) \qquad (5.20)$$

or

$$p = p' \exp\left(\frac{(E - E')\,F}{RT}\right). \qquad (5.21)$$

The index ′ denotes that the values of the quantities involved refer to the stoichiometric point i.e. $\delta = 0$. E′ is virtually equal to the value of the emf found for equilibrium between Ag₂S and liquid sulfur. This was pointed out by C. Wagner [5.1] who compared the titration curves of α-Ag₂S and β-Ag₂S. β-Ag₂S has a much smaller stoichiometric width of existence than α-Ag₂S, i.e. it shows a much smaller deviation from ideal stoichiometry. If β-Ag₂S is heated, transformation into α-Ag₂S occurs which is virtually in equilibrium with sulfur. We may thus conclude that α-Ag₂S in equilibrium with sulfur has a virtually ideal stoichiometry. At this point the concentration p′ of the electron defects is equal to the concentration of the electrons n′ which can be calculated with the aid of Eq. (5.15). For example, at 300 °C one obtains $n' = p' = 2.8 \times 10^{18}$ particles · cm⁻³. Hence, using this

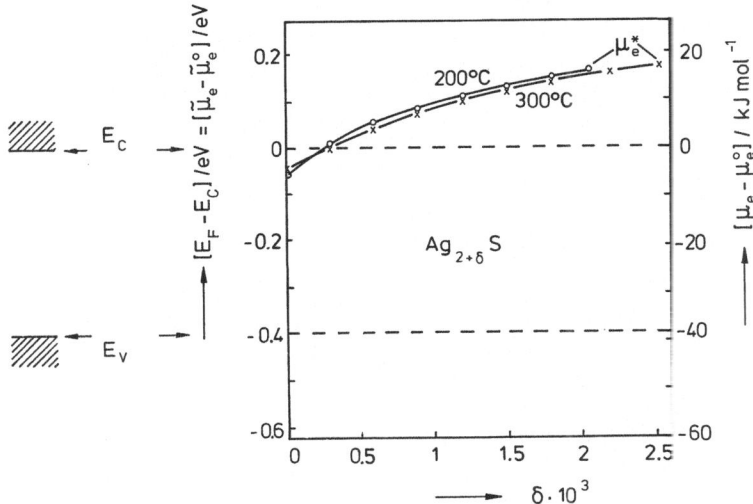

Fig. 5.6. Variation of the chemical potential of the electrons and of the Fermi energy E_F respectively in α-Ag₂S as a function of δ at 200 and 300 °C. E_C, E_V see Fig. 5.5

value and Eq. (5.20) or (5.21), the concentration of the electron defects can be calculated as a function of μ_e or the emf E.

In a semilogarithmic plot shown in Fig. 5.5 we obtain a straight line. With the values of the concentration n of the electrons and p of the electron defects Eq. (5.8) can be used to calculate the deviation δ from ideal stoichiometry. The values obtained in this way agree well with the measured ones. Since the emf E can be related to $\mu_e - \mu_e^0$ or to $E_F - E_C$ with the help of Eqs. (5.14) and (5.10), one can plot δ as a function of these quantities. This is illustrated in Fig. 5.6. The curves shown in this figure reflect the dependence of the Fermi energy of the electrons in Ag$_2$S on the deviation from ideal stoichiometry. The calculation of the value of the band gap (result: $E_G = 0.4$ eV) will be discussed below.

5.3 Entropy and Enthalpy of the Electrons in Ag$_2$S

In addition to the relation between the chemical potential μ_e and the concentration of the electrons, the temperature dependence of the emf of cell (5.I) allows further partial molar quantities of the electrons, e.g. entropy and enthalpy, to be determined as a function of the stoichiometry of α-Ag$_2$S.

But there is one important point to account for if the experimental data are combined. Whereas the chemical potential of the electrons μ_e is virtually the same for all electrons regardless of whether they occupy states in the conduction band or states in the valence band, the partial molar entropy and the partial molar enthalpy are sums of distinct and generally different contributions from electrons in the various bands, the conduction band or the valence band.

We thus speak of the partial molar quantities of the electrons as components (suffix "e") which result from two distinct contributions: that due to the quasi-free electrons in the conduction band (suffix "n") and that due to the electron defects in the valence band (suffix "p"). Both contributions must be taken into account, because the electrons which are generated within the Ag$_2$S via dissolution of silver in Ag$_2$S, either occupy states in the conduction band leading to the formation of free electrons or states in the valence band leading to the annihilation of electron defects.

Returning to our experimental data for the temperature dependence of the emf of cell (5.I), we may first calculate the entropy change corresponding to the dissolution of one mole silver into Ag$_2$S whereby the latter has a definite stoichiometric composition. Deriving Eq. (5.4) with respect to temperature gives the following:

$$\left(\frac{\partial E}{\partial T}\right)_{p,N_J} = -\frac{1}{F}\left[\left(\frac{\partial \mu_{Ag}(Ag_2S)}{\partial T}\right)_{p,N_J} - \left(\frac{\partial \mu_{Ag}^0}{\partial T}\right)_{p,N_J}\right]. \qquad (5.22)$$

The right-hand side of Eq. (5.22) may be rewritten using the fundamental thermodynamic identity $(\partial \mu_i/\partial T)_{p,N_J} = -\bar{S}_i$

$$\left(\frac{\partial E}{\partial T}\right)_{p,N_J} = \frac{1}{F}\left[\bar{S}_{Ag}(Ag_2S) - S_{Ag}^0(s)\right]. \qquad (5.23)$$

Here $\bar{S}_{Ag}(Ag_2S)$ is the partial molar entropy of silver in Ag$_2$S and $S^0_{Ag}(s)$ the molar entropy of pure silver equal to the entropy of silver in the standard state. If, in addition to the chemical potential of silver, the partial molar entropy of Ag is known, we can also calculate the enthalpy change associated with the dissolution of silver into Ag$_2$S:

$$\bar{H}_{Ag}(Ag_2S) - H^0_{Ag}(s) = -EF + T[\bar{S}_{Ag}(Ag_2S) - S^0_{Ag}(s)]. \qquad (5.24)$$

Here, $\bar{H}_{Ag}(Ag_2S)$ is the partial molar enthalpy of silver in Ag$_2$S and $H^0_{Ag}(s)$ the partial molar enthalpy of pure silver which is chosen as standard state. As already mentioned the chemical potential μ_{Ag^+} of the silver ions in Ag$_2$S is virtually independent of small deviations from ideal stoichiometry because of the statistical distribution of the silver ions over a large number of lattice sites. Applying the same argument, we may conclude that a change of the partial molar enthalpy of silver due to a variation in stoichiometry is identical with the change of the partial molar enthalpy $\bar{H}_e$ of the electrons, the latter being considered as a component of the system Ag$_2$S. Therefore, the following relationship is valid

$$d\bar{H}_{Ag} = d\bar{H}_e. \qquad (5.25)$$

Correspondingly, a change in the partial molar entropy $\bar{S}_{Ag}$ of the silver in Ag$_2$S is identical with the change in the partial molar entropy $\bar{S}_e$ of the electrons:

$$d\bar{S}_{Ag} = d\bar{S}_e. \qquad (5.26)$$

In this way it is possible to determine changes of partial molar quantities of the electrons as component as a function of stoichiometry in measuring the partial molar quantities of silver in Ag$_2$S as a function of stoichiometry.

By incorporating dn_{tot} electrons per unit volume into Ag$_2$S a part dn of them may enter the conduction band whereby the number of electron defects may be changed by dp. The infinitesimal enthalpy change $\bar{H}_e \, dn_{tot}$ of the electrons as component contains the contribution $\bar{H}_n \, dn$ resulting from the electrons and the one resulting from the electron defects, i.e. $\bar{H}_p \, dp$. The partial molar enthalpy of the electrons as component is then given as follows:

$$\bar{H}_e = \bar{H}_n \frac{dn}{dn_{tot}} + \bar{H}_p \frac{dp}{dn_{tot}}. \qquad (5.27)$$

According to this expression the bulk property $\bar{H}_e$, which is characteristic of the electrons considered as a component of the sample, is composed of the contributions of the partial molar enthalpy $\bar{H}_n$ of the free electrons and that of $\bar{H}_p$ of the electron defects. A similar expression holds for the partial molar entropy of the electrons as component $\bar{S}_e$, the partial molar entropy of the quasi-free electrons $\bar{S}_n$ and that of the electron defects $\bar{S}_p$

$$\bar{S}_e = \bar{S}_n \frac{dn}{dn_{tot}} + \bar{S}_p \frac{dp}{dn_{tot}}. \qquad (5.28)$$

For the case of ideal behaviour of the quasi-free electrons and electron defects we may apply the law of mass action in the simple form

$$n \cdot p = K(p, T). \qquad (5.29)$$

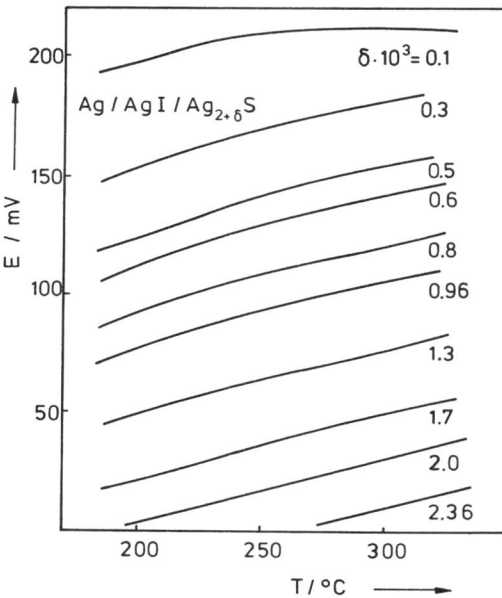

Fig. 5.7. Emf E of the solid-state galvanic cell Ag | AgI | Ag$_2$S as a function of the temperature for different deviations δ from ideal stoichiometry

In this case we get simple expressions for Eq. (5.27) and (5.28), respectively. In Ag$_2$S, however, the quasi-free electrons do not behave ideally because they are degenerate. In this case Eq. (5.27) and (5.28) may be evaluated using a graphical procedure.

The temperature dependence of the emf of cell (5.I) is illustrated in Fig. 5.7. Figures 5.8 and 5.9 show the molar entropies and enthalpies of dissolution of silver into Ag$_2$S. If an additive term of constant magnitude is neglected, these curves represent the partial molar enthalpy and the partial molar entropy of the electrons as components according to Eqs. (5.27) and (5.28). In the following the partial molar enthalpy $\overline{H}_e$ of the electrons as components will be treated in more detail.

The quantities dn/dn_{tot} and dp/dn_{tot} have been evaluated graphically. The partial molar enthalpy $\overline{H}_n$ of the electrons in the conduction band may be determined by differentiation of Eq. (5.9) with respect to T for constant n, taking into account Eq. (5.10) and the relationship which connects $\overline{H}_n$ to the temperature dependence of μ_e/T:

$$\overline{H}_n = -T^2 \frac{\partial}{\partial T}\left(\frac{\mu_e}{T}\right)_{p,N_J} = \frac{3}{2}\frac{F_{1/2}}{F'_{1/2}} \cdot RT. \tag{5.30}$$

Here $F'_{1/2}$ is the derivative of $F_{1/2}$ with respect to μ_e/RT. For determining the partial molar enthalpy $\overline{H}_p$ of the electron defects we can make use of the approximation holding for low concentrations. This is given by

$$\overline{H}_p = E_G + \frac{3}{2}RT. \tag{5.31}$$

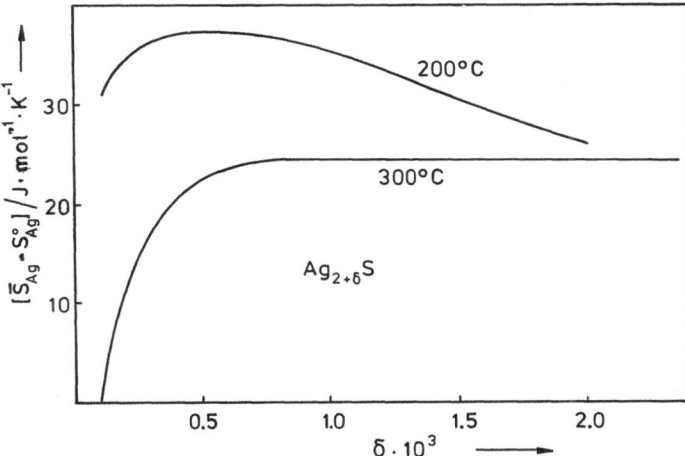

Fig. 5.8. Entropy change due to the dissolution of one mole pure silver into α-Ag$_2$S as a function of δ at 200 and 300 °C

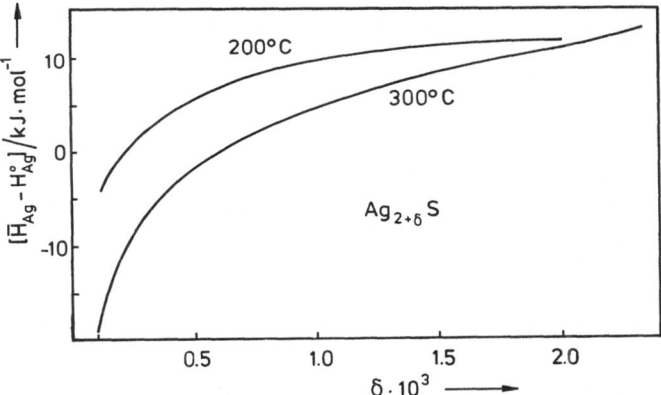

Fig. 5.9. Enthalpy change due to the dissolution of one mole pure silver into α-Ag$_2$S as a function of δ at 200 and 300 °C

The results obtained in this way are shown in Fig. 5.10. The value of 0.4 eV for the band gap E_G in Fig. 5.10 results from fitting the calculated values to the measured curves. A closer look at Fig. 5.10 reveals the degeneracy of the electrons. Since the electrons are confined in a virtually constant volume within the Ag$_2$S-lattice, their enthalpies are nearly identical to their internal energies. It follows from statistical thermodynamics that the internal molar energy of the electrons has the value 3/2 RT if the electrons are assumed to behave like an ideal gas. $\overline{H}_n$ approaches this low value only at the stoichiometric point, i.e. where equilibrium between Ag$_2$S and sulfur is attained.

If one approaches the region of equilibrium with silver, the value of $\overline{\overline{H}}_n$ becomes more than two times as high as that found for the ideal case. According to Eq.

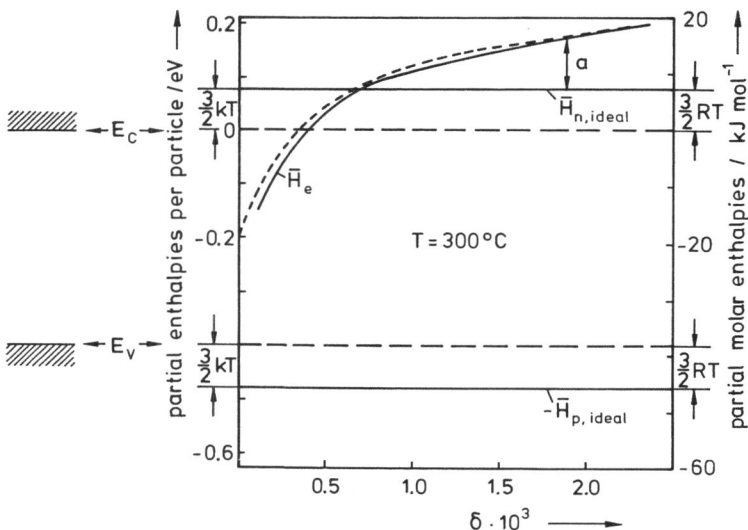

Fig. 5.10. Partial molar enthalpy of the electrons as a component $\overline{H}_e$, of the quasi-free electrons $\overline{H}_n$ and of the electron defects $\overline{H}_p$ respectively in α-Ag$_{2+\delta}$S as a function of δ (E$_C$, E$_V$ see Fig. 5.5). E$_g \approx 0.4$ eV $= 39$ kJ/mol; a, contribution due to degeneracy; —— measured, - - - - calculated values

(4.5.10) the value of E$_G$ may also be determined from the temperature dependence of the intrinsic concentration of the electrons and electron defects respectively. Recently, this has been carried out by Bonnecaze, Lichanot and Gromb [5.9].

The value of E$_G$ resulting from this approach is 0.38 eV which is in quite good agreement with that one chosen in Fig. 5.10. The discussion of the behaviour of the electrons in Ag$_2$S has shown that electrochemical methods, which make use of thermodynamic concepts for the treatment of electrons, may also be important in the field of semiconductor physics.

Further measurements for determining the electronic properties of mixed conducting solid compounds have been carried out by Miyatani [5.10]. For details the reader is referred to the literature.

5.4 References

[5.1]   Wagner, C.: J. chem. Phys. *21*, 1819 (1953)
[5.2]   Miyatani, S.: J. Phys. Soc. Japan *10*, 786 (1955)
[5.3]   Rickert, H., Sattler, V., Wedde, Ch.: Z. phys. Chem. N.F. *98*, 339 (1975)
[5.4]   Tubandt, C., Reinhold, H.: Z. Elektrochem. *37*, 589 (1931)
[5.5]   Rahlfs, P.: Z. phys. Chem. *B31*, 157 (1935)
[5.6]   Rickert, H.: Z. phys. Chem. *23*, 355 (1960)
[5.7]   McDougall, J., Stoner, E. C.: Trans. Roy Soc. (London) *A 237*, 67 (1938)
[5.8]   Geserich, H. P.: Phys. Stat. Sol. *37*, K 85 (1970)
[5.9]   Bonnecaze, G., Lichanot, A., Gromb, S.: J. Phys. Chem. Solids *39*, 813 (1978)
[5.10] Miyatani, S.: J. Phys. Soc. Japan, *24* (2), 328 (1968); *10* (9), 786 (1955)

6 Mobility, Diffusion and Partial Conductivity of Ions and Electrons

In this chapter we shall be concerned with diffusion of ions and electrons in a concentration or activity gradient and with their transport in an electric field. These two processes are often superimposed in which case it is more appropriate to speak of transport in an electrochemical potential gradient. The expressions to be derived are valid for the fluxes of each of the various types of ions and electrons separately. From these expressions the equations for cases in which the transport of various types of particles is interconnected can also be derived; an example is to be found in Chap. 8. Our discussion can be divided into

a) a phenomenological treatment in which the transport processes are described in terms of experimentally measurable quantities such as diffusion coefficients, mobilities, partial conductivities or Onsager coefficients, and
b) a statistical treatment in which the quantities involved in the phenomenological treatment are related to atomistic quantities.

6.1 Phenomenological Treatment of Transport Processes

6.1.1 Diffusion Equations

We shall first consider the special case in which only a concentration or activity gradient of one type of particle is present in the absence of an electrical field. From the microscopic point of view the individual particles move not merely in one but in all possible directions. If no concentration gradient is present, the system is in dynamic equilibrium; the currents flowing in opposite directions are then exactly equal and cancel each other. If, however, a concentration gradient exists, net particle flux results, i.e. the particle current through one unit area in one direction minus the current in the opposite direction is non-zero. The net particle flux is called the diffusion flux j which is the amount of particles flowing through a unit area in a unit time.

When transport processes in solid compounds are considered, one type of particle is often much more mobile than the other. In such cases the partial lattice of the virtually immobile type of particle is chosen as the reference system so that a discussion concerning the choice of the reference system becomes unnecessary. With regard to this point the reader is referred to more detailed treatments which are to be found in the literature [6.1].

In 1855, Fick [6.2] stated the following empirical relationship involving the diffusion flux j_x in a concentration gradient $\partial c / \partial x$

$$j_x = -D \frac{\partial c}{\partial x}. \tag{6.1.1}$$

This is known as Fick's first law; the constant D is called the diffusion coefficient. It will sometimes be referred to as Fick's diffusion coefficient to distinguish it from diffusion coefficients which have been defined in other ways. The concentration gradient in Eq. (6.1.1) is assumed to be only in the x-direction. If there is a concentration gradient in any direction in space, we obtain the more general form of Fick's first law:

$$\vec{j} = -D \text{ grad } c = -D \nabla c \tag{6.1.2}$$

where

$$\text{grad } c = \nabla c = \frac{\partial c}{\partial x} \frac{\vec{x}}{|x|} + \frac{\partial c}{\partial y} \frac{\vec{y}}{|y|} + \frac{\partial c}{\partial z} \frac{\vec{z}}{|z|} = \left(\frac{\partial c}{\partial x}, \frac{\partial c}{\partial y}, \frac{\partial c}{\partial z} \right) \tag{6.1.3}$$

(∇: Nabla-operator). The negative sign is due to the fact that the diffusion current is in a opposite direction to the concentration gradient. In Eq. (6.1.2) we have assumed that the diffusion coefficient D is independent of the direction in the crystal, i.e. D is isotropic and can be written as a scalar quantity. We can only use a direction-independent diffusion coefficient for our calculations when we are dealing with either cubic or regular crystals or samples containing crystals whose orientations are randomly distributed. In the general and anisotropic case D must be expressed as a (second-order) tensor [6.1].

The variation of the concentration with time at a certain point can be obtained from Fick's first law and from continuity considerations if the particles considered are neither generated nor used up (e.g. by recombination); derivations of the resulting equation, known as Fick's second law, can be found in the literature. When we at first assume that there is only a diffusion flux in the x-direction according to Eq. (6.1.1) we obtain the expression

$$\frac{\partial c}{\partial t} = \frac{\partial}{\partial x} \left(D \frac{\partial c}{\partial x} \right). \tag{6.1.4}$$

If the diffusion coefficient D is concentration-independent, it is also independent of the spatial coordinate x and can be written before the differential quotient; we thus obtain

$$\frac{\partial c}{\partial t} = D \frac{\partial^2 c}{\partial x^2}. \tag{6.1.5}$$

If the concentration gradient, and thus the diffusion flux, has any direction in space we obtain instead of Eq. (6.1.4) the more general expression

$$\frac{\partial c}{\partial t} = \text{div } (D \text{ grad } c)$$

$$= \frac{\partial (D \text{ grad } c)_x}{\partial x} + \frac{\partial (D \text{ grad } c)_y}{\partial y} + \frac{\partial (D \text{ grad } c)_z}{\partial z}. \tag{6.1.6}$$

The indices x, y and z refer to the components of the vector (D grad c) in x-, y- and z-direction. If D is independent of the concentration c and the direction of the diffusion, Eq. (6.1.6) becomes

$$\frac{\partial c}{\partial t} = D \text{ div grad } c = D \triangle c, \qquad (6.1.7)$$

where $\triangle$ is the Laplace operator:

$$\triangle = \nabla^2 = \text{div grad} = \frac{\partial^2}{\partial x^2} + \frac{\partial^2}{\partial y^2} + \frac{\partial^2}{\partial z^2}. \qquad (6.1.8)$$

Mathematical solutions of the diffusion equation for various boundary and initial conditions have been given by Jost [6.3], Carslaw and Jaeger [6.4], Wagner [6.5], Crank [6.6], and Hauffe [6.7].

6.1.2 The Mechanical Mobility B_m

The spatial equilibrium for uncharged particles is characterized by the fact that the chemical potential μ is locally constant. If the particles are charged, the electrochemical potential η must be constant. For the case, however, where no electrical field is present the equilibrium condition that the chemical potential be locally constant, i.e. grad $\mu = 0$, is also valid for charged particles. Thus, if in the absence of an electrical field grad μ is not equal to zero, a particle flux is expected, and we assume that this is proportional to the gradient of the chemical potential. The negative gradient of the chemical potential referred to a single particle may be considered as a generalized force $\vec{F}$ acting on the particle.

$$\vec{F} = -\frac{1}{N_A} \text{ grad } \mu, \qquad (6.1.9\,\text{a})$$

N_A is Avogadro's number. When the chemical potential is not referred to one mole but to a single particle and given the symbol $\tilde{\mu}$, Eq. (6.1.9 a) can be expressed as

$$\vec{F} = -\text{grad } \tilde{\mu}. \qquad (6.1.9\,\text{b})$$

Since the particle flux is proportional to grad $\tilde{\mu}$ or to $\vec{F}$ the mean particle velocity $\vec{v}$ (sometimes called the drift velocity) is also proportional to the force $\vec{F}$ acting on a single particle. The proportionality constant is called the mechanical mobility B_m:

$$\vec{v} = B_m \vec{F}. \qquad (6.1.10)$$

Thus, the mechanical mobility B_m is defined as the ratio of the mean stationary particle velocity $\vec{v}$ to the force $\vec{F}$ which acts on the particle

$$B_m = \frac{|\vec{v}|}{|\vec{F}|}. \qquad (6.1.11)$$

Since the flux of a particular type of particle is equal to the product of the concentration c of these particles and their mean velocity $\vec{v}$,

$$\vec{j} = c\vec{v}, \tag{6.1.12}$$

we obtain the following expression from Eqs. (6.1.9a), (6.1.10) and (6.1.12)

$$\vec{j} = -\frac{cB_m}{N_A}\,\text{grad}\,\mu. \tag{6.1.13}$$

Equation (6.1.13) expresses the particle flux in terms of the gradient of the chemical potential μ and the mechanical mobility B_m.

6.1.3 The Nernst-Einstein Equation [6.8]

We shall first make the limiting assumption that, just as in the case of an ideal solution, the relation

$$\mu = \mu^0 + RT\ln\frac{c}{c^0} \tag{6.1.14}$$

is valid for the chemical potential; μ^0 is the chemical potential in the standard state, c the concentration of the particles, c^0 their concentration in the standard state, R the general gas constant and T the absolute temperature. The relation (6.1.14) is not expected to be valid for the components of a solid compound, but should be true for the defects, e.g. the vacancies. We shall further assume that there is a concentration gradient only in the x-direction; this does not cause a further limitation of the validity of our treatment, but merely makes the situation clearer. According to Eq. (6.1.14), $d\mu$ can be replaced by RT d ln c.
Thus, in the one-dimensional case, Eq. (6.1.13) can be expressed as follows

$$j_x = -\frac{cB_mRT}{N_A}\cdot\frac{d\ln c}{dx}. \tag{6.1.15}$$

Since c d ln c is equal to dc and R/N_A is equal to Boltzmann's constant k, we can rewrite Eq. (6.1.15) as follows

$$j_x = -B_mkT\frac{dc}{dx}. \tag{6.1.16}$$

By comparing Eq. (6.1.16) with Fick's first law (6.1.1) a relationship between the diffusion coefficient D for ideal behaviour of the diffusing particles (according to Eq. (6.1.14)) and the mechanical mobility B_m is obtained:

$$D = B_mkT \quad \text{(ideal behaviour)}. \tag{6.1.17}$$

This is the so-called Nernst-Einstein equation; it relates the diffusion coefficient D with the mechanical mobility for the case of ideal behaviour, i.e. if Eq. (6.1.14) for the chemical potential is valid.

The Nernst-Einstein equation (6.1.17) could also have been derived by rewriting Fick's first law (6.1.1), incorporating the expression for the chemical potential for the case of ideal behaviour (6.1.14) by replacement of dc by the expression $\frac{c\,d\mu}{RT}$. We should then have obtained the following equation

$$j_x = -\frac{Dc}{RT}\frac{d\mu}{dx} \quad \text{(ideal behaviour)}. \tag{6.1.18}$$

Equation (6.1.17) would then result from Eqs. (6.1.18) and (6.1.13).

6.1.4 The Component Diffusion Coefficient D_K

In the case of non-ideal behaviour of the system under study the more general expression

$$\mu = \mu^0 + RT \ln a \tag{6.1.19}$$

must be used for the chemical potential of the diffusing species, where a is the activity. The component diffusion coefficient D_K is defined in such a manner that an expression analogous to Eq. (6.1.18) is obtained:

$$j_x = -\frac{D_K c}{RT}\frac{d\mu}{dx}, \tag{6.1.20}$$

or, when the activity gradient is not only in the x-direction,

$$\vec{j} = -\frac{D_K c}{RT}\,\text{grad}\,\mu. \tag{6.1.21}$$

A comparison of Eqs. (6.1.21) and (6.1.13) again yields the Nernst-Einstein equation

$$D_K = B_m kT. \tag{6.1.22}$$

The Nernst-Einstein equation involving the component diffusion coefficient D_K is valid for ideal and non-ideal behaviour. We can also say that D_K is defined by the Nernst-Einstein equation. For the case of ideal behaviour, the component diffusion coefficient and the Fick diffusion coefficient are identical.

6.1.5 Relationship Between D and D_K

Since dc is equal to c d ln c, Fick's first law (Eq. (6.1.1)) can be expressed as follows:

$$j_x = -Dc\,\frac{d\ln c}{dx}. \tag{6.1.23}$$

From Eqs. (6.1.20) and (6.1.19) we can obtain an expression for the particle flux which contains the component diffusion coefficient D_K:

$$j_x = -D_K c\,\frac{d\ln a}{dx}. \tag{6.1.24}$$

A comparison of Eqs. (6.1.23) and (6.1.24) gives the following relationship between D_K and Fick's diffusion coefficient D:

$$D = D_K \frac{d \ln a}{d \ln c}.\tag{6.1.25}$$

As can be seen from Eq. (6.1.25), $D = D_K$ if the activity a is proportional to concentration c. The factor $d \ln a / d \ln c$ was introduced by Darken [6.9] and is called the "thermodynamic factor".

6.1.6 Relationship Between the Component Diffusion Coefficient D_K and the Tracer Diffusion Coefficient D_{Tr}

The definition of the component diffusion coefficient D_K is an appropriate one from two points of view. Firstly, as shown in Eq. (6.1.22), the Nernst-Einstein equation is universally valid, i.e. there is a general connection with the mechanical mobility B_m; secondly, there is a close connection with the tracer or self-diffusion coefficient D_{Tr}, which can be measured at electrochemical equilibrium by using a radioactive isotope. Under certain conditions (the reader is referred to the specialist literature [6.10]), e.g. for vacancy diffusion, the tracer diffusion coefficient D_{Tr} is related with the mechanical mobility B_m by

$$D_{Tr} = f B_m kT,\tag{6.1.26}$$

where f is the so-called correlation factor, sometimes also referred to as the Haven ratio, denoted as H_R [6.10]. The relationship between D_{Tr} and D_K can be obtained from Eqs. (6.1.22) and (6.1.26)

$$D_{Tr} = f D_K.\tag{6.1.27}$$

The correlation factor f is of the order of one for most cases.

6.1.7 Particle Flux in an Electrical Field

As long as the electrical field is not too strong, i.e. if the product of the electrical potential difference per atomic distance in the lattice and particle charge is much smaller than kT we can assume that there is a linear relationship between the flux of the charged particles or the electrical current density $\vec{i}$ and the electrical field $\vec{E}$; Ohm's law is then valid:

$$\vec{i} = \sigma \vec{E},\tag{6.1.28}$$

σ is the specific electrical conductivity and — for simplicity — is assumed to be independent of the direction in the crystal. Equation (6.1.28) can also be written in terms of the electrical potential:

$$\vec{i} = -\sigma \, \mathrm{grad} \, \varphi.\tag{6.1.29}$$

The total conductivity σ is the sum of the partial conductivities σ_i of the various types of particles i if the different charge carriers are transported independently of one another. This is true in most cases:

$$\sigma = \sum_i \sigma_i. \tag{6.1.30a}$$

The ratio of the partial conductivity σ_i of one type of particle to the total conductivity $\sigma = \sum_i \sigma_i$ is called the transference number t_i of this type of particle:

$$t_i = \frac{\sigma_i}{\sum_i \sigma_i}. \tag{6.1.30b}$$

For the following discussion we shall assume that only one type of particle is mobile, i.e. σ is equal to the σ_i of the mobile particles. The relation between the particle flux in units of mol cm^{-2} s^{-1} and the electrical current density $\vec{i}$ involves Faraday's constant F and the charge number z:

$$\vec{j} \,(\text{mol/cm}^2\,\text{s}) = \frac{\vec{i}}{zF}. \tag{6.1.31}$$

From Eqs. (6.1.29) and (6.1.31) we can obtain the relationship between the particle flux $\vec{j}$, the partial conductivity $\sigma_i = \sigma$ and the gradient of the electrical potential:

$$\vec{j} = - \frac{\sigma}{zF} \,\text{grad}\,\varphi. \tag{6.1.32}$$

In Eq. (6.1.32) the particle flux is written in terms of the partial conductivity σ. It is, however, also possible to express the particle flux in terms of relationships analogous to Eqs. (6.1.12) and (6.1.10) involving the mechanical mobility and the force $\vec{F}$ which acts on the particles. When only an electrical field is present the latter can be expressed as follows:

$$\vec{F} = -z\,\text{e}\,\text{grad}\,\varphi \tag{6.1.33}$$

where z is the valency of the particles and e the elementary charge; z is positive for cations and negative for anions and electrons. Using Eqs. (6.1.33), (6.1.12) and (6.1.10) we obtain:

$$\vec{j} = -cB_m z\,\text{e}\,\text{grad}\,\varphi \tag{6.1.34}$$

which is a description of the particle flux in terms of the mechanical mobility B_m and the electrical field $-\text{grad}\,\varphi$. Using the Nernst-Einstein equation (6.1.22), the mechanical mobility may be replaced by the component diffusion coefficient D_K and we obtain from Eq. (6.1.34)

$$\vec{j} = -\frac{cD_K z\,\text{e}}{kT} \,\text{grad}\,\varphi, \tag{6.1.35}$$

i.e. the particle flux is described in terms of the component diffusion coefficient D_K and the electrical field $-\text{grad } \varphi$. A comparison of Eqs. (6.1.32), (6.1.34) and (6.1.35) provides the relationship between the partial electrical conductivity σ, the mechanical mobility B_m and the component diffusion coefficient D_K:

$$\sigma = cB_m z^2\, eF, \tag{6.1.36}$$

$$\sigma = \frac{cD_K z^2 F^2}{RT}. \tag{6.1.37}$$

The relationship between the partial electrical conductivity of one type of particle i and the mechanical mobility is expressed by Eq. (6.1.36). The mechanical mobility B_m is defined as the ratio of the velocity of a particle to the force acting on that particle. The connection between σ and the component diffusion coefficient D_K is given by Eq. (6.1.37). The concentration must always be expressed in mol cm^{-3}.

In this section, when dealing with the electrical particle flux in an electrical field, we have made the assumption that the mechanical mobility used here is identical with that which was introduced when dealing with particle flux in an activity gradient. We shall show in the next section that this assumption is valid.

The so-called electrical mobility u is also often used; this is defined as the ratio of the magnitude of the mean particle velocity $\vec{v}$ to the magnitude of the electrical field $\vec{E}$:

$$u = \frac{|\vec{v}|}{|\vec{E}|}. \tag{6.1.38}$$

From Eqs. (6.1.10), (6.1.33) and (6.1.38) we can obtain the following relationship between electrical and mechanical mobility:

$$|z|\, eB_m = u. \tag{6.1.39}$$

6.1.8 Particle Flux in the Presence of a Concentration or Activity Gradient and an Electrical Field

If a concentration or activity gradient and an electrical field are both present we assume that the particle fluxes caused by the two types of forces can be linearly added to give the resulting particle flux. In this case, since Faraday's constant is equal to the product of e and N_A, Eqs. (6.1.13) and (6.1.34) give the expression

$$\vec{j} = -\frac{cB_m}{N_A} (\text{grad } \mu + zF \text{ grad } \varphi). \tag{6.1.40}$$

In Eq. (6.1.40) we have again assumed that the mechanical mobility B_m for the particle flux in a concentration gradient is identical with that used to describe the particle flux in an electrical field. If these mobilities were unequal and symbolized as B'_m and B''_m, it would be necessary to write expression (6.1.40) as follows:

$$\vec{j} = -\frac{cB'_m}{N_A} \, \text{grad} \, \mu - \frac{cB''_m}{N_A} \, zF \, \text{grad} \, \varphi. \tag{6.1.41}$$

It can be shown from Eq. (6.1.41) that a zero-particle flux for equilibrium conditions, i.e. the gradient of the electrochemical potential $\eta = \mu + zF\varphi$ is zero, can only result if B'_m is equal to B''_m. Using the definition $\eta = \mu + zF\varphi$, Eq. (6.1.40) can be expressed as follows:

$$\vec{j} = -\frac{cB_m}{N_A} \, \text{grad} \, \eta. \tag{6.1.42}$$

Equation (6.1.42) relates the particle flux $\vec{j}$, the mechanical mobility and the gradient of the electrochemical potential, i.e. Eq. (6.1.42) gives the particle flux for the general case. As a special case Eq. (6.1.42) involves the particle flux in a concentration or activity gradient and the particle flux in an electrical field. Using the Nernst-Einstein equation (6.1.22), Eq. (6.1.42) becomes

$$\vec{j} = -\frac{cD_K}{RT} \, \text{grad} \, \eta \tag{6.1.43}$$

and, using Eq. (6.1.37)

$$\vec{j} = -\frac{\sigma}{z^2 F^2} \, \text{grad} \, \eta. \tag{6.1.44}$$

Equations (6.1.42—44) are the general expressions for particle fluxes in the presence of a gradient of the electrochemical potential η. They contain the mechanical mobility B_m, the component diffusion coefficient D_K or the partial electrical conductivity σ. The particle flux $\vec{j}$ is expressed in Eq. (6.1.44) in units of mol $\times \text{cm}^{-2} \, \text{s}^{-1}$; in Eqs. (6.1.42) and (6.1.43) the dimensions of the particle flux depend on the dimensions chosen for the concentration.

Equations (6.1.42—44) are the basic relationships for the description of transport processes of particles in solids. We shall use these equations in Chap. 10 to discuss diffusion-controlled reactions in solids.

6.1.9 Description of Particle Fluxes Using the General Transport Equations of Irreversible Thermodynamics

The phenomenological equations of irreversible thermodynamics are used to describe irreversible processes (see Ref. [6.12]). These include transport processes of particles. The forces and fluxes involved in the thermodynamic treatment must be so chosen that the sum of the products of forces and fluxes is equal to the rate of entropy production per unit volume in the process considered. For this reason, the forces suitable for the description of particle fluxes are the gradients of η/T, where η_i is the electrochemical potential of the type of particle considered and T the absolute temperature. Considering the fluxes of two types of particles 1 and 2 we obtain the equations

$$\vec{j}_1 = L_{11} \operatorname{grad} \frac{\eta_1}{T} + L_{12} \operatorname{grad} \frac{\eta_2}{T},$$
(6.1.45)

$$\vec{j}_2 = L_{21} \operatorname{grad} \frac{\eta_1}{T} + L_{22} \operatorname{grad} \frac{\eta_2}{T},$$
(6.1.46)

L_{11}, L_{12}, L_{21} and L_{22} are the so-called Onsager coefficients. The particle flux $\vec{j}_1$ may be for example the flux of a particular type of ions and $\vec{j}_2$ that of electrons. The so-called reciprocity relation applies to the cross-coefficients

$$L_{12} = L_{21}.$$
(6.1.47)

In many cases where investigations have been carried out involving simultaneous ion and electron flux, it has been shown that L_{12} is virtually equal to zero, i.e. in processes of this type a coupling between electron and ion fluxes can be neglected. However, this statement does not always apply, though it is often true for ions which have a constant valency. In this case it can be assumed that

$$L_{12} = 0.$$
(6.1.48)

In principle, this assumption must be checked experimentally for each system considered. It is for example no longer valid if large electron currents are caused to flow. In this case coupling between the motion of electrons and ions is observed; this must be treated using coupling terms. Using Eq. (6.1.48), the expression or the ion flux $\vec{j}_1$ in Eq. (6.1.45) becomes

$$\vec{j}_1 = L_{11} \operatorname{grad} \frac{\eta_1}{T}.$$
(6.1.49)

A comparison of Eq. (6.1.49) with Eqs. (6.1.42—44) gives a relationship between the Onsager coefficient L_{11}, the mechanical mobility B_m, the component diffusion coefficient D_k, and the partial conductivity σ of the particles considered:

$$L_{11} = -\frac{cB_mT}{N_A},$$
(6.1.50)

$$L_{11} = -\frac{cD_kT}{RT},$$
(6.1.51)

$$L_{11} = -\frac{\sigma T}{z^2F^2}.$$
(6.1.52)

These three equations connect the phenomenological coefficient L_{11} of Eq. (6.1.45) with B_m, D_k and σ for the case that coupling terms (i.e. L_{12} and L_{21}) may be neglected. From the opposite point of view the above treatment shows that in Sects. (6.1.1—3) dealing with the transport of particles it was assumed that coupling terms can be neglected. If this were not true it would not be possible to obtain the total conductivity from the sum of the partial conductivities or to treat the fluxes as independent of one another.

6.1.10 Chemical Diffusion

For processes involving changes of stoichiometry in solids it is necessary on the ground of electrical neutrality that, apart from ions, electrons or electron defects must migrate simultaneously, i.e. the fluxes of ions and electrons are related to one another. The diffusion coefficient which describes such processes is called the chemical diffusion coefficient $\tilde{D}$. Darken [6.13] and Wagner [6.14] have treated the chemical diffusion coefficient $\tilde{D}$ theoretically. In the following discussion we shall establish the relationship between $\tilde{D}$ and the component diffusion coefficient or the partial conductivity of the mobile ion for a simple case. The experimental determination of the chemical diffusion coefficient for the model substances $Fe_{1-\delta}O$ and $Ag_{2+\delta}S$ is discussed in Sect. 6.6. For these compounds the partial conductivity σ_{X-} of non-metal ions is negligible in comparison with that of metal ions, and at the same time their electron partial conductivity is much larger than that of metal ions. Thus

$$\sigma_e \gg \sigma_{Me^+} \gg \sigma_{X-}. \tag{6.1.53}$$

An expression analogous to Fick's first law can be used to describe the flux of the metal j_{Me} relative to the virtually constant anion partial lattice:

$$j_{Me} = -\,\tilde{D}_{Me}\,\frac{\partial c_{Me}}{\partial x}, \tag{6.1.54}$$

where c_{Me} is the concentration of the metal in mol/cm^3. j_{Me} can be considered as the sum of the fluxes of the metal ions and the electrons. The latter can be expressed as follows in terms of partial conductivities σ_i and electrochemical potentials η_i according to Eq. (6.1.44):

$$j_{Me^+} = -\frac{\sigma_{Me^+}}{z_{Me}^2 F^2}\,\frac{\partial \eta_{Me^+}}{\partial x} \tag{6.1.55}$$

and

$$j_e = -\frac{\sigma_e}{F^2}\,\frac{\partial \eta_e}{\partial x}. \tag{6.1.56}$$

Since, on the grounds of electrical neutrality, the fluxes of metal ions, electrons and neutral metal must be equivalent,

$$j_{Me^+} = \frac{j_e}{z_{Me^+}} = j_{Me} \tag{6.1.57}$$

and since the sum of the electrochemical potentials η_{Me^+} and η_e is equal to the chemical potential μ_{Me} of the neutral metal, Eqs. (6.1.55—57) yield the following expression for j_{Me^+}:

$$j_{Me^+} = -\frac{\sigma_{Me^+}\sigma_e}{z_{Me^+}^2 F^2(\sigma_{Me^+} + \sigma_e)}\,\frac{\partial \mu_{Me}}{\partial x}. \tag{6.1.58}$$

In the special case considered where $\sigma_e \gg \sigma_{Me^+}$ we can obtain the simpler relationship

$$\dot{j}_{Me} = -\frac{\sigma_{Me^+}}{z_{Me^+}^2 F^2} \frac{\partial \mu_{Me}}{\partial x} \tag{6.1.59}$$

or, using the expression

$$\mu_{Me} = \mu_{Me}^0 + RT \ln a_{Me}$$

together with Eq. (6.1.37):

$$\dot{j}_{Me} = -c_{Me} D_{K,Me^+} \frac{\partial \ln a_{Me}}{\partial x} \tag{6.1.60}$$

or

$$\dot{j}_{Me} = -D_{K,Me^+} \frac{\partial \ln a_{Me}}{\partial \ln c_{Me}} \frac{\partial c_{Me}}{\partial x}. \tag{6.1.61}$$

From this equation and Eq. (6.1.54) we can obtain the relationship

$$\tilde{D}_{Me} = D_{K,Me^+} \frac{\partial \ln a_{Me}}{\partial \ln c_{Me}} \tag{6.1.62}$$

between the chemical diffusion coefficient $\tilde{D}$ and the component diffusion coefficient D_K.

6.2 Statistical Treatment of Transport Quantities

In the previous section the motion of particles was discussed using phenomenological equations and phenomenological coefficients such as the partial conductivity σ, the mobility B_m and the diffusion coefficients D or D_K. In this section we relate the phenomenological quantities D_K, B_m, σ etc. to microscopic (that is atomistic) ones.

In atomistic terms the movements of particles in solids — in this section we are mainly concerned with material particles, i.e. ions or atoms — consist of successive jumps from one lattice or interstitial lattice site to another site in the lattice. The mean distance travelled per jump is given the symbol r and the mean jump frequency the symbol ν whereby this jump frequency is defined as the number of jumps made by one particle per unit time (ν = n/t). We shall see that the jump distance r and the jump frequency ν at equilibrium are two very important quantities for the atomistic treatment. A further quantity to be considered is the so-called correlation factor f, whose microscopic significance will be discussed below. In order to be able to relate the phenomenological and atomistic quantities to one another, we shall calculate the mean square displacement of a particle in different ways; for simplicity, we shall discuss the one-dimensional case. We shall consider a cylindrical sample which will be first cut into two halves in order to introduce N_0 tracer atoms into the cut surface; the two halves will then be re-united. At time t = 0 the sample is heated to the temperature of the experiment;

the tracer atoms diffuse into both halves of the sample, which are considered as being infinitely long, in the positive and negative x-directions. The diffusion coefficient describing this diffusion is given the sympol D_{Tr}, since the experiment is carried out with tracer atoms. The tracer diffusion is in fact a counter-diffusion, i.e. tracer atoms (or ions) and the corresponding normal isotopic particles diffuse in opposite directions. To obtain the change of the concentration of tracer atoms with time and in space we must solve Fick's second law according to the following equation:

$$\frac{\partial c_{Tr}}{\partial t} = D_{Tr} \frac{\partial c_{Tr}^2}{\partial x^2},$$ (6.2.1)

using the initial and boundary conditions given below

$$c_{Tr} = 0 \quad \text{for} \quad |x| > 0, \quad t = 0,$$

$$c_{Tr} = \infty \quad \text{for} \quad x = 0, \quad t = 0.$$

If both halves of the sample are infinitely long, the solution of this equation is as follows [15]

$$c_{Tr}(x, t) = \frac{N_0}{(4\pi D_{Tr}t)^{1/2}} \exp\left(-\frac{x^2}{4D_{Tr}t}\right).$$ (6.2.2)

Figure 6.2.1 shows the concentration profile as a function of x at different times t. We shall now assume that the side-by-side existence of tracer atoms in space and in time leads to the same result as the diffusion of the particles one after the other from the point x = 0 into the sample. The result of the diffusion experiment and the solution of Eq. (6.2.1) then allow to obtain the probability $w(x, t) \cdot dx$ that, at time t a particle is to be found between x and x + dx. This

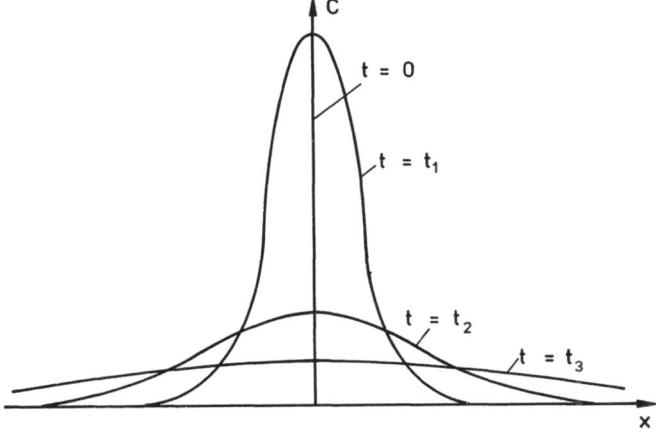

Fig. 6.2.1. Schematic representation of the concentration as a function of the x-coordinate in the case of one-dimensional diffusion in the positive and negative x-directions at different times t

probability $w(x, t) \cdot dx$ is related to the concentration $c(x, t)$ and the total number N_0 of the tracer atoms used in the experiment according to the equation

$$w(x, t) \cdot dx = \frac{1}{N_0} c(x, t), \tag{6.2.3}$$

since multiplication of $w(x, t) \cdot dx$ by N_0 must afford the result of the diffusion experiment carried out simultaneously with N_0 particles.

Comparison of Eq. (6.2.2) with (6.2.3) leads to the expression

$$w(x, t) = (4\pi D_{Tr}t)^{-1/2} \exp\left(-\frac{x^2}{4D_{Tr}t}\right). \tag{6.2.4}$$

From the definition of the mean square displacement $\overline{X^2}(t)$,

$$\overline{X^2}(t) = \int\limits_{-\infty}^{+\infty} x^2 w(x, t) \, dx, \tag{6.2.5}$$

and with the aid of Eq. (6.2.4) it follows that

$$\overline{X^2}(t) = \int\limits_{-\infty}^{+\infty} \frac{x^2}{(4\pi D_{Tr}t)^{1/2}} \exp\left(-\frac{x^2}{4D_{Tr}t}\right) dx \tag{6.2.6}$$

The solution of this integral (see e.g. [6.16]) for the one-dimensional case is

$$\overline{X^2}(t) = 2D_{Tr}t. \tag{6.2.7}$$

Similarly, we can obtain an expression for the mean square displacement $\overline{\vec{R}^2}(t)$ for the three-dimensional case

$$\overline{\vec{R}^2}(t) = 6D_{Tr}t. \tag{6.2.8}$$

On the other hand, it is also possible to calculate the mean square displacement from the various jumps occurring at time t. We shall denote the i-th jump by the symbol $\vec{r}_i$ taking into account its vector character, i.e. every jump is properly

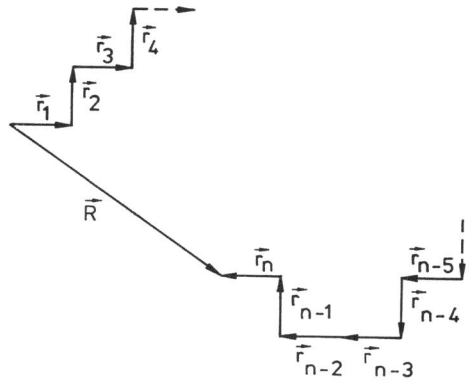

Fig. 6.2.2. Displacement vector $\vec{R}$ resulting from the displacement of a diffusing particle after n diffusion jumps with the single jump vector $\vec{r}_i$

specified if its direction and distance are known. After time t in which the particle has carried out n jumps it has been shifted by the vector $\vec{R}$ (see Fig. 6.2.2).

This displacement vector $\vec{R}$ is the sum of the individual jumps $\vec{r}_i$

$$\vec{R} = \vec{r}_1 + \vec{r}_2 + \dots + \vec{r}_n = \sum_{i=1}^{n} \vec{r}_i. \tag{6.2.9}$$

The square of the displacement $\vec{R}^2(t)$ can be calculated from $\vec{R}$:

$$\vec{R}^2(t) = R^2(t) = (\vec{r}_1 + \vec{r}_2 + \dots \vec{r}_n)^2 = \left(\sum_{i=1}^{n} \vec{r}_i \right)^2 \tag{6.2.10}$$

or in detail

$$\begin{aligned} \vec{R}^2(t) = \ &\vec{r}_1 \cdot \vec{r}_1 + \vec{r}_1 \cdot \vec{r}_2 + \dots + \vec{r}_1 \cdot \vec{r}_n \\ &\vec{r}_2 \cdot \vec{r}_1 + \vec{r}_2 \cdot \vec{r}_2 + \dots + \vec{r}_2 \cdot \vec{r}_n \\ &\qquad\qquad\vdots \\ &\vec{r}_n \cdot \vec{r}_1 = \vec{r}_n \cdot \vec{r}_2 + \dots + \vec{r}_n \cdot \vec{r}_n \\ = \ &\sum_{i=1}^{n} \sum_{j=1}^{n} \vec{r}_i \vec{r}_j. \end{aligned} \tag{6.2.11}$$

If we assume that all individual jump vectors have the same magnitude r, then, using the formula

$$\vec{r}_i \cdot \vec{r}_j = r^2 \cos \alpha_{ij} \tag{6.2.12}$$

for the scalar product of $\vec{r}_i$ and $\vec{r}_j$, where α_{ij} denotes the angle between $\vec{r}_i$ and $\vec{r}_j$, we obtain the following expression for $\vec{R}^2(t)$

$$\vec{R}^2(t) = nr^2 + 2 \sum_{i=j+1}^{n} \sum_{j=1}^{n-1} r^2 \cos \alpha_{ij}. \tag{6.2.13}$$

In Eq. (6.2.13) the diagonal elements in the matrix representation of $\vec{R}^2$ given above, i.e. the products of equal jump vectors are written separately as $n \cdot r^2$, since for such jumps $\cos \alpha = 1$. To obtain the mean square of the displacement $\overline{\vec{R}^2}(t) = \overline{R^2}(t)$ we can make use of the fact that the mean value of a sum is equal to the sum of mean values of the terms to be added. Thus, Eq. (6.2.13) can be written as follows

$$\overline{\vec{R}^2}(t) = \overline{R^2}(t) = nr^2 + 2 \sum_{i=j+1}^{n} \sum_{j=1}^{n-1} r^2 \overline{\cos \alpha_{ij}}. \tag{6.2.14}$$

If the directions of the different jumps, e.g. of successive jumps, are independent of one another, α_{ij} is equally distributed among all angles so that

$$\overline{\cos \alpha_{ij}} = 0. \tag{6.2.15}$$

It then follows from Eq. (6.2.15) and (6.2.14) that

$$\overline{\vec{R}^2}(t) = \overline{R^2}(t) = nr^2 \quad \text{(uncorrelated)}. \tag{6.2.16}$$

Equation (6.2.16) is valid only for the case in which the directions of the various jumps are not related one to another, i.e. there is no correlation between the directions of the different jumps and in particular between those of successive jumps. Under these conditions, the mean square of the displacement is equal to the product of the number of jumps and the square of the distance travelled per jump.

If the directions of successive jumps are not independent of one another, i.e. if there is a correlation between the directions of the different jumps made by a particle, then

$$\overline{\cos \alpha_{ij}} \neq 0 . \tag{6.2.17}$$

In this case the so-called correlation factor f is introduced into Eq. (6.2.16), since the second term in Eq. (6.2.14) is no longer equal to zero. Equation (6.2.16) becomes

$$\overline{\vec{R}^2}(t) = \overline{R^2}(t) = fnr^2 . \tag{6.2.18}$$

The correlation factor depends on the geometrical arrangement of the particles in the lattice and on the diffusion mechanism (compare [6.17]). For example, for the vacancy mechanism in a simple cubic lattice f = 0.65, in the NaCl-lattice f = 0.78, in the CsCl lattice f = 0.72, for the interstitial mechanism f = 1 and for the indirect interstitial mechanism (interstitialcy mechanism) with collinear jumps f = 0.67.

The vacancy mechanism involves movements of atoms through the crystal in which lattice atoms jump to neighbouring unoccupied lattice sites and from there, if there is a vacancy in their vicinity to further sites (see Fig. 6.2.3a). The interstitial mechanism involves jumps of particles in the interstitial lattice directly from one interstitial site to another (see Fig. 6.2.3b). An interstitialcy mechanism involves the displacement of a particle located on a regular lattice site into the interstitial lattice by an interstitial particle, which itself then occupies the regular site (see Fig. 6.2.3c). The new interstitial particle then repeats the process.

It is quite simple to show qualitatively that the jump directions for successive jumps of a particle are correlated with one another if the vacancy mechanism applies. When a particle makes a jump, the vacancy jumps in the opposite direction and thus occupies the position formerly occupied by the particle. Since the site formerly occupied by the particle is now vacant, the particle has a pre-

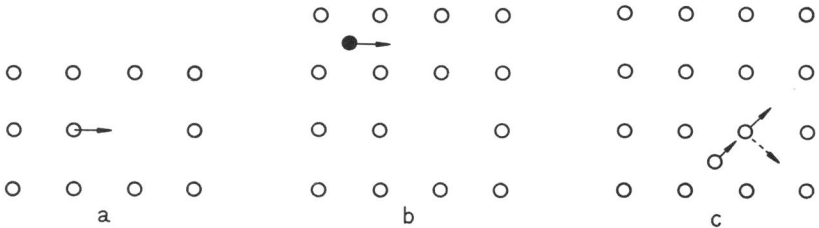

Fig. 6.2.3a—c. Site exchange mechanism for diffusion in solid crystals. **a** Vacancy mechanism; **b** interstitial mechanism; **c** interstitialcy mechanism

ferred tendency to jump back to this old site. Thus, the directions of successive jumps are correlated; the same is valid for all further jumps. The determination of correlation factors has become important for resolving diffusion mechanisms.

If we now wish to relate the tracer diffusion coefficient to atomistic quantities, we must compare (6.2.8) and (6.2.18) which both contain expressions for the mean square displacement. This gives us the following relationship

$$D_{Tr} = \frac{1}{6} f \frac{n}{t} r^2 \qquad (6.2.19)$$

or, in terms of the jump frequency $\nu = n/t$,

$$D_{Tr} = \frac{1}{6} f\nu r^2. \qquad (6.2.20)$$

Equation (6.2.20) shows the relationship between the macroscopic tracer diffusion coefficient and the microscopic quantities, jump distance r and jump frequency ν, and correlation factor f. This correlation factor must also be included in the Nernst-Einstein equation if the latter is written in terms of D_{Tr}:

$$D_{Tr} = fB_m kT. \qquad (6.2.21)$$

A comparison of Eqs. (6.2.21) and (6.1.22) leads to the relationship

$$D_{Tr} = fD_K \qquad (6.2.22)$$

while a comparison of Eqs. (6.2.22) and (6.2.20) yields the following expression for the component diffusion coefficient D_K

$$D_K = \frac{1}{6} \nu r^2, \qquad (6.2.23)$$

i.e. the component diffusion coefficient D_K is related in an even simpler manner — no correlation factor appears — to the jump frequency ν and the square of the jump distance of a single jump r^2. Using the relationship between the component diffusion coefficient D_K and the mobilities or partial conductivities of the various types of particles according to Eqs. (6.1.22) and (6.1.37), the latter may also be calculated from jump distance r and the jump frequency ν at equilibrium. The calculation of absolute values for transport quantities such as partial conductivities and diffusion coefficients would therefore be possible if the jump distance of a single jump and the jump frequency at thermodynamic equilibrium are known. The jump distance r can be obtained from the geometrical parameters of the lattice which in their turn can be obtained using methods such as X-ray diffraction. However, until now no calculations of the absolute value for the jump frequency ν have been carried out. In principle, the jump frequency of a single particle can be calculated using the transition state method; the reader is referred to Jost's detailed treatment [6.18]. Calculations of this type involve the enthalpy and the entropy of activation for a single jump. In simple cases absolute calculations of the activation enthalpy have been carried out; these are in agreement with the experimental results (see Barr and Lidiard [6.19]).

It is possible to estimate an upper limit of the jump frequency ν. The maximum jump frequency ν_{max}, i.e. the highest frequency of change of a lattice site results, if the particles move with thermal speed v between the lattice sites without performing oscillations on such a site. For the maximum jump frequency then applies:

$$\nu_{max} = \frac{v}{r}.\tag{6.2.24}$$

For a given temperature this natural limit cannot be exceeded. The maximum diffusion coefficient, denoted as $D_K(max)$, is then given according to Eq. (6.2.23) and (6.2.24)

$$D_K(max) \cdot = \frac{1}{6}\, v \cdot r.\tag{6.2.25}$$

Inserting values for the jump distance and for the upper limit of the jump frequency into Eq. (6.2.23), the diffusion coefficients obtained lie in the range of 10^{-5} cm^2/s at temperatures above the room temperature. Here, a value of 10^{-10} m = 1 Å has been used for r, corresponding to the order of magnitude of lattice parameters. Liquids and also the so-called superionic conductors like silver iodide or β-alumina are found to exhibit such high diffusion coefficients.

Recent investigations [6.20] on AgI based upon microwave measurements and neutron scattering have established that this very crude model must only slightly be adjusted; the values of the diffusion coefficients are, however, not affected by this adjustment.

6.3 Experimental Determination of Partial Conductivities

We have shown in Sect. 6.1 that the total conductivity σ of a solid in which ionic and electronic conduction take place is the sum of the partial conductivities σ_i of all the types of particles i present in the solid:

$$\sigma = \sum_i \sigma_i.\tag{6.3.1}$$

Experimentally, it is often found that one type of particle, label with index 1, has a much higher partial conductivity than all others; in this case

$$\sigma \cong \sigma_1, \quad \text{if} \quad \sigma_1 \gg \sigma_2, \sigma_3, \ldots,\tag{6.3.2}$$

i.e. the transference number of the particles 1 is virtually equal to unity

$$t_1 = \frac{\sigma_1}{\sigma} \cong 1, \quad \text{if} \quad \sigma_1 \gg \sigma_2, \sigma_3, \ldots\tag{6.3.3}$$

while other transference numbers are virtually zero.

In this case, it is possible to obtain the partial conductivity σ_1 simply by measuring the total conductivity σ. For medium conductivities of the material to be investigated this can be done by carrying out a.c. measurements on single crystals or for polycrystalline materials on pellets prepared by pressing powdered material; as electrodes one can for example use platinum leads. The a.c. measurements should be carried out as a function of frequency using an impedance bridge. It is often useful to take then complex impedance diagrams. Since a.c. impedance measurements have found widespread use in recent times in order to study impedances and electrode processes, a discussion of this method with special emphasis on complex impedance diagrams — also called complex plane plots — will be given later in this chapter. If the resistance is very high or very low it is often more advisable to carry out d.c. measurements and to use probes for determining the potential drop in order to eliminate contact resistances at the electrodes and polarization effects, which will probably occur here. It is also necessary to ensure that those particles, responsible for the total conductivity, e.g. electrons or ions, can be exchanged between the electrodes and the sample. If electrodes are used which do not permit the exchange of one or several types of particles, the contribution of the latter to the charge transport and to the total conductivity will be suppressed during d.c. measurements. In the early years of solid-state electrochemistry, this effect has sometimes led to errors and to inaccurate experimental results. At present, this phenomenon is used to obtain small contributions from one or more particles to the total conductivity, i.e. to determine small transference numbers by means of polarization measurements which will be discussed below.

The following three methods can be shown to be particularly suited for the measurement of partial conductivities or transference numbers (e.g. the partial conductivity of the electrons in mixed conductors with predominant ion conduction and vice versa):

a) transference measurements,
b) emf measurements using galvanic cells,
c) polarization measurements.

These three methods will be discussed below. The following holds for all measurements: it is necessary to take into account that the partial conductivities σ_i may depend on external parameters, e.g. on the chemical potentials of the components. This is also true for measurements of the total conductivity discussed above. For example, in the case of binary oxides, the conductivity should only be measured if the partial pressure of oxygen for which equilibrium with the oxide is attained is known. In some cases the ionic conductivity is virtually independent of changes in the chemical potential of a component of the solid. This depends on the type of disorder in the compound and is found to be true for many important electrolytes.

6.3.1 Transference Measurements

To apply this method an electrical current is caused to flow through the solid to be investigated, e.g. a solid electrolyte; the amount of substance transported is compared with the charge which has flown through the solid. This method is sometimes called the electrolysis method.

The first transference measurements were carried out in the last century, and were used to calculate the ionic and electronic partial conductivities of solid substances. Warburg [6.21] and Tegetmeier [6.22], Haber and Tolloczko [6.23] and Bruni and Scarpa [6.24] were able to confirm Faraday's law for solid compounds and thus to show that pure ionic conduction occurs in the substances which they investigated. In about 1930, Tubandt and his co-workers [6.25] carried out extended investigations on the conductivity behaviour of solid compounds. In the course of these studies the transference number, which equals the ratio of partial conductivity to total conductivity, was determined from changes in the weight of the electrodes in the galvanic cells. As an example, we shall consider silver iodide, whose conductivity behaviour above 149 °C was studied by Tubandt. He used the following galvanic cell consisting of five pellets:

$$+ \text{ Ag } | \text{ AgI } | \text{ AgI } | \text{ AgI } | \text{ Ag } - \tag{6.3.I}$$

$^*\Delta m$ for 10^{-3} F: $-$ 108 mg   0   0 $+$108 mg

$$\xleftarrow{\hspace{1cm}} \quad \xrightarrow{\hspace{2cm}} \quad \xleftarrow{\hspace{1cm}}$$
$$\text{e}^- \qquad\qquad \text{Ag}^+ \qquad\quad \text{e}^-$$

$$\text{Ag} = \text{e}^-(\leftarrow) + \text{Ag}^+(\rightarrow) \qquad \text{Ag}^+(\rightarrow) + \text{e}^-(\leftarrow) = \text{Ag}$$

$^*\Delta m = $ mass change

The two silver pellets serve as electrodes, and the electrolyte to be studied is divided into three pellets. After a charge transport of 10^{-3} Faraday through the cell the changes in weight Δm shown in the figure were measured; the right-hand silver electrode was grown together to the silver iodide pellet next to it because of the deposition of silver. The interpretation of this measurement led to the conclusion that the transference number of silver ions $t_{\text{Ag}^+} = 1$, while the transference numbers of iodide ions and electrons are much smaller than unity.

For measurements of this type the condition that the electrodes of the galvanic cell permit the flow of ions and electrons must be fulfilled. As already mentioned, errors have in some cases been made in the interpretation of the results obtained because this requirement was not obeyed. The result of the measurements carried out on the cell (6.3.I) is only valid for silver iodide in equilibrium with silver, since silver iodide must be in equilibrium with silver as a consequence of the fact that the silver pellets were pressed at both sides of the silver iodide. In fact, it has been observed that silver iodide at equilibrium with iodine vapour shows notable electronic conduction [6.26] which is due to electron defects.

6.3.2 Measurements of Transference Numbers by Means of Emf Measurements Using Galvanic Cells

In this method the solid conductor to be studied is used as an electrolyte in a galvanic cell. If the electrolyte is a pure ionic conductor, the emf E of the cell is given by the equation

$$\Delta G = -nFE \tag{6.3.4}$$

(see Chap. 8); ΔG denotes the Gibbs reaction energy of the cell reaction, F the Faraday constant and n the number of Faradays which must flow through the cell corresponding to a per formula conversion of the cell reaction. We shall mainly be concerned here with binary electrolytes which may also be doped; exceptions are possible for ternary electrolytes. If the ionic conduction in the solid electrolyte is accompanied by an electronic one, the absolute value of the emf of the cell is smaller than that resulting from Eq. (6.3.4). This decrease in the emf of the cell was first calculated in general terms by Wagner [6.27]; under certain circumstances, the value of this voltage reduction permits predictions to be made about the transference number of the ions. For the discussion of the method following below we shall take as an example a solid electrolyte whose conduction is mainly due to oxygen ions but also to some extent to electrons. In such a case it must always be taken into account that the partial conductivity of the electrons is in general a function of the oxygen partial pressure or the chemical potential of the oxygen.

Using such an oxygen ion conductor as solid electrolyte for the construction of a galvanic cell with different oxygen pressures or different chemical potentials of oxygen μ'_{O_2} and μ''_{O_2} on both sides of the solid electrolyte, the following expression applies to the emf E; the explicit derivation of this equation is given below:

$$E = \frac{1}{4F} \int_{\mu'_{O_2}}^{\mu''_{O_2}} t_{ion} \, d\mu_{O_2}, \tag{6.3.5}$$

F denotes Faraday's constant and t_{ion} the transference number of the ions, which must be regarded as a function of the chemical potential μ_{O_2} of oxygen in the solid electrolyte. For the case of pure ionic conduction ($t_{ion} = 1$) and remembering that $\mu_{O_2} = \mu^0_{O_2} + RT \ln \frac{p_{O_2}}{p^0_{O_2}}$ we obtain the following relationship from Eq. (6.3.5)

$$E = \frac{RT}{4F} \ln \frac{p''_{O_2}}{p'_{O_2}} \quad \text{if } t_{ion} = 1, \tag{6.3.6}$$

R denotes the gas constant, T the absolute temperature, p''_{O_2} and p'_{O_2} the oxygen partial pressures on both sides of the solid electrolyte.

If conduction by electrons can take place in the solid electrolyte in addition to ionic conduction involving oxygen ions, oxygen (in the form of ions and electrons) flows through the electrolyte from the electrode with higher oxygen partial pressure to that with lower one even if the circuit is open. The electrical partial current density i_k for each type of particle k is given by the equation (see also Eqs. (6.1.44) and (6.1.31))

$$i_k = - \frac{\sigma_k}{z_k F} \frac{d\eta_k}{dx} \tag{6.3.7}$$

where $\sigma_k = c_k z_k u_k F$. z_k is the valency of particle species k and η_k is its electrochemical potential, which can be expressed in terms of the chemical potential μ_k and the electrical potential φ:

$$\eta_k = \mu_k + z_k F \varphi. \tag{6.3.8}$$

At open circuit the partial current densities of oxygen ions $i_{O^{2-}}$ and electrons i_{e^-} must, on the grounds of electrical neutrality, be equal with respect to their absolute values, if the mobility of the metal ions is negligibly small:

$$i_{O^{2-}} = -i_{e^-}. \tag{6.3.9}$$

Thus, using Eq. (6.3.7), it follows that

$$\frac{1}{2}\, \sigma_{O^{2-}} \frac{d\eta_{O^{2-}}}{dx} = -\sigma_e \frac{d\eta_e}{dx} \tag{6.3.10}$$

or

$$\frac{1}{2}\, d\eta_{O^{2-}} = -\frac{\sigma_e}{\sigma_{O^{2-}}}\, d\eta_e. \tag{6.3.11}$$

The chemical potential of oxygen is related to the electrochemical potential of the oxygen ions and that of the electrons according to the following equation written with differentials for the quantities involved:

$$2d\eta_{O^{2-}} - 4d\eta_e = d\mu_{O_2}. \tag{6.3.12}$$

Equations (6.3.11) and (6.3.12) lead to the expression

$$\frac{\sigma_e + \sigma_{O^{2-}}}{\sigma_{O^{2-}}}\, d\eta_e = -\frac{1}{4}\, d\mu_{O_2} \tag{6.3.13}$$

or

$$d\eta_e = -\frac{1}{4}\, t_{O^{2-}}\, d\mu_{O_2}. \tag{6.3.14}$$

Integration over the thickness of the electrolyte gives

$$\eta_e'' - \eta_e' = -\frac{1}{4} \int_{\mu_{O_2}'}^{\mu_{O_2}''} t_{O^{2-}}\, d\mu_{O_2}. \tag{6.3.15}$$

Since the difference in the electrochemical potential of the electrons itself is related to the measurable emf by the equation

$$\eta_e'' - \eta_e' = -EF, \tag{6.3.16}$$

Eqs. (6.3.15) and (6.3.16) yield

$$E = \frac{1}{4F} \int_{\mu_{O_2}'}^{\mu_{O_2}''} t_{O^{2-}}\, d\mu_{O_2}, \tag{6.3.17a}$$

i.e. Eq. (6.3.5), which we have thus derived.

It is possible in principle, if $t_{ion} \neq 1$, to use the measured value of the emf E to obtain information about t_{ion}. This method was applied by Schmalzried [6.28] to doped zirconium dioxide. Working with Eq. (6.3.17a) it must, however, be noticed that t_{ion} is a function of the chemical potential of oxygen. This dependence, which follows from the disorder model, must be included in the evaluation to be carried out. If the disorder model is not known, the transference number corresponding to a certain chemical potential of oxygen can be obtained by measuring the emf E as a function of the oxygen partial pressure or of the chemical potential of oxygen at the second electrode. Differentiation of Eq. (6.3.17a) with respect to the upper limit gives the following expression (see C. Wagner [6.29]):

$$t_{O^{2-}}(\mu''_{O_2}) = 4F \left[\frac{\partial E(\mu'_{O_2}, \mu''_{O_2})}{\partial \mu''_{O_2}} \right]_{\mu'_{O_2}} . \tag{6.3.17b}$$

A definite statement can always be obtained if $t_{ion} = 1$ over the total range of the chemical potential considered: this is an important condition imposed on an electrolyte used in galvanic cells for the determination of ΔG-values as described in Chap. 8. In this case, Eqs. (6.3.4) or (6.3.6) apply.

6.3.3 Determination of Partial Conductivities by Means of Stationary Polarization Measurements

In this method either the ionic or the electronic current is suppressed by appropriate choice of electrodes and polarity. If for example the ionic current in a substance which exhibits predominant ionic conduction is suppressed, it is possible to carry out direct and exact measurements of the electronic partial conductivity, even though this may be very small in comparison with the ionic conductivity. This method is due to Hebb [6.30] and C. Wagner [6.31].

In the following discussion we shall deal with three important electrode combinations. The discussion of the third electrode combination will include results obtained for zirconium dioxide and thorium dioxide.

a) Mixed Conductor Between Two Ionic Conductors

As an example of this case we shall discuss the measurement of the ionic partial conductivity σ_{Ag^+} in Ag_2S. The following cell has been used for such investigations [6.32]:

$$Ag''| \; AgI \; | \; Ag_2S \; | \; AgI \, |' Ag. \tag{6.3.II}$$

AgI is a pure ionic conductor for silver ions at high temperatures. When current flows through the cell a pure ionic current flows through Ag_2S since the transport of electrons is suppressed by the AgI pellets. The potential difference E applied to induce current flow is a measure of the electrochemical potential difference $\eta'_{Ag^+} - \eta''_{Ag^+}$ of the silver ions. In order to eliminate polarization phenomena which may occur at the current-carrying electrodes, separate probes, as shown in Fig. 6.3.1, are used to measure the difference of the electrochemical potential of the silver ions in Ag_2S.

This can be derived as follows: the measured electrical potential difference E_{12} is primarily the difference of the electrical potential between lead Pt 1 and lead Pt 2. Since the Galvani potential differences occurring at the phase boundaries Ag 1/Pt 1 and Ag 2/Pt 2 respectively are equal, one can conclude, that E_{12} is also given by the difference of the electrical potential between the silver probes

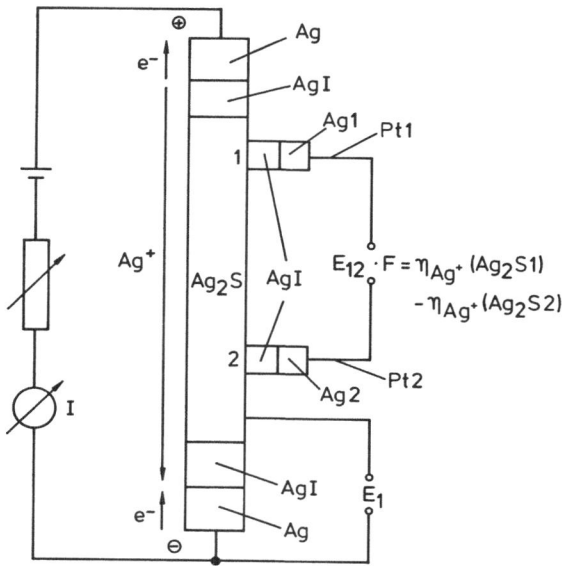

Fig. 6.3.1. Experimental set-up used for the electrochemical measurement of the ionic partial conductivity in Ag_2S

Ag1 and Ag2. Thus, the following is valid:

$$E_{12} = \varphi(Pt\,1) - \varphi(Pt\,2) = \varphi(Ag\,1) - \varphi(Ag\,2). \qquad (6.3.18\,a)$$

Multiplying Eq. (6.3.18a) by F and adding the chemical potential difference of the silver ions $\mu_{Ag^+}(Ag\,1) - \mu_{Ag^+}(Ag\,2)$, which is equal to zero, on the right-hand side of Eq. (6.3.18a) and rearrangement leads to

$$E_{12}F = [\mu_{Ag^+}(Ag\,1) + F\varphi(Ag\,1)] - [\mu_{Ag^+}(Ag\,2) + F\varphi(Ag\,2)]. \qquad (6.3.18\,b)$$

Introducing the electrochemical potential of the silver ions $\eta_{Ag^+} = \mu_{Ag^+} + F\varphi$ Eq. (6.3.18b) can be written as

$$E_{12}F = \eta_{Ag^+}(Ag\,1) - \eta_{Ag^+}(Ag\,2). \qquad (6.3.18\,c)$$

It can thus be seen that the electrical potential difference E_{12} is a measure of the electrochemical potential difference of the silver ions between probe Ag1 and probe Ag2. In silver iodide only silver ions are mobile. Hence, as long as no current flows through the probes the electrochemical potential of the silver ions is constant within each AgI phase. Since each AgI probe is in electrochemical equilibrium both with the corresponding silver probe at one side and with the Ag_2S sample at the other side, the following equations apply:

$$\eta_{Ag^+}(Ag\,1) = \eta_{Ag^+}(Ag_2S\,1), \qquad (6.3.18\,d)$$

$$\eta_{Ag^+}(Ag\,2) = \eta_{Ag^+}(Ag_2S\,2). \qquad (6.3.18\,e)$$

From Eqs. (6.3.18 c—e) it thus follows

$$E_{12}F = \eta_{Ag^+}(Ag_2S\,1) - \eta_{Ag^+}(Ag_2S\,2). \tag{6.3.18f}$$

The electrochemical potential of the electrons in Ag_2S is constant in this experiment, since the electrons, although mobile, do not flow. By means of electrochemical titration it is possible, using the cell

$$Ag \mid AgI \mid Ag_2S, \tag{6.3.III}$$

to change the chemical potential of silver in Ag_2S which corresponds to a change in the stoichiometry of Ag_2S.

b) Mixed Conductor Between Two Electronic Conductors

As an example of a measurement of this type we shall discuss the determination of the electronic partial conductivity in Ag_2S, which was carried out by Miyatani [6.33]. The cell shown in Fig. 6.3.2 can be used for this purpose; its most important part is the cell

$$Pt \mid Ag_2S \mid Pt \tag{6.3.IV}$$

which is used for determining the electronic partial conductivity of Ag_2S. To measure the latter quantity the two platinum wires $Pt1$ and $Pt2$ are used as current leads. Since the platinum leads can only exchange electrons with the Ag_2S at the various phase boundaries of the type Pt/Ag_2S, only electrons flow through the Ag_2S under stationary state conditions. The electrochemical potential gradient of the electrons in Ag_2S is obtained by measuring the emf E_{34} between the platinum probes $Pt3$ and $Pt4$. This can be derived as follows: the potential difference E_{34} is primarily the difference of the electrical potential between lead $Pt3$ and lead $Pt4$. Hence, we can write

$$E_{34} = \varphi(Pt3) - \varphi(Pt4). \tag{6.3.19a}$$

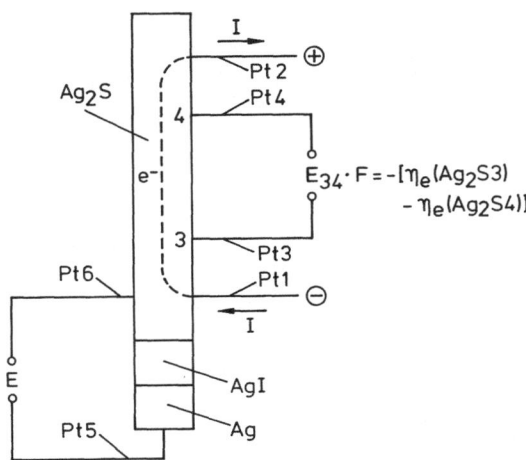

Fig. 6.3.2. Experimental set-up used for the measurement of the partial conductivity of the electrons in Ag_2S according to Hebb [6.30] and Miyatani [6.33]

Multiplying Eq. (6.3.19a) by F and adding the chemical potential difference of the electrons $\mu_e(\text{Pt}4) - \mu_e(\text{Pt}3)$ — this term is equal to zero, because both probes consist of the same material, here Pt — on the right-hand side of Eq. (6.3.19a) gives:

$$E_{34}F = [\mu_e(\text{Pt}4) - F\varphi(\text{Pt}4)] - [\mu_e(\text{Pt}3) - \varphi(\text{Pt}3)]. \qquad (6.3.19\,\text{b})$$

Introducing the electrochemical potential of the electrons according to $\eta_e = \mu_e - F\varphi$ Eq. (6.3.19b) can be written as follows

$$E_{34}F = \eta_e(\text{Pt}4) - \eta_e(\text{Pt}3) = -[\eta_e(\text{Pt}3) - \eta_e(\text{Pt}4)]. \qquad (6.3.19\,\text{c})$$

Since both probes Pt3 and Pt4 are in electrochemical equilibrium with the Ag_2S sample at the corresponding sites — note that E_{34} is measured under conditions of zero current through the potential measuring circuit — the following equations are valid.

$$\eta_e(\text{Pt}3) = \eta_e(Ag_2S\,3) \qquad (6.3.19\,\text{d})$$

and

$$\eta_e(\text{Pt}4) = \eta_e(Ag_2S\,4). \qquad (6.3.19\,\text{e})$$

Comparing Eq. (6.3.19c) with Eqs. (6.3.19d) and (6.3.19e) thus gives

$$E_{34}F = -[\eta_e(Ag_2S\,3) - \eta_e(Ag_2S\,4)]. \qquad (6.3.19\,\text{f})$$

Using the cell

$$\text{Pt}5 \mid Ag \mid AgI \mid Ag_2S \mid \text{Pt}6 \qquad (6.3.\text{V})$$

which forms the lower part of the experimental set-up shown in Fig. 6.3.2 it is possible to fix, vary or measure the activity of silver in Ag_2S (see Sect. 8.2).

c) Mixed Conductor Between a Reversible Electrode and an Electronic Conductor

In this case the polarity chosen permits only electrons to flow through the mixed conductor in the stationary state. The current is carried by an electron

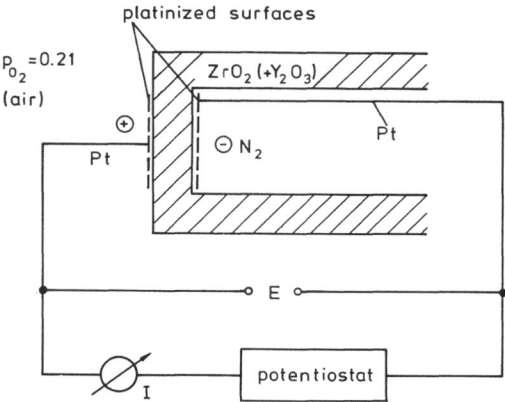

Fig. 6.3.3. Solid state galvanic cell used for investigating the partial conductivities and mobilities of electrons e and electron defects h in doped ZrO_2

conductor on one side of the cell and on the other side by a "reversible" electrode which can exchange electrons and ions; due to the chosen polarity, however, the ions cannot migrate within the mixed conductor under steady state conditions. The reversible electrode determines the chemical potential of one of the components on one side of the sample. On the other side of the sample, the chemical potentials vary with the voltage applied across the cell. The current flowing through the cell is measured as a function of the voltage. Analysis of the current-voltage curve gives the electronic partial conductivity as a function of the chemical potential of one component of the compound. Measurements of this type have been carried out among others for AgBr [6.26] and doped zirconium dioxide and thorium dioxide [6.34, 6.35], CuCl [6.36] and PbCl$_2$ [6.37].

As an example of the method we shall discuss the measurement of the electronic partial conductivity as a function of the oxygen partial pressure in doped zirconium and thorium dioxides. These oxides are particularly important for studies in solid-state electrochemistry, since they are virtually pure oxygen ion conductors over a wide range of oxygen partial pressures, and can thus be used in galvanic cells as solid electrolytes. Doped zirconium dioxide exhibits a considerable electronic conduction at very low oxygen partial pressures while doped thorium dioxide is an electron-defect conductor at higher oxygen partial pressures. The experimental arrangement used is shown in Fig. 6.3.4; the basic part is the galvanic cell

$$p'_{O_2}, \text{Pt} \mid \text{investigated solid electrolyte} \mid \text{Pt}, N_2 \qquad\qquad (6.3.\text{VI})$$

The left-hand electrode consists of porous platinum surrounded by oxygen at a certain partial pressure p'_{O_2}, e.g. air. The right-hand electrode also consists of porous platinum, in this case in contact with nitrogen. As shown in Figs. 6.3.3 and 6.3.4, the two electrodes are separated from each other in such a manner that no gas exchange is possible. If an electric current flows through this cell —

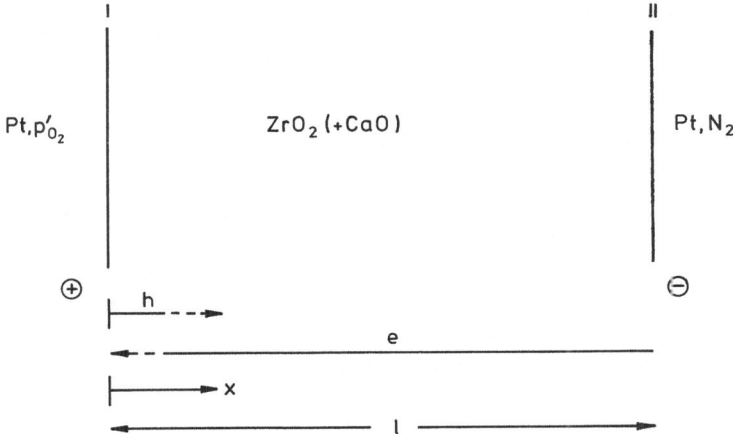

Fig. 6.3.4. Arrangement (schematically) for measuring the total electronic partial conductivity $\sigma_e = \sigma_n + \sigma_p$ (polarization measurement)

with the negative pole taken on the right-hand side — only electrons can flow through the cell, since oxygen ions cannot be replaced on the nitrogen side of the cell. The nitrogen electrode is thus called a blocking electrode for oxygen ions.

The oxygen electrode on the left-hand side can exchange electrons and oxygen ions; by means of this electrode a certain oxygen partial pressure is established on the left-hand side of the electrolyte. As a further condition for the flow of a pure electron current the potential difference E applied must be smaller than the decomposition voltage of the electrolyte. Since the oxygen ions in the electrolyte are mobile, yet do not flow, there exists no gradient in the electrochemical potential $\eta_{O^{2-}}$ of the oxygen ions in the electrolyte:

$$\frac{d\eta_{O^{2-}}}{dx} = 0. \tag{6.3.20}$$

The distance from the left-hand electrode is given the symbol x. Since the current flowing through the cell can only be carried by electrons or electron defects, i.e. it is of purely electronic nature, the current density i is given by the equation

$$i = i_e = \frac{\sigma_e}{F} \frac{d\eta_e}{dx} \tag{6.3.21a}$$

where σ_e denotes the electronic partial conductivity, which contains the contribution of the quasi-free electrons σ_n and that of the electron defects σ_p. F is Faraday's constant and η_e is the electrochemical potential of the electrons. Integration over the thickness l of the solid electrolyte leads to the equation

$$i_e = \frac{1}{lF} \int_{\eta_e'}^{\eta_{e]}''} \sigma_e \, d\eta_e. \tag{6.3.21b}$$

η_e' and η_e'' are the electrochemical potentials of the electrons at the two phase boundaries of the solid electrolyte. The difference of the electrochemical potential of the electrons on both sides of the solid electrolyte is related to the difference in the chemical potential of oxygen:

$$\mu_{O_2}'' - \mu_{O_2}' = 2(\eta_{O^{2-}}'' - \eta_{O^{2-}}') - 4(\eta_e'' - \eta_e'). \tag{6.3.22}$$

Because of the relation stated in Eq. (6.3.20) the first term on the right-hand side of Eq. (6.3.22) is zero so that the difference in the electrochemical potential of the electrons can be expressed in terms of the difference in the chemical potential of oxygen:

$$\mu_{O_2}'' - \mu_{O_2}' = -4(\eta_e'' - \eta_e'). \tag{6.3.23}$$

The measured electrical potential difference E between the two electrodes is always a measure of the difference in the electrochemical potential of the electrons. We can thus write

$$-FE = \eta_e'' - \eta_e'. \tag{6.3.24}$$

It then follows from Eqs. (6.3.23) and (6.3.24) that

$$\mu''_{O_2} - \mu'_{O_2} = 4\,FE, \tag{6.3.25}$$

i.e. for stationary polarization measurements the potential difference E is a measure of the difference in the chemical potential of oxygen on both sides of the electrolyte, despite some additional partial electronic conductivity, just as is the case with a cell containing a pure ionic conductor through which no current flows. Since the chemical potential of the oxygen at the left-hand electrode is fixed, the applied potential difference E is thus a measure of the chemical potential of oxygen which is set up at the right-hand electrode of cell (6.3.VI). It follows from Eqs. (6.3.21b) and (6.3.23) that

$$i_e = -\frac{1}{4lF} \int_{\mu'_{O_2}}^{\mu''_{O_2}} \sigma_e \, d\mu_{O_2}. \tag{6.3.26}$$

The partial electronic conductivity σ_e is a function of the chemical potential of oxygen. Differentiation of Eq. (6.3.26) with respect to μ''_{O_2} leads to the following expression

$$\sigma_e(\mu_{O_2} = \mu''_{O_2}) = -4Fl \left(\frac{di_e}{d\mu''_{O_2}}\right) \tag{6.3.27}$$

or using Eq. (6.3.25)

$$\sigma_e(\mu_{O_2} = \mu''_{O_2}) = -l \left(\frac{di_e}{dE}\right). \tag{6.3.28}$$

One can use Eqs. (6.3.27) or (6.3.28) to obtain the electronic partial conductivity σ_e as a function of the chemical potential of oxygen. If it is possible to make assumptions regarding the disorder of ions and electrons, a more detailed evaluation is possible; this is the case with doped zirconium dioxide and thorium dioxide, the disorder in which has been discussed in Sect. 3.1. We were able to show that, as expressed by Eqs. (3.1.9) and (3.1.10), the partial conductivities of the electrons σ_n and electron defects σ_p are proportional to $p_{O_2}^{-1/4}$ or $p_{O_2}^{1/4}$ respectively where p_{O_2} is the partial pressure of oxygen in equilibrium with the oxide. Using the equation $\mu_{O_2} = \mu^0_{O_2} + RT \ln p_{O_2}$ for the chemical potential of oxygen, it follows from Eqs. (3.1.9) and (3.1.10) that

$$\sigma_n = \sigma'_n \exp\left[-(\mu_{O_2} - \mu'_{O_2})/4RT\right] \tag{6.3.29}$$

or

$$\sigma_p = \sigma'_p \exp\left[+(\mu_{O_2} - \mu'_{O_2})/4RT\right]. \tag{6.3.30}$$

The electronic partial conductivity is given by the expression

$$\sigma_e = \sigma_n + \sigma_p. \tag{6.3.31}$$

If this expression is inserted into Eq. (6.3.26) we can, noting the validity of Eqs. (6.3.29), (6.3.30) and (6.3.25), obtain the following expression for i_e:

$$i_e = \frac{RT}{lF} \left[\sigma'_n \left(\exp\frac{-EF}{RT} - 1\right) + \sigma'_p \left(1 - \exp\frac{EF}{RT}\right)\right]. \tag{6.3.32}$$

This equation contains two limiting cases for negative E-values (E is always negative for the chosen experimental arrangement):

a) If the partial conductivity of the electron defects σ'_p at the left-hand electrode is larger than that of the quasi-free electrons σ'_n, the current first attains a limiting value

$$i_e \cong \frac{RT}{Fl} \sigma'_p. \tag{6.3.33}$$

If the absolute value of the voltage applied increases further, the current of the quasi-free electrons increases to a point at which it exceeds by far that of the electron defects; in this case

$$i_e \cong \frac{RT}{Fl} \sigma'_n \exp\left(-\frac{EF}{RT}\right) \tag{6.3.34}$$

or

$$\ln i_e \cong \ln\left(\frac{RT}{Fl} \sigma'_n\right) - \frac{EF}{RT}. \tag{6.3.35}$$

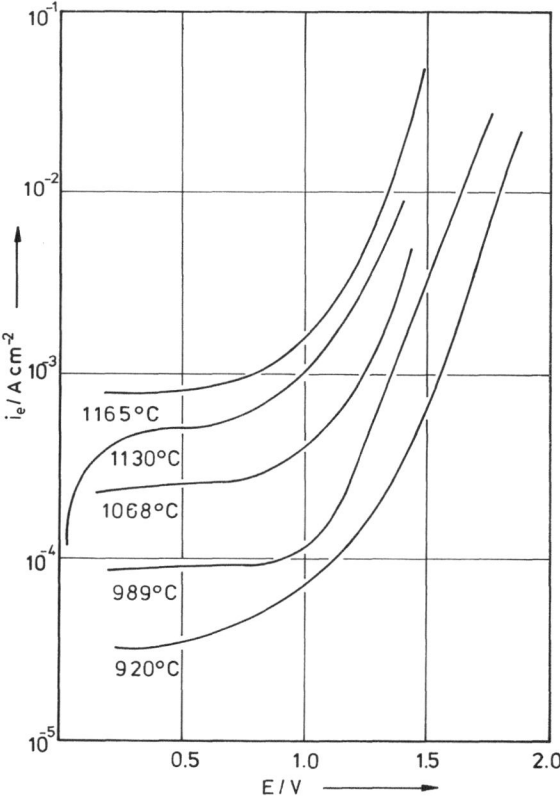

Fig. 6.3.5. Logarithmic plot of the total electronic current density i_e as a function of the applied potential difference E for stationary polarization measurements carried out on $Zr_{0.9} Y_{0.1} O_{1.95}$. The reference electrode is atmospheric air ($p_{O_2} = 0.21$ atm)

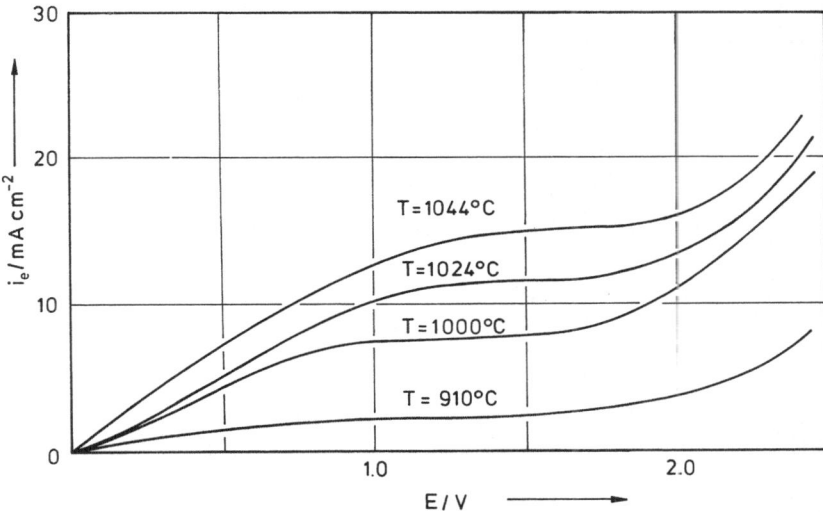

Fig. 6.3.6. Total electronic current density i_e as a function of the applied potential difference E at different temperatures for stationary polarization measurements carried out on $Th_{0.9} Y_{0.1} O_{1.95}$. Reference electrode: $p_{O_2} = 0.21$ atm

Equation (6.3.35) states that at $1000\,°C$ the current density i_e rises by an order of magnitude if E increases by an absolute value of 252.2 mV.

Equations (6.3.33) or (6.3.34) and (6.3.35) can be used for the determination of σ_n' or σ_p' and thus also, by means of Eqs. (6.3.29) and (6.3.30), for the determination of σ_p and σ_n as a function of the chemical potential of oxygen μ_{O_2}. Figure 6.3.5 shows the stationary current density i_e for measurements carried out with zirconium dioxide (plus 10 Mol% Y_2O_3) as a function of the potential difference E at various temperatures [6.34]. It can be seen that there is a plateau between about 250 mV and 700 mV and an exponential increase of the current density at larger E-values; this behaviour is in agreement with theoretical considerations and with Eqs. (6.3.33) and (6.3.34).

Figure 6.3.6 shows the results of corresponding measurements carried out with thorium dioxide, doped with yttrium oxide. Because of the much larger electron defect partial conductivity the plateaus lie much higher. Figure 6.3.7 shows the dependence of the partial conductivities σ_n and σ_p on the partial pressure of oxygen calculated from the measurements on doped ZrO_2 discussed above.

Apart from partial conductivities evaluation of non-stationary measurements allows one to determine diffusion coefficients and thus electrical mobilities of electrons and electron defects. It is then possible using the mobility u_n of the electrons to calculate their concentration c_n according to the following equation

$$\sigma_n = c_n\, e\, u_n. \tag{6.3.36}$$

Further complications arise in the case of instationary measurements, if both electrons and electron defects contribute to the conductivity. An approximate solution for this case has been given by Weppner [6.38].

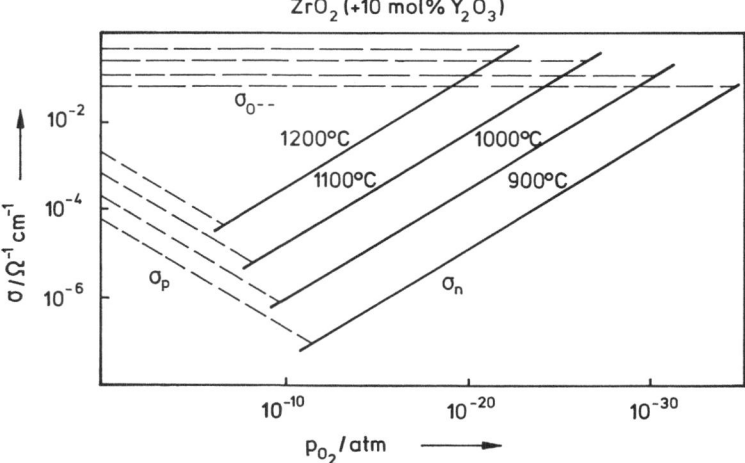

Fig. 6.3.7. Partial conductivities of electrons, electron defects and oxygen ions in ZrO_2 (+ 10 mol-% Y_2O_3) at different temperatures as a function of the oxygen partial pressure

Polarization measurements of other solid compounds with dominating ionic conductivity have been combined with instationary measurements for determining diffusion coefficients. Such studies have been carried out on AgBr by Weiss [6.39] and Mizusaki, Fueki and Mukaibo [6.40], on CuCl by Joshi and Wagner [6.41] and on CuI by Jow and Wagner [6.42].

6.4 Alternating Current Measurements. Complex Impedance Diagrams

For determining the conductivity of a solid electrolyte or a solid mixed conductor it may be preferable in some cases to carry out alternating current (a.c.) measurements. The impedance is determined as a function of frequency. This frequency dependence of the impedance is characteristic of the various types of cells and electrodes. It is useful to represent the results in a complex impedance diagram, for example imaginary versus real part of the impedance with the frequency as a parameter (so-called complex-plane representations or Argand diagrams).

Often, though not in all cases, the results can be interpreted in terms of simple equivalent circuits consisting of combinations of resistive and capacitive elements. A blocking electrode attached to a solid electrolyte for example gives a contribution to the total cell impedance, that can be represented by a capacitance in series with the bulk resistance of the electrolyte (see Fig. 6.4.1). In the case of a reversible electrode one can often use the equivalent circuit given in Fig. 6.4.2, consisting of a capacity parallel to a resistance. Real cases may be much more complicated than simple equivalent circuits, which are introduced to account for the most significant contributions present in a certain cell. For further details we refer to the literature [6.43].

In the following some principles of such a.c. measurements will be represented.

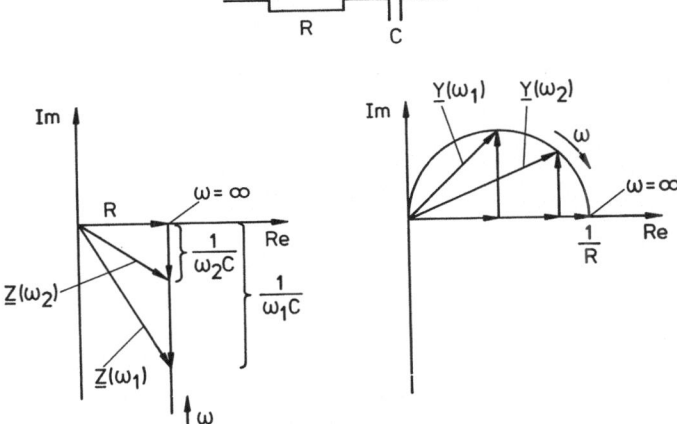

Fig. 6.4.1. Complex plane plots schematically representing the impedance and the admittance respectively of a series connection of a resistance and a capacity

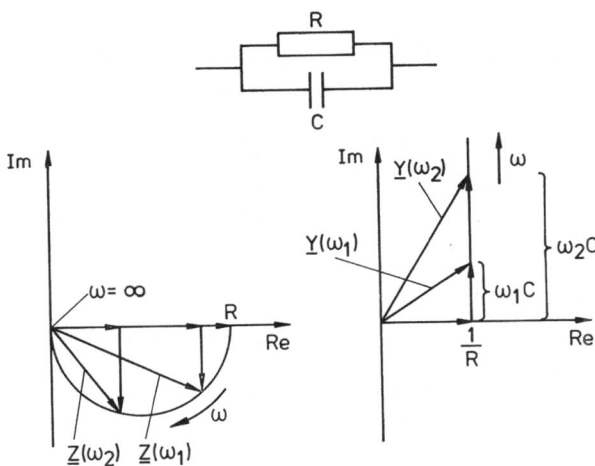

Fig. 6.4.2. Complex plane plots schematically representing the impedance and the admittance respectively of a parallel connection of a resistance and a capacity

The task in general will be to determine the current I(t) flowing through an electrochemical cell while applying a sinusoidally varying voltage, as a function of frequency. The sine-shaped voltage U(t) can be written in the form

$$U(t) = U_{max} \cos \omega t, \qquad (6.4.1)$$

U_{max} denotes the maximum value of the sinusoidally varying voltage, i.e. the amplitude, and ω the circular frequency which is related to the frequency ν and the duration of period τ respectively as follows

$$\omega = 2\pi\nu = \frac{2\pi}{\tau}. \qquad (6.4.2)$$

Such sinusoidal voltages may also be represented in complex exponential form according to

$$\underline{U}(t) = U_{max} e^{i\omega t}. \tag{6.4.3}$$

Here the symbol $\underline{U}$ is used to emphasize the fact that the voltage is a vector quantity. It is determined by the two scalar quantities U_{max} and the instantaneous phase angle $\omega \cdot t$.

The two representations are related by Euler's formula:

$$e^{i\omega t} = \cos \omega t + i \sin \omega t. \tag{6.4.4}$$

One can see that Eq. (6.4.1) is equivalent to the real part of $\underline{U}(t)$ in Eq. (6.4.3).

Applying a voltage $\underline{U}(t)$ that is given by Eq. (6.4.3) to an electrochemical cell, the resulting current I(t) may be written in complex form as follows:

$$\underline{I}(t) = I_{max} e^{i(\omega t - \varphi)}. \tag{6.4.5}$$

As one can see the current is assumed to change with time like the voltage, i.e. it has the same circular frequency ω. The phase angle φ occurring in Eq. (6.4.5) gives the phase shift of the current with respect to the voltage, i.e. it describes the fact, that the current "vector" lags behind the voltage "vector" by φ radians.

Now a complex resistance or simply impedance $\underline{Z}(\omega)$ is defined according to the following equation:

$$\underline{U} = \underline{Z} \cdot \underline{I}. \tag{6.4.6}$$

This equation states Ohm's law in a complex form. In general, the impedance is a function of the frequency ω but is assumed to be independent of the voltage $\underline{U}$. In this case one speaks of a linear circuit. Therefore, the following considerations are restricted to small amplitudes of the voltage $\underline{U}$, where the electrochemical systems are assumed to exhibit such a linear behaviour; otherwise, one has to take into account nonlinear response which results in complicated current voltage relationships.

The reciprocal of the impedance, i.e. the quantity $\underline{Y} = \underline{Z}^{-1}$, is referred to as the complex conductance or admittance. According to Eq. (6.4.6) we may rewrite Ohm's law in the following form

$$\underline{Y} \cdot \underline{U} = \underline{I}. \tag{6.4.7}$$

Noting the vector character of complex impedances and admittances, these quantities may be diagramatically represented in the complex plane.

In many cases it is necessary to transform impedances into admittances and vice versa. Since impedances and admittances are related to each other by

$$\underline{Y} \cdot \underline{Z} = 1 \tag{6.4.8}$$

as can be seen by comparing Eqs. (6.4.6) and (6.4.7) the following is valid if the impedance vector $\underline{Z}$ is written in exponential form according to $\underline{Z} = |\underline{Z}| e^{i\varphi}$:

$$\underline{Y} = \underline{Z}^{-1} = |\underline{Z}|^{-1} e^{-i\varphi}. \tag{6.4.9}$$

In the complex plane the vector $\underline{Y}$ is obtained from $\underline{Z}$ by inversion at the unit circle $|\underline{Z}| = 1$ (a special conformal mapping). The vector $\underline{Y} = \underline{Z}^{-1}$ has the negative phase angle $-\varphi$ and a phase angle -2φ relative to the vector $\underline{Z}$. Its magnitude $|\underline{Y}|$ is the reciprocal of the magnitude of $\underline{Z}$.

If impedances are represented in terms of their real and imaginary parts according to

$$\underline{Z} = a + bi \tag{6.4.10}$$

it follows that the corresponding admittance $\underline{Y}$ is given by

$$\underline{Y} = (a + bi)^{-1}. \tag{6.4.11}$$

Extension of the right-hand side of Eq. (6.4.11) by (a — bi) yields

$$\underline{Y} = \frac{a - bi}{a^2 + b^2} \tag{6.4.12}$$

or

$$\underline{Y} = \frac{a}{a^2 + b^2} - \frac{b}{a^2 + b^2} \cdot i. \tag{6.4.13}$$

The transformation of admittances into impedances may be performed in a similar way.

Let us now consider a series connection of several impedances. Each of the individual impedances $\underline{Z}_1$, $\underline{Z}_2$, $\underline{Z}_3$ etc. may in its own be composed of various components. The following is valid for such a series connection:

$$\underline{Z}_{ser} = \sum_k \underline{Z}_k. \tag{6.4.14}$$

For a parallel connection of impedances we get the following equation:

$$\underline{Z}_{par}^{-1} = \sum_k \underline{Z}_k^{-1}. \tag{6.4.15}$$

The inverse of the total impedance $\underline{Z}_{par}$ is equal to the sum of all reciprocal terms $\underline{Z}_k^{-1}$. The total admittance for the parallel connection is given simply by the sum of the individual admittances $\underline{Y}_k = \underline{Z}_k^{-1}$:

$$\underline{Y}_{par} = \sum_k \underline{Y}_k. \tag{6.4.16}$$

In the following we discuss three simple examples of equivalent circuits that are important in discussing the behaviour of electrochemical systems. We consider combinations of one or more ohmic resistances R and a capacitance C, which have by their own the following impedances:

$$\underline{Z}_R = R, \tag{6.4.17}$$

$$\underline{Z}_C = -\frac{i}{\omega C} = \frac{1}{i\omega C}. \tag{6.4.18}$$

Equation (6.4.18) can be derived starting from the well-known expression holding for the current-voltage relationship of a capacitance according to

$$I(t) = \frac{dQ(t)}{dt} = C\frac{dU(t)}{dt} \tag{6.4.19}$$

where Q(t) is the charge on the capacitor at time t. The current I(t) is given by the time derivative of the voltage U(t) times the capacitance C. This relationship is also valid in the case of complex quantities. Insertion of the expression for $\underline{U}(t)$ according to

$$\underline{U}(t) = U_{max} e^{i\omega t} \tag{6.4.20}$$

which is equal to Eq. (6.4.3), into Eq. (6.4.19) results in the following:

$$\underline{I}(t) = i\omega C U_{max} e^{i\omega t} \tag{6.4.21}$$

or

$$\underline{U}(t) = \frac{\underline{I}(t)}{i\omega C}. \tag{6.4.22}$$

Comparing Eq. (6.4.22) with the general definition (6.4.6) for the impedance $\underline{Z}$ leads to the expression $\underline{Z}_C = (i\omega C)^{-1}$.

Complex Impedance Diagrams of Simple Equivalent Circuits

Example 1: Let us first consider the series connection of an ohmic resistance R and an ideal capacitance C as shown in Fig. 6.4.1. It follows from Eqs. (6.4.14), (6.4.17) and (6.4.18) that the impedance $\underline{Z}$ of this circuit is given by

$$\underline{Z} = R - i(\omega C)^{-1} \tag{6.4.23}$$

where R is the real part of $\underline{Z}$ and $-(\omega C)^{-1}$ the imaginary one. To calculate the corresponding admittance $\underline{Y}$ we apply the transformation formula according to Eq. (6.4.13). Hence, the following is valid for $\underline{Y}$

$$\underline{Y} = \frac{\omega^2 R C^2}{1 + \omega^2 R^2 C^2} + \frac{\omega C}{1 + \omega^2 R^2 C^2} \cdot i. \tag{6.4.24}$$

The variation of $\underline{Z}$ and $\underline{Y}$ respectively as a function of ω is schematically shown in Fig. 6.4.1 where we have made use of the fact that both quantities have vectorial character which allows them to be represented graphically in the complex plane. The sets of all terminal points of the arrows, which correspond to the impedances $\underline{Z}$ and $\underline{Y}$ respectively, are the $\underline{Z}$- and $\underline{Y}$-loci respectively.

Example 2: We now consider the parallel connection of an ohmic resistance R and an ideal capacitance C. According to Eq. (6.4.15) the reciprocal of the total impedance of a parallel circuit is equal to the sum of the reciprocal of the individual impedances. Thus, the following holds for $\underline{Z}$

$$\frac{1}{\underline{Z}} = \left(\frac{1}{R} + i\omega C\right). \tag{6.4.25}$$

The right-hand side of Eq. (6.4.25) may be transformed by applying the same procedure that has been introduced for converting impedances into equivalent admittances (see Eqs. (6.4.11−13). We then have

$$\underline{Z} = \frac{R}{1 + \omega^2 R^2 C^2} - \frac{\omega R^2 C}{1 + \omega^2 R^2 C^2} \cdot i. \tag{6.4.26}$$

Since the total admittance of a parallel connection simply results from the addition of the various individual admittances (see Eq. (6.4.16)) we can write

$$\underline{Y} = \frac{1}{R} + i\omega C. \tag{6.4.27}$$

Figure 6.4.2 schematically shows the loci of $\underline{Z}$ and $\underline{Y}$ respectively for this circuit.

Example 3: As shown in Fig. 6.4.3 this example involves a circuit, where a parallel connection of an ohmic resistance R and an ideal capacitance C is in series with a second resistance r. Referring to the considerations given above the following can be written for the impedance of the whole circuit:

$$\underline{Z} = r + \left(\frac{1}{R} + i\omega C\right)^{-1}. \tag{6.4.28}$$

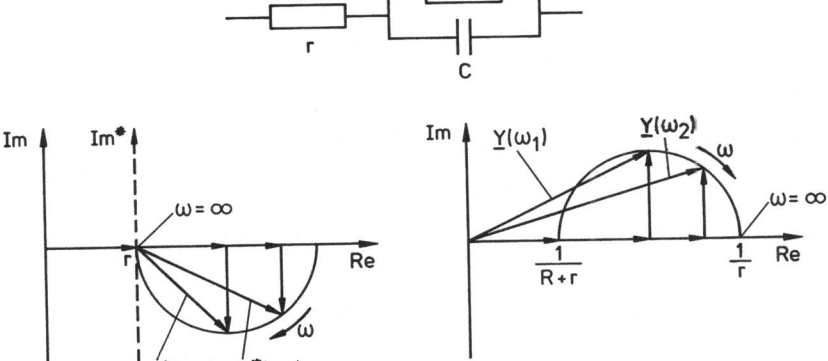

Fig. 6.4.3. Complex plane plots schematically representing the impedance and the admittance respectively of a parallel connection of a resistance and a capacity in series with another resistance

According to Eq. (6.4.28) this impedance may be split into two contributions, one of them equal to r, the other one-denoted as $\underline{Z}^*$ in the following — equal to the second term on the right-hand side of Eq. (6.4.28).

The contribution $\underline{Z}^*$ is identical with the impedance $\underline{Z}$ of the circuit discussed in Example 2. Replacing $\underline{Z}^*$ by the expression (6.4.26), we can rewrite Eq. (6.4.28) to give

$$\underline{Z} - r = \underline{Z}^* = \frac{R}{1 + \omega^2 R^2 C^2} - \frac{\omega R^2 C}{1 + \omega^2 R^2 C^2} \cdot i. \tag{6.4.29}$$

To calculate the total admittance $\underline{Y}$ of the circuit considered we start from the relationship

$$\underline{Y} = \underline{Z}^{-1}. \tag{6.4.30}$$

Using Eq. (6.4.28) we can insert the expression for $\underline{Z}$, whereby the following results:

$$\underline{Y} = \left[r + \left(\frac{1}{R} + \omega Ci \right)^{-1} \right]^{-1} \tag{6.4.31}$$

or

$$\underline{Y} = \left[r + \frac{R}{1 + \omega RCi} \right]^{-1}. \tag{6.4.32}$$

Further calculation finally yields the rather awkward looking expression

$$\underline{Y} = \frac{R + r + \omega^2 R^2 C^2 r}{(r + R)^2 + \omega^2 R^2 C^2 \cdot r^2} + \frac{\omega R^2 C}{(r + R)^2 + \omega^2 R^2 C^2 \cdot r^2} i. \tag{6.4.33}$$

Figure 6.4.3 schematically shows the frequency dependence of the vectors $\underline{Z}$ and $\underline{Y}$ respectively in the complex plane for this circuit. The curves result from the various positions of the tips of the vectors $\underline{Z}$ and $\underline{Y}$ corresponding to different values of the frequency.

6.5 References

[6.1]   Barr, L. W., Lidiard, A. B.: Physical Chemistry, An Advanced Treatise. Eyring, H.
        et al. (ed.), Vol. X, p. 152. New York, London: Academic Press 1970
        Darken, L. S.: Trans. AIME, *175*, 184 (1948)
        Hauffe, K.: Reaktionen in und an festen Stoffen, 2nd edit., Berlin—Heidelberg—
        New York: Springer 1966
        Jost, W.: Diffusion in Solids, Liquids, Gases. London: Academic Press 1969
        Lidiard, A. B.: Handbuch der Physik. Flügge, S., (ed.), Berlin—Göttingen—Heidel-
        berg: Springer 1957
        Manning, J. R.: Diffusion Kinetics for Atoms in Crystals, Princeton: van Nostrand
        1968
        Shewmon, P. G.: Diffusion in Solids. New York: McGraw-Hill 1963
[6.2]   Fick, A.: Pogg. Ann. *94*, 59 (1855)
[6.3]   Jost, W.: Diffusion in Solids, Liquids, Gases, London: Academic Press 1969
[6.4]   Carslaw, H. S., Jaeger, J. C.: Conduction of Heat in Solids. Oxford University
        Press 1959
[6.5]   Dünwald, H., Wagner, C.: Z. phys. Chem. *B24*, 53 (1934)
[6.6]   Crank, J.: The Mathematics of Diffusion. London, Oxford: University Press 1967
[6.7]   Hauffe, K.: Reaktionen in und an festen Stoffen, 2nd edit., Berlin—Heidelberg—
        New York: Springer 1966
[6.8]   Einstein, A.: Ann. Phys. (4) *17*, 549 (1905)
        Nernst, W.: Z. Phys. Chem. *2*, 613 (1888)
[6.9]   Darken, L. S.: Trans. AIME *175*, 184 (1948)
[6.10] Bardeen, J., Herring, C.: Atom Movements, Am. Soc. for Metals p. 87. Cleveland:
        1951
        Bardeen, J., Herring, C.: Imperfection in Nearly Perfect Crystals. New York:
        Wiley 1952
        Compaan, K., Haven, Y.: Trans. Faraday Soc. *52*, 786 (1956), Trans. Faraday Soc.
        54, 1498 (1958)
        LeClaire, A. D., Lidiard, A. B.: Philos. Mag (8) *1*, 518 (1956)
        Manning, J. R.: Phys. Rev. *116*, 819 (1959)
[6.11] LeClaire, A. D.: Physical Chemistry, An Advanced Treatise Eyring, H. et al. (ed.),
        Vol. X, p. 261. New York, London: Academic Press 1970
[6.12] DeGroot, S. R.: Thermodynamics of Irreversible Processes. Amsterdam: North-
        Holland Publ. Comp. 1951
        Wagner, C.: Prog. Solid State Chem. *10*, 3—16 (1975)
        DeGroot, S. R., Mazur, P.: Non-equilibrium Thermodynamics. Amsterdam: North-
        Holland Publ. Comp. 1962
        Denbigh, K. G.: The thermodynamics of the steady state. London: Methuen 1951
        Prigogine, I.: Etude thermodynamique des phénomènes irréversibles. Paris: Dunod
        1947
[6.13] Darken, L. S.: Trans. AIME *175*, 184 (1948)
[6.14] Wagner, C.: Atom Movements. Am. Soc. f. Metals, p. 153. Cleveland, Ohio, 1951
[6.15] Carslaw, H. S., Jaeger, J. C.: Conduction of Heat in Solids. Oxford: University
        Press 1959
        Shewmon, P. G.: Diffusion in Solids. New York: McGraw-Hill 1963
[6.16] Bronstein, I. N., Semendjajew, K. A.: Taschenbuch der Mathematik. Zürich,
        Frankfurt/M.: Deutsch 1980
[6.17] Hauffe, K.: Reaktionen in und an festen Stoffen, 2nd edit. Berlin—Heidelberg—
        New York: Springer 1966
        Manning, J. R.: Diffusion Kinetics for Atoms in Crystals. Princeton: van Nostrand
        1968
        Shewmon, P. G.: Diffusion in Solids. New York: McGraw-Hill 1963
[6.18] Jost, W.: Halbleiterprobleme Vol. II. Schottky, W., (ed.), Braunschweig: Vieweg
        1955

[6.19] Barr, L. W., Lidiard, A. B.: Physical Chemistry, An Advanced Treatise Eyring, H., et al. (ed.), Vol. X, p. 152. New York, London: Academic Press 1970

[6.20] Funke, K., Hackenberg, R.: Ber. Bunsenges. Phys. Chem. *76*, 885 (1972)
Funke, K., Kàlus, J., Lechner, R.: Solid State Commun. *14*, 1021 (1974)

[6.21] Warburg, E.: Wiedemann. Ann. Phys. *21*, 622 (1884)

[6.22] Warburg, E., Tegetmeier, F.: Wiedemann. Ann. Phys. *32*, 455 (1888)

[6.23] Haber, F., Tolloczko, A.: Z. anorg. Chem. *41*, 407 (1904)

[6.24] Bruni, E., Scarpa, O.: Rend. reale accad. naz. Lincei *22*, 438 (1913)

[6.25] Tubandt, C., Lorenz, F.: Z. phys. Chem. *87*, 543 (1913)
Tubandt, C., Eggert, S.: Z. anorg. allgem. Chem. *110*, 196 (1920)
Tubandt, C., Reinhold, H.: Z. Elektrochem. *29*, 313 (1923)
Tubandt, C.: Handbuch der Experimentalphysik Wien, W., Harms, F. (eds.), Vol. 12, Part 1, p. 383. Leipzig: Akad. Verlagsges. 1932

[6.26] Ilschner, B.: J. chem. Phys. *28*, 1109 (1958)

[6.27] Wagner, C.: Z. phys. Chem. *B21*, 42 (1933)

[6.28] Schmalzried, H.: Z. phys. Chem. N.F. *38*, 87 (1963)

[6.29] Wagner, C.: Adv. Electrochem. Eng. Delahay, P. (ed.), Vol. 4, p. 40. New York: Wiley 1966

[6.30] Hebb, M. H.: J. chem. Phys. *20*, 185 (1952)

[6.31] Wagner, C.: Proc. of the 7th. Meeting of the International Committee on Electrochemical Thermodynamics and Kinetics, p. 361 ff., Lindau 1955

[6.32] Hebb, M. H.: J. Chem. Phys. *20*, 185 (1952)
Rickert, H.: Z. phys. Chem. N.F. *23*, 355 (1960)
Valverde, N.: Z. phys. Chem. N.F. *75*, 1 (1971)

[6.33] Miyatani, S.: J. Phys. Soc. Japan *10*, 786 (1955)

[6.34] Patterson, J. W., Bogren, E. C., Rapp, R. A.: J. Electrochem. Soc. *114*, 752 (1967)

[6.35] Burke, L. D., Rickert, H., Steiner, R.: Z. phys. Chem. N.F. *74*, 146 (1971)

[6.36] Wagner, J. B., Wagner, C.: J. Chem. Phys. *26*, 1597 (1957)

[6.37] Wagner, J. B., Wagner, C.: J. Electrochem. Soc. *104*, 509 (1957)

[6.38] Weppner, W.: J. Solid State Chem. *20*, 305 (1977)

[6.39] Weiss, K.: Z. Phys. Chem., N.F. *59*, 242 (1968)
Weiss, K.: Electrochim. Acta *16*, 201 (1971)

[6.40] Mizusaki, J., Fueki, K., Mukaibo, T.: Bull. Chem. Soc. Jpn., *52*, 1890 (1979)

[6.41] Joshi, A. V., Wagner, J. B., Jr.: J. Electrochem. Soc., *122*, 1071 (1975)

[6.42] Jow, T., Wagner, J. B., Jr.: J. Electrochem. Soc. *125*, 613 (1978)

[6.43] Bottelberghs, P. H.: Low-Frequency Measurements on Solid Electrolytes and Their Interpretations, in: Solid Electrolytes, Hagenmüller, P., van Gool, W., (eds.). New York: Academic Press 1978
Bauerle, J. E.: J. Phys. Chem. Solids *30*, 2657 (1969)
McDonald, J. R.: Interpretation of AC Impedance Measurements in Solids, in: Superionic Conductors. Mahan, G. D., Roth, W. L. (eds.). New York: Plenum Press 1976

7 Solid Ionic Conductors, Solid Electrolytes and Solid-Solution Electrodes

The first part of this chapter contains a compilation of solid electrolytes — solid ionic conductors — according to the type of mobile ion which causes the ionic conductivity characteristic of the particular class of solid materials. The second part gives an account of mixed conductors which can be used as electrodes, the so-called solid-solution electrodes.

7.1 Solid Ionic Conductors, Solid Electrolytes

The following compilation of solid electrolytes does not pretend to be complete, since reports of new electrolytes are currently being published, at frequent intervals. Figure 7.1.1 shows the conductivities of some of the most important solid electrolytes as a function of temperature or reciprocal temperature respectively. The conductivity of liquid sulfuric acid is included as a basis for comparison. As will be noticed the conductivity of solid electrolytes may reach the same order of magnitude as that of liquid electrolytes.

Several important electrolytes will be treated below and some of these will be discussed in more detail.

7.1.1 Oxygen Ion Conductors

In Table 7.1.1 are compiled oxygen ion conductors. The most important one is doped zirconium dioxide, used at temperatures above 600 °C. As early as 1899 Nernst [7.1] suggested that ZrO_2 is a solid conductor for oxygen ions. A satisfactory explanation of the conduction mechanism, however, was first given by C. Wagner [7.2] in 1943. Doping of ZrO_2 with various metal oxides like CaO Y_2O_3 or MgO results in the generation of oxygen ion vacancies. Calcium ions for example are incorporated on zirconium sites in the ZrO_2 lattice; since, however, only one oxygen ion is introduced per calcium ion, one oxygen site remains unoccupied due to the addition of one molecule of CaO. These disorder sites, i.e. oxygen ion vacancies, are responsible for the ionic conductivity of zirconium dioxide. The amount of doping is of the order of 10 mol-%. It thus becomes clear why doping produces a very large number of vacancies in zirconium dioxide. The disorder in ZrO_2 and the dependence of the conductivity on the oxygen partial pressure have already been discussed in more detail in Sect. 3.1. It can be concluded from the disorder model of ZrO_2 that electron conductivity occurs at low oxygen partial pressures while electron defect conductivity is present at high

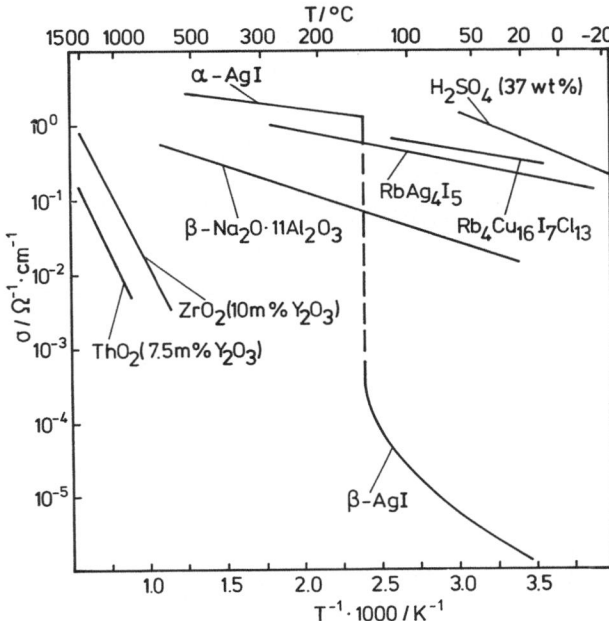

Fig. 7.1.1. Conductivity of several solid electrolytes as a function of temperature.

$ZrO_2(Y_2O_3)$	[7.26]
$ThO_2(Y_2O_3)$	[7.27]
$\beta\text{-}Na_2O \cdot 11\,Al_2O_3$	[7.28]
AgI	[7.11]
$RbAg_4I_5$	[7.15]
$Rb_4Cu_{16}I_7Cl_{13}$	[7.21]
H_2SO_4	[7.34]

Table 7.1.1. Conductivity of several oxygen ion conductors for the temperature range shown

Compound	Temp./°C	$\sigma \cdot 10^2/\Omega^{-1}\text{cm}^{-1}$	Ref.
ZrO_2 (10 mol-% Sc_2O_3)	600—1400	2—100	[7.26]
ZrO_2 (10 mol-% Y_2O_3)	600—1400	0.3—50	[7.26]
ZrO_2 (13 mol-% CaO)	640—1400	0.2—30	[7.27]
ThO_2 (7.5 mol-% Y_2O_3)	1000—1500	1.3—12	[7.27]
Bi_2O_3 (20 mol-% Er_2O_3)	270— 730	0.001—45	[7.28]

oxygen partial pressures. Because of this zirconium dioxide can only be used as a solid electrolyte for "medium" ranges of oxygen partial pressures. This range of ionic conductivity does not only depend on the oxygen partial pressure but also on the temperature. The limits of this range are reached if the conductivity of the electrons or electron defects respectively is higher than 1% of the total conductivity. The range of ionic conductivity in doped ZrO_2 is shown in Fig. 7.1.2.

From the practical point of view doped zirconium dioxide is used as a solid electrolyte in sensors for measuring oxygen partial pressures and in fuel cells and electrolyzers. These topics are discussed in Chap. 9.

Another important oxygen ion conductor is doped thorium dioxide: here pure ionic conductivity also occurs at lower oxygen partial pressures than are required for ZrO_2. ThO_2 can therefore still be employed as solid electrolyte when ZrO_2

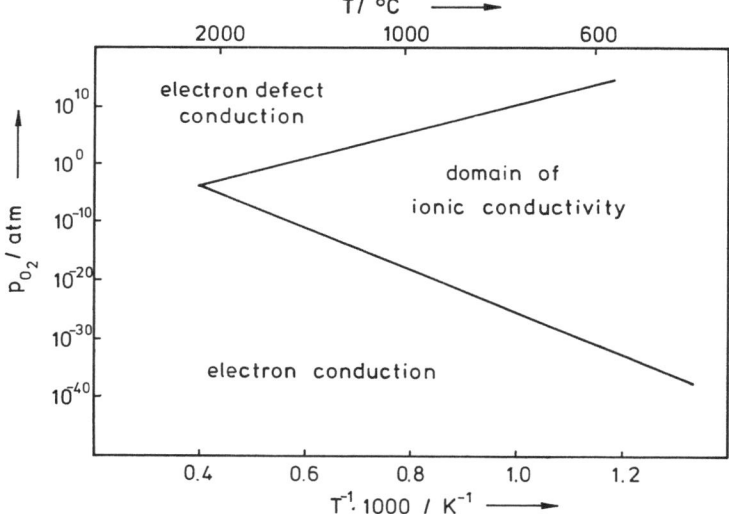

Fig. 7.1.2. Schematic representation of the conductivity behaviour of doped ZrO_2 as a function of the oxygen partial pressure and temperature

already exhibits considerable electron conduction. The use of doped cerium dioxide [7.38] and bismuth oxide for certain applications has recently also been discussed.

The temperature dependence of the conductivity of some common solid oxygen ion conductors is shown in Fig. 7.1.3. More details can be found in various review articles [7.3].

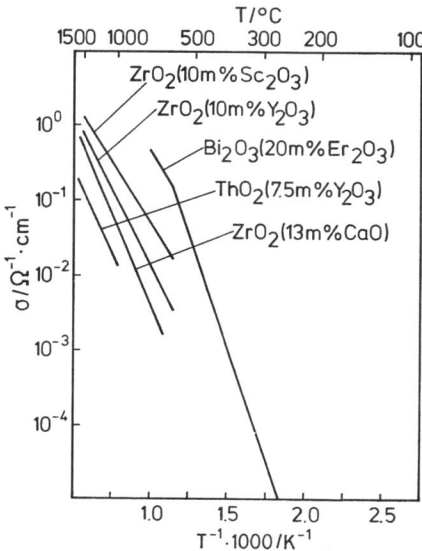

Fig. 7.1.3. Conductivity of several solid oxygen ion conductors as a function of temperature

$ZrO_2(Y_2O_3)$	[7.26]
$ZrO_2(Sc_2O_3)$	[7.26]
$ZrO_2(CaO)$	[7.26]
$ThO_2(Y_2O_3)$	[7.27]
$Bi_2O_3(Er_2O_3)$	[7.28]

7.1.2 Sodium Ion Conductors

Sodium ion conductors and their conductivity σ in a certain temperature range are listed in Table 7.1.2; Fig. 7.1.4 shows the temperature dependence of the conductivity of some of them. The most important is β-alumina with the formula $Na_2O \cdot 11\,Al_2O_3$ discovered by Kummer, Weber and Yao in 1967 [7.4] which is normally used at about $300-400\,°C$. It follows from investigations of this compound, carried out by Bettman and Peters [7.5] that β-alumina crystallizes in several phases. The most important of these with respect to practical applications are β- and β''-alumina. Since the β''-phase exhibits a higher conductivity and can be stabilized by adding large amounts of Na_2O to Al_2O_3 and further by doping with suitable oxides such as MgO and Li_2O, it is possible to optimize the conductivity as a function of the composition. The highest conductivities obtained in this way are found in polycrystalline phases of the ternary system Al_2O_3/Na_2O/MgO [7.6] and Al_2O_3/Na_2O/Li_2O [7.7]. The conductivity may reach values of about $0.3\ \Omega^{-1}\,cm^{-1}$. The mobile sodium ions are incorporated into lattice sheets located between non-conducting spinel blocks; hence, movements of these ions are only possible in two dimensions. Polycrystalline material must therefore

Table 7.1.2. Conductivity of several sodium ion conductors for the temperature range shown

Compound	Temp./°C	$\sigma \cdot 10^2/\Omega^{-1}\,cm^{-1}$	Ref.
β-$NaAl_{11}O_{17}$	$20-640$	$1.5-55$	[7.29]
$Na_5GdSi_4O_{12}$	$40-210$	$0.3-10$	[7.10]
$Na_3Zr_2Si_2PO_{12}$	$40-220$	$0.3-7.5$	[7.9]
$NaSbO_3 \cdot \frac{1}{6}\,NaF$	$40-220$	$0.001-0.7$	[7.30]

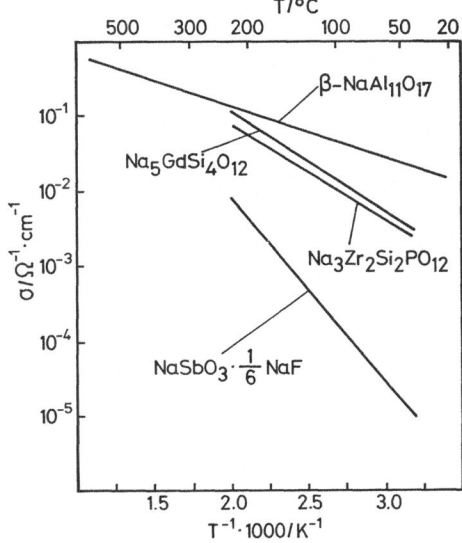

Fig. 7.1.4. Conductivity of some sodium ion conductors as a function of temperature

β-$NaAl_{11}O_{17}$	[7.29]
$Na_5GdSi_4O_{12}$	[7.10]
$Na_3Zr_2Si_2PO_{12}$	[7.9]
$NaSbO_3 \cdot \frac{1}{6}\,NaF$	[7.30]

be used to obtain a three-dimensional conductor. β-alumina has become very interesting as a solid electrolyte for rechargeable sodium sulfur batteries: this will be discussed in more detail in Chap. 9. If the sodium ions in the Na-β-alumina are replaced by for example Ag^+-, K^+- or Rb^+-ions, other β-aluminas are produced [7.8].

Another good sodium ion conductor with a conductivity comparable to that of β-alumina is the so-called Nasicon [7.9] whose composition is given by the empirical formula $Na_{1+x}Zr_2Si_xP_{3-x}O_{12}$ where x is about 2. In contrast to β-alumina, Nasicon is a three-dimensional conductor.

Other sodium ion conductors, for example sodium-ytterbium-silicates [7.10], are also under investigation.

7.1.3 Silver Ion Conductors

α-AgI was one of the first solid electrolytes to be found to exhibit a very high ionic conductivity. In 1914, Tubandt and Lorenz [7.11] observed that the conducting α-phase was stable above 146 °C. The first attempt to explain the conductivity was made by Strock [7.12] about 20 years later. He carried out investigations on the structure of α-AgI and pointed out that the iodide sublattice of crystalline AgI is fixed while the silver ions are randomly distributed over 42 lattice-sites available to them within a unit cell. As has been pointed out in Chap. 3, this picture has somewhat been modified: in contrast to distinct lattice positions one mereley speaks of spatial regions, which the Ag^+-ions are allowed to occupy in a statistical manner. The high mobility of the silver ions is due to this structural feature. This so-called structural disorder, already discussed in Chap. 3, causes the high ionic conductivity of this electrolyte.

α-Ag_3SI, an electrolyte similar to AgI, was studied in 1961 by Reuter and Hardel [7.13]. Since then several silver ion-conducting solid electrolytes of high conductivity have been discovered [7.14]. The main disadvantage of all these compounds lies in the fact that they attain their high conductivity only at higher temperatures. The first known exception was silver rubidium iodide, $RbAg_4I_5$, discovered in 1966 by Bradley and Greene and Owens and Argue in 1967 [7.15]. This electrolyte has attracted particular attention because it exhibits the highest silver ion conductivity at room temperature at present known. A number of other solid electrolytes showing silver ion conduction at room temperature have

Table 7.1.3. Conductivity of several silver ion conductors for the temperature range shown

Compound	Temp./°C	$\sigma \cdot 10^2/\Omega^{-1}\,cm^{-1}$	Ref.
AgI	150−500	130−260	[7.11]
$RbAg_4I_5$	−15−270	14−100	[7.15]
Ag_3SI	250−440	84−100	[7.13]
Ag_3SBr	20−300	0.1−8.2	[7.13]
Ag_2HgI_4	50−90	0.001−0.4	[7.15]
$C_5H_5NHAg_5I_6$	−20−480	0.1−440	[7.31]

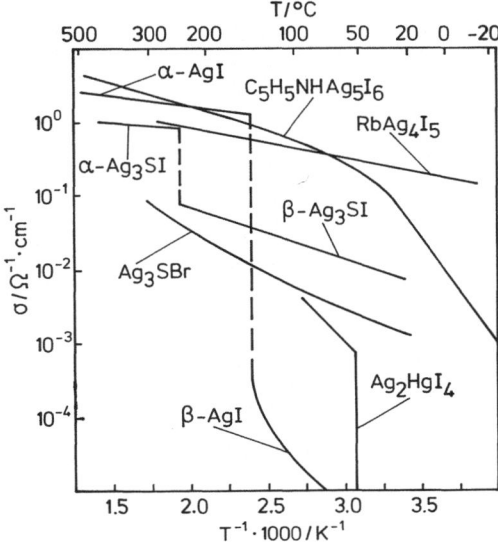

Fig. 7.1.5. Conductivity of some silver ion conductors as a function of temperature

AgI	[7.11]
RbAg$_4$I$_5$	[7.15]
Ag$_2$HgI$_4$	[7.15]
Ag$_3$SI	[7.13]
Ag$_3$SBr	[7.13]
C$_5$H$_5$NHAg$_5$I$_6$	[7.31]

recently been developed: these include Ag$_{19}$I$_{15}$P$_2$O$_7$ [7.16], Ag$_7$I$_4$PO$_4$ [7.17] Ag$_6$I$_4$WO$_4$ [7.18], Ag$_6$I$_4$CrO$_4$ [7.19] and also some which contain organic molecules in the crystal. Table 7.1.3 lists some silver ion conductors and their conductivities σ for the given temperature ranges. Figure 7.1.5 shows the temperature dependence of the conductivity of some of these conductors.

7.1.4 Copper Ion Conductors

The first solid electrolytes with high copper ion conductivity at room temperature were discovered in 1973 by Takahashi, Yamamoto and Ikeda [7.20]. These are n-alkyl(or hydro-)-hexamethylenetetraminehalide-copper(I)halide double salts. An example is 7CuBr·C$_6$H$_{12}$N$_4$CH$_3$Br, whose conductivity at room temperature is 0.017 Ω^{-1} cm^{-1}.

Several other copper ion conductors have since been described, e.g. the ternary system CuCl/CuI/RbCl. One of these conductors represented by the formula Rb$_4$Cu$_{16}$I$_7$Cl$_{13}$ has a conductivity of 0.34 Ω^{-1} cm^{-1} at 25°C. This is the solid electrolyte with the highest conductivity at room temperature known at present

Table 7.1.4. Conductivity of several copper ion conductors for the temperature range shown

Compound	Temp./°C	$\sigma \cdot 10^2/\Omega^{-1}$ cm^{-1}	Ref.
Rb$_4$Cu$_{16}$I$_7$Cl$_{13}$	10—110	28—62	[7.21]
7CuBr · C$_6$H$_{12}$N$_4$CH$_3$Br	10—130	1.5—14	[7.20]
7CuCl · C$_6$H$_{12}$N$_4$HCl	20—110	0.4—5	[7.20]
17CuI·3C$_6$H$_{12}$N$_4$CH$_3$I	20—140	0.1—2.2	[7.20]

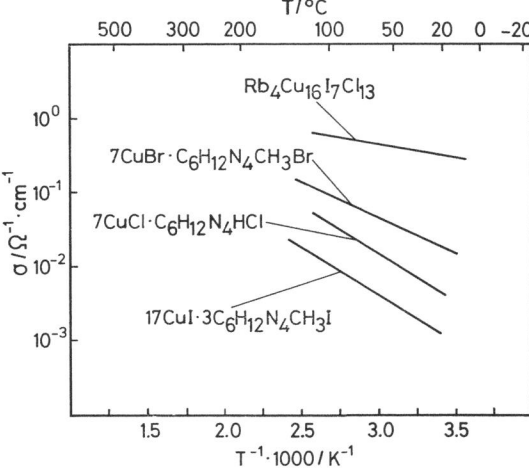

Fig. 7.1.6. Conductivity of some copper ion conductors as a function of temperature.

$Rb_4Cu_{16}I_7Cl_{13}$	[7.21]
$7\,CuBr \cdot C_6H_{12}N_4CH_3Br$	[7.20]
$7\,CuCl \cdot C_6H_{12}N_4HCl$	[7.20]
$17\,CuI \cdot 3\,C_6H_{12}N_4CH_3I$	[7.20]

[7.21]. Copper ion conductors and their conductivities σ at certain temperatures are listed in Table 7.1.4 and in Fig. 7.1.6 the temperature dependence of the conductivity of some of these is shown.

7.1.5 Lithium Ion Conductors

Among the lithium halides lithium iodide exhibits the highest conductivity, $10^{-7}\ \Omega^{-1}\ cm^{-1}$ at 25 °C [7.22]. This relatively low conductivity should increase upon incorporation of cations of other valencies [7.23], e.g. upon addition of Ca^{2+}-ions. The increase in conductivity resulting from this doping, however, was not sufficient. In 1973, Liang [7.24] discovered that the ionic conductivity of lithium iodide could be increased by a factor of about 100 at 25 °C by preparing a mechanical mixture of LiI and 35 mol-% Al_2O_3; this electrolyte was shown to be very stable. In 1977, Owens and Hanson [7.25] were able to establish that the conductivity of LiI as well as that of LiBr, LiCl and LiF could be increased by mixing these compounds with Al_2O_3 and H_2O. Some typical examples for such mixtures are contained in Table 7.1.5, together with other lithium ion

Table 7.1.5. Conductivity of several lithium ion conductors at 25 °C

Compound	Temp./°C	$\sigma \cdot 10^5/\Omega^{-1}\ cm^{-1}$	Ref.
LiI	25	0.01	[7.22]
LiI(+ 1 mol-% CaI_2)	25	1.2—0.2	[7.23]
LiI(+ 35 mol-% Al_2O_3)	25	1	[7.24]
LiI(+ 25 wt-% Al_2O_3 + 11 wt-% H_2O)	25	4	[7.25]
LiF(+ 34 wt-% Al_2O_3 + 10 wt-% H_2O)	25	10	[7.25]
$LiAlCl_4$	25	0.1	[7.33]

conductors. Lithium ion conductors find their main applications in primary high-energy density batteries with long life-times where only low currents are to be drawn, i.e. in batteries required for pace makers and so on.

7.1.6 Fluorine Ion Conductors

Several solid electrolytes exist which exhibit high fluorine ion conductivities; they are predominantly of the fluorite or tysonite type. Fluorine ion conductors of the fluorite type are for example alkaline-earth metal fluorites and lead fluorite. Tysonite-type fluorine ion conductors include lanthanum or cerium fluoride doped with for example calcium or strontium fluoride. Uses of the fluorine ion conductors include their incorporation into solid-state batteries with high power density. Table 7.1.6 lists some fluorine ion conductors and their conductivities σ at certain temperatures, while Fig. 7.1.7 shows the temperature dependence of the conductivity of some of these.

Table 7.1.6. Conductivity of several fluorine ion conductors for the temperature range shown [7.32]

Compound	Temp./°C	$\sigma \cdot 10^2/\Omega^{-1}\,cm^{-1}$
CaF_2	440 — 1160	0.0001 — 45
SrF_2	580 — 980	0.00014 — 1.4
BaF_2	430 — 1080	0.0001 — 25
$Ca_{0.75}Y_{0.25}F_{2.25}$	230 — 600	0.0001 — 0.5
LaF_3	130 — 670	0.003 — 1.2
$La_{0.95}Sr_{0.05}F_{2.95}$	130 — 510	0.05 — 1.4

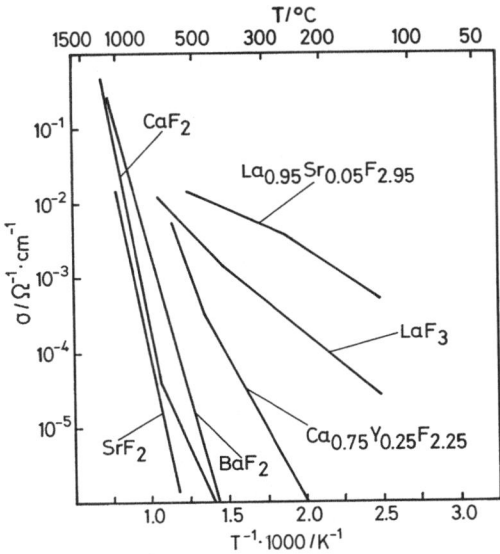

Fig. 7.1.7. Conductivity of some fluorine ion conductors as a function of temperature [7.32]

7.2 Solid-Solution Electrodes

A second class of important solids next to solid electrolytes is that of mixed conducting solids exhibiting both fast ion transport and electronic conductivity. If, in addition to these properties, a range of stoichiometry is present, which is large enough, these mixed conducting solids can be used as electrodes for batteries, the so-called solid-solution electrodes. These batteries may contain either liquid electrolytes or solid electrolytes. One of the most promising solid-solution electrodes is based on titanium disulfide, TiS_2, where it is possible to dissolve relatively large amounts of lithium metal in the TiS_2-phase. There is a continuous range of non-stoichiometry from TiS_2 to $LiTiS_2$. The structure of TiS_2 and other similar chalcogenides of transition metals can be described as a sequence of layers held together by van der Waals forces only. The lithium is dissolved between the layers whereby the distance between the layers is slightly increased. However, other changes in the crystal structure do not occur. Solutions of this kind are sometimes called insertion or intercalation compounds. The partial molar Gibbs energy of solution of lithium is relatively constant over the whole range of non-stoichiometry. Therefore, the emf of a cell containing a Li_xTiS_2 electrode is nearly constant during the discharging process. Figure 7.2.1 shows the emf of the cell

$$Li \mid Li^+\text{-electrolyte} \mid Li_xTiS_2 \tag{7.2.I}$$

as a function of x in Li_xTiS_2 (see for example [7.35]). The emf times Faraday's constant is equivalent to the difference of the chemical potentials of lithium

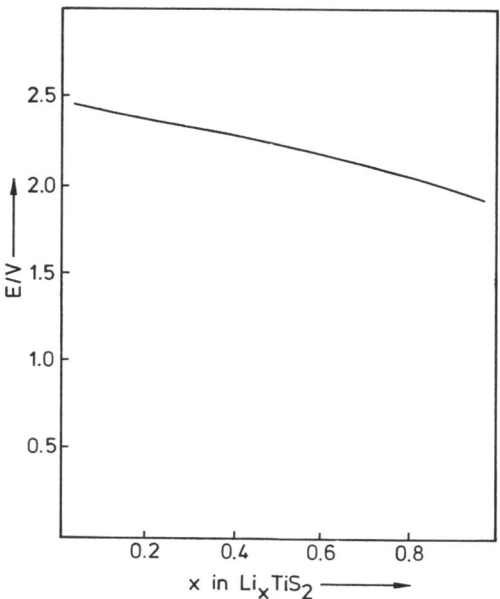

Fig. 7.2.1. Emf of cell (7.2.I) as a function of the stoichiometric composition x of Li_xTiS_2 [7.35]

μ_{Li} in the compound Li_xTiS_2 and μ_{Li}^0 in the pure metal:

$$\mu_{Li} - \mu_{Li}^0 = -EF. \tag{7.2.1}$$

This difference corresponds to the Gibbs energy of solution of lithium in Li_xTiS_2 and shows a relatively slow change over a large range of possible stoichiometric compositions x as can be seen in Fig. 7.2.1.

Up to now, many other solid-solution electrodes or, to be more general, compounds exhibiting good ionic and electronic conductivity have been investigated. Reviews on this topic can be found in the literature [7.36, 7.37]. Examples of other layer compounds besides TiS_2, that are able to dissolve certain alkali metals and in some cases Cu^+- or Ag^+-ions are represented by $TiSe_2$, MoS_2, WS_2, TaS_2, ZrS_2, NbS_2, VSe_2, $MoSe_2$, WSe_2, CrS_2. A well-known example of a material with a layered structure is graphite which can dissolve alkali metals between the carbon layers whereby alkali intercalates are generated as, for example, C_xK. However, the formation of intercalates based on graphite is accompanied by very low Gibbs energies of solution of the metals. Hence, these compounds represent no suitable candidates for good solid-solution electrodes. All these compounds show high mobilities of the metal ions in the region between the layers of the host compound.

In this connection one should also mention a related series of certain transition metal oxides which do not exhibit layered structures but are able to insert alkali metals or copper or silver, for example TiO_2, MnO_2, MoO_3, WO_{3-y}, V_2O_5, Ta_2O_5.

7.3 References

[7.1]   Nernst, W.: Z. Elektrochem. *6*, 41 (1899)
[7.2]   Wagner, C.: Naturwissenschaften *31*, 265 (1943)
[7.3]   Hladik, J.: Physics of Electrolytes, Academic Press, New York 1972
        van Gool, W.: Fast Ion Transport in Solids — Solid-State Batteries and Devices, North-Holland, Amsterdam 1973
        Kleitz, M., Dupuy, J.: Electrode Processes in Solid-State Ionics, Reidel, Dordrecht 1976
        Alcock, C. B.: EMF-Measurements in High-Temperature Systems, Inst. Mining Metallurgy, London 1968
        Fischer, W. A., Jahnke, D.: Metallurgische Elektrochemie, Springer, Berlin, Heidelberg, New York 1975
        Mahan, G. D., Roth, W. L.: Superionic Conductors, Plenum Press, New York, London 1976
        Geller, S.: Solid Electrolytes, Springer Verlag Berlin, Heidelberg, New York 1977
        Hagenmuller, P., van Gool, W.: Solid Electrolytes, Academic Press, New York, San Francisco, London 1978
        Vashishta, P., Mundy, J. N., Shenoy, G. K., (eds.): Fast Ion Transport in Solids Electrolytes North-Holland, N.Y., Amsterdam, Oxford 1979
        Subbarao, E. C. (ed.): Solid Electrolytes and Their Applications, Plenum Press, New York, London 1980
        Etsell, T. H., Flengas, S. N.: Chem. Rev. *70*, 339 (1970)
[7.4]   Weber, N., Kummer, J. T.: Proc. Ann. Power Sources Conf. *21*, 37 (1967)
[7.5]   Bettman, M., Peters, C. R.: J. Phys. Chem. *73*, 1774 (1969)
        Peters, C. R., et al.: Acta Cryst. *B27*, 1826 (1971)

[7.6] Haar, W., et al.: High-Temperature Batterie Workshop, Argonne National Laboratory, USA, 1967

[7.7] Weiner, S. A., Ford Motor Co.: Research on Electrodes and Electrolytes for the Ford Sodium Sulfur Batterie; Annual Report, Contract NSF-C 805 July 1975, 1976, 1977

[7.8] Yao, Y., Kummer, J. T.: J. Inorg. Nucl. Chem. *29*, 2453 (1967)
Kummer, J. T.: Solid-State Chem. *17*, 141 (1972)

[7.9] Hong, H. Y.-P.: Mat. Res. Bull. *11*, 173 (1976)
Goodenough, J. B., Hong, H. Y.-P., Kafalas, J. A.: Mat. Res. Bull. *11*, 203 (1976)

[7.10] Shannon, R. D., Chen, H.-Y., Berzins, T.: Mat. Res. Bull. *12*, 969 (1977)

[7.11] Tubandt, C., Lorenz, F.: Z. phys. Chem. *87*, 543 (1914)

[7.12] Strock, L. W.: Z. phys. Chem. *B25*, 441 (1934)
Strock, L. W.: Z. phys. Chem. *B31*, 132, (1935)

[7.13] Reuter, B., Hardel, K.: Naturwissenschaften *48*, 161 (1961)
Reuter, B., Hardel, K.: Z. anorg. allg. Chem. *340*, 158 (1965)

[7.14] Geller, S., (ed.): Solid Electrolytes, Springer-Verlag Berlin, Heidelberg, New York 1977

[7.15] Bradley, J. N., Greene, P. D.: Trans. Faraday Soc. *62*, 2069 (1966)
Bradley, J. N., Greene, P. D.: Trans. Faraday Soc. *63*, 424 (1967)
Owens, B. B., Argue, G. R.: Science *157*, 308 (1967)

[7.16] Takahashi, T., Ikeda, S., Yamamoto, O.: J. Electrochem. Soc. *119*, 447 (1972)

[7.17] see Ref. [7.16]

[7.18] Takahashi, T., Ikeda, S., Yamamoto, O.: J. Electrochem. Soc. *120*, 647 (1973)

[7.19] Chiodelli, G., Magistris, A., Schiraldi, A.: Electrochim. Acta *19*, 655 (1974)

[7.20] Takahashi, T., Yamamoto, O., Ikeda, S.: J. Electrochem. Soc. *120*, 1431 (1973)

[7.21] Takahashi, T. et al.: J. Electrochem. Soc. *126*, 1654 (1979)

[7.22] Ginnings, D. C., Phipps, T. E.: J. Am. Chem. Soc. *52*, 1340 (1930)

[7.23] Schlaitkjer, C. R., Liang, C. C.: J. Electrochem. Soc., *118*, 1447 (1971)

[7.24] Liang, C. C.: J. Electrochem. Soc. *120*, 1289 (1973)

[7.25] Owens, B. B., Hanson, H. J.: US Patent 4 007/22 (1977)

[7.26] Etsell, T. H., Flengas, S. N.: Chem. Rev. *70*, 339 (1970)

[7.27] Wimmer, J. M., Bidwell, L. R., Tallan, N. M.: J. Amer. Ceram. Soc. *50*, 198 (1967)

[7.28] Verkerk, M. J., Keizer, K., Burggraaf, A. J.: J. Appl. Electrochem. *10*, 81 (1980)

[7.29] Wittingham, M. S., Huggins, R. A.: J. Chem. Phys. *54*, 414 (1971)

[7.30] Singer, J., et al.: J. Electrochem. Soc. *123*, 614 (1976)

[7.31] Geller, S., Owens, B. B.: J. Phys. Chem. Solids *33*, 1241 (1972)

[7.32] Nagel, L. E., O'Keeffe, M.: Fast Ion Transport in Solids, Solid State Batteries and Devices. van Gool, W. (ed.). North-Holland Publ. Co., Amsterdam, London: Elsevier Publ. Co., New York 1973, p. 165

[7.33] Weppner, W., Huggins, R. A.: J. Electrochem. Soc. *124*, 35 (1977)

[7.34] Drotschmann, C.: Batterien *19*, 761 (1965)

[7.35] Whittingham, M. S.: J. Electrochem. Soc. *123*, 315 (1976)

[7.36] Steele, B. C. H.: Electrochemical injection of ions into non-stoichiometric electrodes, in: Trends in Electrochemistry. Bockris, J. O'. M., Rand, D. A. J., Welch, B. J., (eds.), p. 145 Plenum Press, New York 1977

[7.37] Whittingham, M. S.: Chemistry of intercalation compounds, metal guests in chalkogenide hosts, in: Progress in Solid-State Chemistry, Rosenblatt, G. M., Worrell, W. L., (eds.), Vol. 12, p. 41 Pergamon Press, Oxford, New York 1978

[7.38] Tuller, H. L., Nowick, A. S.: J. Electrochem. Soc. *122*, 255 (1975)
Kudo, T., Obayashi, H.: J. Electrochem. Soc. *122*, 142 (1975)

8 Galvanic Cells with Solid Electrolytes for Thermodynamic Investigations

In analogy to galvanic cells with liquid electrolytes, those with solid electrolytes consist of at least two electrodes separated by an electrolyte, which in this case is a solid ionic conductor. In a simple case both electrodes and the electrolyte consist of powdered chemical compounds which have been pressed to pellets; these, in turn, are pressed together and provided with two leads. Thermodynamic investigations involving such galvanic cells with solid electrolytes permit the determination of

a) Gibbs reaction energies,
b) chemical potentials, activities or partial pressures,
c) reaction enthalpies and reaction entropies obtained from investigations of temperature dependencies.

In this section we shall only consider galvanic cells with solid electrolytes which are pure ionic conductors, i.e. the electronic conductivity is negligible compared to the ionic conductivity.

Just as in the electrochemistry of liquid electrolytes, galvanic cells with solid electrolytes may be treated from two different points of view:

a) that due to Helmholtz, in which the maximum available work done by the cell during the cell reaction is considered
b) that due to Nernst, in which single electrode potentials are considered, these then being summed to give the total emf of the galvanic cell.

The Helmholtz approach provides direct information about the emf of a cell, but few data on the factors causing the emf to assume this particular value and on the various physical processes taking place in the cell. We can obtain more information of this type by using the Nernst approach whereby electrochemical and electrical potentials are considered; we shall discuss this approach when dealing with measurements of chemical potentials.

8.1 Determination of Gibbs Reaction Energies

From the Helmholtz point of view, we are interested in the cell reaction proceeding in a galvanic cell accompanied by a definite charge flow, e.g. n Faradays. The maximum electrical work done by the galvanic cell is equal to nFE, neglecting polarization effects, where E is the open circuit electrical potential difference, i.e. the electromotive force (emf) of the cell. According to the Stockholm convention [8.1] E is taken as positive if the right-hand electrode of the galvanic cell is positive. The flow of charge through the cell is called positive if a positive electrical current flows from left to right through the cell.

The value of n is generally chosen to correspond to one formula conversion of the cell reaction. In practice, however, the potential difference is not measured while the cell reaction is proceeding, since a finite current flow through the cell, that is present under these conditions, is almost always accompanied by polarization effects; instead, the potential difference is measured at open circuit to obtain the emf which is then multiplied by nF to give the maximum electrical work nFE. This quantity is equal to the negative Gibbs energy ΔG of the cell reaction. Thus, the following equation applies:

$$\Delta G = -nFE. \tag{8.1.1}$$

Equation (8.1.1) can be derived as follows: According to the first law of thermodynamics the state function U, i.e. the internal energy of the system, obeys the following relation:

$$dU = \delta Q + \delta W, \tag{8.1.2}$$

i.e. an infinitesimal change in the internal energy U of a system results because of exchange of heat δQ and work δW between the system considered and the surroundings.

If we only consider mechanical work involving changes in volume and electrical work δW_{el} we can write for δW

$$\delta W = -p \, dV + \delta W_{el} \tag{8.1.3}$$

where the equality holds for reversible processes since only for such processes is the mechanical work identical with $-pdV$. Regarding only reversible processes we can relate the heat exchanged between the system and the surroundings to the corresponding entropy change with the aid of the second law of thermodynamics according to the equation

$$\delta Q_{rev} = T \, dS. \tag{8.1.4}$$

Combining Eqs. (8.1.2−4) and solving for δW_{el} we obtain

$$\delta W_{el} = dU - T \, dS + p \, dV. \tag{8.1.5}$$

The expression on the right-hand side of Eq. (8.1.5) is equal to the differential change dG of the Gibbs energy for p and T constant as will be shown in the following. The Gibbs energy is defined in thermodynamic terms as follows

$$G = U - TS + pV \tag{8.1.6}$$

with the total differential

$$dG = dU - T \, dS - S \, dT + p \, dV + V \, dp \tag{8.1.7}$$

If pressure p and temperature T are held constant, Eq. (8.1.7) reduces to

$$dG = dU - T \, dS + p \, dV \quad \text{for} \quad p, T = \text{const.} \tag{8.1.8}$$

Combining Eqs. (8.1.8) and (8.1.5) we see

$$\delta W_{el} = dG \quad \text{for} \quad p, T = \text{const.,} \tag{8.1.9}$$

i.e. the maximum (reversible) electrical work done on or by the system is equal to the change in the Gibbs energy G of the system. Regarding one formula conversion of the cell reaction we write instead of δW_{el} and dG W_{el} and ΔG respectively according to finite changes of these quantities. The following equation for the electrical work applies taking into account the sign convention adopted in thermodynamics

$$W_{el} = -nFE \quad \text{for} \quad p, T = \text{const.} \tag{8.1.10}$$

Combination of Eqs. (8.1.9) and (8.1.10) gives Eq. (8.1.1).

The cell reaction and the corresponding Gibbs reaction energy are referred to the chemical substances as they are present in the galvanic cell. If all of these are in their standard state, one measures ΔG^0, the standard value of the Gibbs reaction energy.

We shall now discuss several examples to demonstrate the determination of thermodynamic quantities with the aid of solid-state galvanic cells.

a) Determination of the Gibbs energy of formation $\Delta_f G_{AgCl}$ of silver chloride from silver and gaseous chlorine at a temperature of 400 °C.

Since solid AgCl is a virtually pure ionic conductor for Ag^+ ions as long as the chlorine partial pressure is not too high, AgCl itself can be used as a solid electrolyte in the cell (8.1.I). The vertical lines signify phase boundaries, the chemical symbols denote the phases which are arranged in series to constitute the galvanic cell.

$$C \mid Ag \mid AgCl \mid Cl_2(g), C \qquad (8.1.I)$$

1F:
$$
\begin{array}{cc}
e^- \quad Ag^+ & e^- \\
\end{array}
$$

$$
\begin{aligned}
Ag &= && Ag^+(\rightarrow) \\
Ag^+(\rightarrow) && &+ \frac{1}{2}\, Cl_2(g) \\
+\, e^-(\leftarrow) && &+ e^-(\leftarrow) \\
&= && AgCl
\end{aligned}
$$

The total cell reaction proceeding under current flow follows from a consideration of the single electrode reactions. If a charge of 1 Faraday flows through the cell, one mole of silver ions enters the AgCl at the left-hand electrode, while the corresponding amount of electrons flows off to the left. At the right-hand side the silver ions transported through the AgCl react with gaseous chlorine and with the electrons entering the AgCl-phase via the external circuit. The arrows indicate the different directions with respect to the migration of the reacting particles. The summation of the so called half-reactions proceeding at the electrodes gives the total cell reaction

$$Ag + \frac{1}{2}\, Cl_2 = AgCl \cdots \Delta_f G_{AgCl}. \qquad (8.1.11)$$

The Gibbs energy of formation $\Delta_f G_{AgCl}$ is related to the emf E of the cell by the equation

$$\Delta_f G_{AgCl} = -EF. \qquad (8.1.12)$$

Reinhold [8.2] used this cell in 1928 to determine the $\Delta_f G$ value for the formation of AgCl. The experimental arrangement is shown in Fig. 8.1.1. Care must be taken to ensure that the electrodes are separated from each other in a gas-tight manner and particularly that no direct reaction between chlorine and silver is possible. The cell (8.1.I) is a particularly simple one, since the reaction product AgCl of the cell reaction is at the same time the solid electrolyte. If the solid reaction product, in this case AgCl, were not a pure ionic conductor, it would

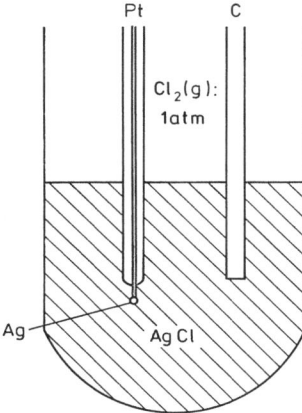

Fig. 8.1.1. Experimental set-up (schematic) of the galvanic cell (8.1.I) for the determination of $\Delta_f G^0_{AgCl}$

be necessary to use a pure ion-conducting auxiliary electrolyte. This is for example the case when we wish to determine the $\Delta_f G^0$-value of Ag_2S.

b) Determination of the standard Gibbs energy of formation of silver sulfide $\Delta_f G^0_{Ag_2S}$ in the temperature range 200—400 °C [8.3.].

The following cell (8.1.II) may be used with silver iodide being a virtually pure ionic conductor for silver ions:

$$Pt \mid Ag \mid AgI \mid Ag_2S \mid S(l) \mid Pt \tag{8.1.II}$$

$$\begin{array}{ccc} \longleftarrow & \longrightarrow & \longleftarrow \\ 2\,F: \qquad 2\,e^- & 2\,Ag^+ & 2\,e^- \end{array}$$

$$\begin{aligned} 2\,Ag &= \quad 2\,Ag^+(\rightarrow) \\ 2\,Ag^+(\rightarrow) & \quad + 2\,e^-(\leftarrow) \\ + 2\,e^-(\leftarrow) & \quad + S(l) \\ & \quad = Ag_2S \end{aligned}$$

$$2\,Ag + S(l) = Ag_2S \cdots \Delta_f G^0_{Ag_2S} \tag{8.1.13}$$

The cell reaction accompanied by the transport of two Faradays through the cell corresponds to the formation of one mole of Ag_2S from solid silver and liquid sulfur. Here again it is necessary to make sure that the electrodes are separated from one another in a gas-tight manner. Figure 8.1.2 shows the construction of such a cell. The standard Gibbs energy of formation may be obtained from the equation

$$\Delta_f G^0_{Ag_2S} = -2\,EF. \tag{8.1.14}$$

The presence of liquid sulfur at the right-hand electrode of the cell (8.1.II) and in similar cells causes difficulties in carrying out the experimental measurements, since particular care must be taken that the solid electrolyte separates the two electrodes in a gas-tight manner. Construction of the cell becomes simpler if the vapour pressures of the substances at the electrodes are sufficiently low. In this case it is for example possible to use electrodes and solid electrolytes in the form of pellets pressed together in an inert atmosphere; the emf measure-

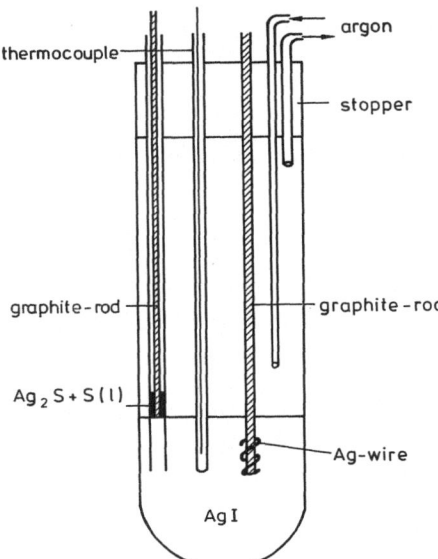

thermocouple

argon

stopper

graphite-rod — graphite-rod

$Ag_2S + S(l)$

Ag-wire

Ag I

Fig. 8.1.2. Experimental set-up of the galvanic cell (8.1.II) for the determination of $\Delta_f G^0_{Ag_2S}$

ments may then be carried out in an oven at the desired temperature. The presence of liquid sulfur may in fact be avoided if the Gibbs energy of formation of the sulfides is not measured directly but indirectly, using a method involving a displacement reaction in the galvanic cell. This may be achieved, as shown in the following example.

c) Determination of the standard Gibbs energy of formation of lead sulfide $\Delta_f G^0_{PbS}$ at a temperature of 350 °C.

The following cell can be used:

$$C \mid Pb(l) \mid PbCl_2 \mid PbS, \mid C \qquad (8.1.III)$$
$$(+ KCl) \quad Ag_2S, Ag$$

$$2F: \quad 2e^- \quad 2Cl^- \quad 2e^-$$

$$\begin{array}{ll}
Pb(l) & PbCl_2 \\
+ 2Cl^-(\leftarrow) & + Ag_2S \\
= PbCl_2 & + 2e^-(\leftarrow) \\
+ 2e^-(\leftarrow) & = PbS \\
& + 2Ag \\
& + 2Cl^-(\leftarrow)
\end{array}$$

$$Pb(l) + Ag_2S = PbS + 2Ag \cdots \Delta G^0 \qquad (8.1.15)$$

The right-hand electrode is a mixture of the substances PbS, Ag_2S and Ag. The cell reaction proceeding in cell (8.1.III) is the displacement reaction given by Eq. (8.1.15), i.e. lead sulfide and silver are formed from lead and silver sulfide. The standard Gibbs energy ΔG^0 of the displacement reaction (8.1.15), which can be determined from measurements of the emf E of cell (8.1.III) is equal to the difference of the standard Gibbs energies of formation of PbS and Ag_2S, respectively,

$$\Delta G^0 = -2\,EF = \Delta_f G^0_{PbS} - \Delta_f G^0_{Ag_2S}. \tag{8.1.16}$$

Rearrangement of Eq. (8.1.16) gives the following expression for the standard Gibbs energy of formation of lead sulfide $\Delta_f G^0_{PbS}$

$$\Delta_f G^0_{PbS} = -2\,EF + \Delta_f G^0_{Ag_2S}, \tag{8.1.17}$$

i.e. $\Delta_f G^0_{PbS}$ may be calculated from the emf values of cell (8.1.III), if $\Delta_f G^0_{Ag_2S}$ is known. This type of galvanic cell has been used by Kiukkola and Wagner [8.4].
d) Determination of the standard Gibbs energy of formation $\Delta_f G^0_{Cu_2O}$ of Cu_2O at higher temperatures.

A solid electrolyte suitable for this purpose is zirconium dioxide, doped with calcium oxide, magnesium oxide or yttrium oxide. The doping leads to the formation of oxygen ion vacancies in the zirconium dioxide so that zirconia exhibits a good oxygen ion conductivity. The following galvanic cell (8.1.IV) may be used; the cell reaction is the formation of Cu_2O:

$$
\begin{array}{ccc}
\text{Pt} \quad | \quad \text{Cu,} & | \quad \text{ZrO}_2 & | \quad \text{Pt,} \\
\text{Cu}_2\text{O} & (+\text{CaO}) & \text{O}_2(\text{g}) \ (1 \ \text{atm})
\end{array} \tag{8.1.IV}
$$

$2F: \qquad \overset{\longleftarrow}{2e^-} \quad \overset{\longleftarrow}{O^{2-}} \quad \overset{\longleftarrow}{2e^-}$

$$
\begin{array}{cc}
2\,\text{Cu} & \dfrac{1}{2}\,\text{O}_2 \\
+\,\text{O}^{2-}(\leftarrow) & +\,2e^-(\leftarrow) \\
=\,\text{Cu}_2\text{O} & =\,\text{O}^{2-}(\leftarrow) \\
+\,2e^-(\leftarrow) &
\end{array}
$$

$$2\,\text{Cu} + \frac{1}{2}\,\text{O}_2 = \text{Cu}_2\text{O} \cdots \Delta_f G^0_{Cu_2O} \tag{8.1.18}$$

$\Delta_f G^0_{Cu_2O}$ is related to the emf of the cell according to:

$$\Delta_f G^0_{Cu_2O} = -2\,EF. \tag{8.1.19}$$

Since gaseous oxygen is present at the right-hand electrode at a pressure of one atmosphere, it is again necessary to ensure a gas-tight separation of the electrodes. In practice, as shown in Fig. 8.1.3, it is possible to use an electrolyte consisting

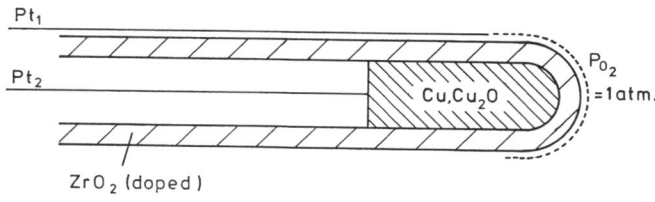

Fig. 8.1.3. Experimental arrangement (schematic) of the galvanic cell (8.1.IV) used for the electrochemical determination of $\Delta_f G^0_{Cu_2O}$

of a tube closed at one end and containing a mixture of Cu and Cu_2O; porous platinum, surrounded by oxygen at a pressure of 1 atmosphere, at the outside of the tube forms the other electrode. It is possible using similar cells to determine $\Delta_f G^0$-values of oxides whose $\Delta_f G^0$-values are not more negative than that of FeO. For oxygen activities lower than those corresponding to the equilibrium between Fe and FeO zirconium dioxide becomes electron conducting and therefore it is no more usable as a solid electrolyte. The Gibbs energy of formation can also be determined indirectly, starting from displacement reactions as shown in the following example.

e) Determination of the standard Gibbs energy of formation of FeO, $\Delta_f G^0_{FeO}$ at higher temperatures.

We shall consider the following cell:

$$Pt \mid \begin{matrix} Fe, \\ FeO \end{matrix} \mid \begin{matrix} ZrO_2 \\ (+CaO) \end{matrix} \mid \begin{matrix} Cu, \\ Cu_2O \end{matrix} \mid Pt \qquad (8.1.V)$$

$$2F: \qquad \overset{\leftarrow}{2e^-} \quad \overset{\leftarrow}{O^{2-}} \quad \overset{\leftarrow}{2e^-}$$

$$\begin{matrix} Fe & 2e^-(\leftarrow) \\ +O^{2-}(\leftarrow) & +Cu_2O \\ = FeO & = 2Cu \\ +2e^-(\leftarrow) & +O^{2-}(\leftarrow) \end{matrix}$$

$$Fe + Cu_2O = FeO + 2Cu \ldots \Delta G^0 \qquad (8.1.20)$$

The cell reaction is the displacement reaction given in Eq. (8.1.20). The standard Gibbs reaction energy ΔG^0 can be obtained from the emf E of cell (8.1.V) and is related to the standard Gibbs energies of formation of FeO, $\Delta_f G^0_{FeO}$ and to the corresponding quantity of Cu_2O, i.e. $\Delta_f G^0_{Cu_2O}$, according to the equation

$$\Delta G^0 = -2EF = \Delta_f G^0_{FeO} - \Delta_f G^0_{Cu_2O} \qquad (8.1.21)$$

Rearrangement yields:

$$\Delta_f G^0_{FeO} = -2EF + \Delta_f G^0_{Cu_2O}, \qquad (8.1.22)$$

i.e. the standard Gibbs energy of formation $\Delta_f G^0_{FeO}$ may be calculated, if $\Delta_f G^0_{Cu_2O}$ is known and the emf E of cell (8.1.V) is measured.

The cell (8.1.V) can be constructed in a simpler way than (8.1.IV); as shown in Fig. 8.1.4, it is only necessary to press together the corresponding pellets and to connect these by means of platinum leads and to heat the cell in a nitrogen atmosphere. An analogous example is given by the following cell (8.1.VI):

$$Pt \mid Fe, FeO \mid ZrO_2(+CaO) \mid Ni, NiO \mid Pt \qquad (8.1.VI)$$

$$\overset{\leftarrow}{2e^-} \qquad \overset{\leftarrow}{O^{2-}} \qquad \overset{\leftarrow}{2e^-}$$

$$\begin{matrix} 2Fe & 2e^-(\leftarrow) \\ +O^{2-}(\leftarrow) & +NiO \\ = FeO & = Ni \\ +2e^-(\leftarrow) & +O^{2-}(\leftarrow) \end{matrix}$$

$$Fe + NiO = FeO + Ni \ldots \Delta G^0 \qquad (8.1.23)$$

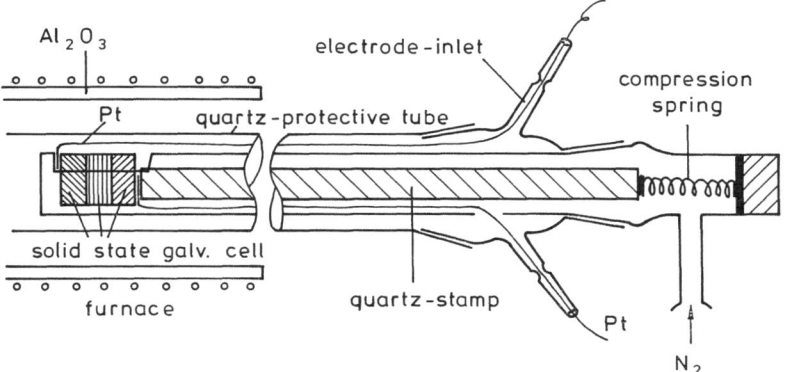

Fig. 8.1.4. Experimental set-up of the solid-state galvanic cell (8.1.V)

The cell reaction is the displacement reaction given by Eq. (8.1.23). The standard Gibbs reaction energy ΔG^0 of the cell reaction is related to the emf E of cell (8.1.VI) and also to the $\Delta_f G^0$ values of the formation of NiO and FeO respectively:

$$\Delta G^0 = -2\,EF = \Delta_f G^0_{FeO} - \Delta_f G^0_{NiO}. \tag{8.1.24}$$

It follows from Eq. (8.1.24) that

$$\Delta_f G^0_{NiO} = \Delta_f G^0_{FeO} + 2\,EF, \tag{8.1.25}$$

i.e. the $\Delta_f G^0_{NiO}$ value of NiO may be obtained if the corresponding value of FeO is known and the emf of cell (8.1.VI) is measured. Kiukkola and Wagner [8.4] have carried out investigations on cell (8.1.V) and (8.1.VI); their results are presented in Table 8.1.1.

The Gibbs energies of formation of more complex compounds may also be obtained by using suitable galvanic cells, e.g. the $\Delta_f G^0$ values for the formation of ternary sulfides or ternary oxides from those of the simple sulfides or oxides.

Table 8.1.1. Emf of the galvanic cells (8.1.V) and (8.1.VI) together with $\Delta_f G^0_{NiO}$ and $\Delta_f G^0_{Cu_2O}$ at different temperatures [8.4]

T/°C	emf/mV cell (8.1.VI)	$\Delta_f G^0_{NiO}$/kJ	T/°C	emf/mV cell (8.1.V)	$\Delta_f G^0_{Cu_2O}$/kJ
750	261 ± 2	−147.35	800	532	−91.92
800	266 ± 1	−143.29	900	539	−83.59
850	271 ± 1	−139.35	1000	543	−76.56
900	276 ± 1	−134.58	1050	545	−72.92
950	281 ± 1	−130.56			
1000	286 ± 2	−126.21			
1050	291 ± 2	−121.85			
1100	296 ± 2	−117.67			
1140	300 ± 1	−113.94			

As an example, we shall discuss the determination of the Gibbs energy of formation of the nickel aluminium spinel $NiAl_2O_4$ from the simple oxides NiO and Al_2O_3 at a temperature of $1000\,°C$. The following cell can be used:

$$Pt \ | \ \underset{Al_2O_3}{Ni, \ NiAl_2O_4} \ | \ \underset{(+Y_2O_3)}{ZrO_2} \ | \ \underset{NiO}{Ni,} \ | \ Pt \qquad (8.1.\text{VII})$$

$$2\,F: \qquad\qquad \overset{\longleftarrow}{2\,e^-} \qquad \overset{\longleftarrow}{O^{2-}} \qquad \overset{\longleftarrow}{2\,e^-}$$

$$
\begin{array}{ll}
Ni & NiO \\
+O^{2-}(\leftarrow) & +2\,e^-(\leftarrow) \\
+Al_2O_3 & = Ni \\
= NiAl_2O_4 & +O^{2-}(\leftarrow) \\
+2\,e^-(\leftarrow) &
\end{array}
$$

$$NiO + Al_2O_3 = NiAl_2O_4 \ldots \Delta G^0 \qquad (8.1.26)$$

The left-hand electrode is a mixture of Ni, $NiAl_2O_4$ and Al_2O_3 whereas the right-hand electrode consists of a mixture of Ni and NiO. As may be seen by considering the electrode reactions the cell reaction proceeding during a transport of two Faradays through the cell is the formation of one mole of the spinel $NiAl_2O_4$ from NiO and Al_2O_3 according to Eq. (8.1.26). The corresponding ΔG^0 value can thus be obtained directly from the emf of the cell (8.1.VII):

$$\Delta G^0 = -2\,EF. \qquad (8.1.27)$$

Measurements of this type have been carried out by Schmalzried [8.5]. He obtained a ΔG^0 value of 21 kJ per mole at $1000\,°C$ from the emf of the above cell. A number of thermodynamic measurements have also been carried out using the solid electrolyte CaF_2, in particular by Egan and his co-workers [8.6, 8.7]. With the aid of the following galvanic cells

$$Mg, MgF_2 \ | \ CaF_2 \ | \ ThF_4, Th, \qquad (8.1.\text{VIII})$$

$$Th, ThF_4 \ | \ CaF_2 \ | \ AlF_3, Al, \qquad (8.1.\text{IX})$$

$$U, UF_3 \ | \ CaF_2 \ | \ AlF_3, Al, \qquad (8.1.\text{X})$$

$$Th, ThF_4 \ | \ CaF_2 \ | \ NiF_2, Ni, \qquad (8.1.\text{XI})$$

$$Al, AlF_3 \ | \ CaF_2 \ | \ PbF_2, Pb, \qquad (8.1.\text{XII})$$

$$Al, AlF_3 \ | \ CaF_2 \ | \ CoF_2, Co \qquad (8.1.\text{XIII})$$

they obtained the standard Gibbs energy of formation of ThF_4, AlF_3, NiF_2, PbF_2, CoF_2 and UF_3. At $600\,°C$ the values are as follows: $\Delta_f G^0_{ThF_4} = -1.935$ MJ, $\Delta_f G^0_{AlF_3} = -1.280$ MJ, $\Delta_f G^0_{NiF_2} = -530.7$ kJ, $\Delta_f G^0_{PbF_2} = -547.9$ kJ, $\Delta_f G^0_{CoF_2} = -542.4$ kJ, $\Delta_f G^0_{UF_3} = -1.23$ MJ.

The cells discussed in this section are merely a selection of typical examples. Many more cells of this type are described in the literature. Some results obtained by different authors are compiled in Table 8.1.2.

Table 8.1.2. ΔG^0 values of different reactions obtained by various authors

Reaction $p^0_{O_2} = 1$ atm	T/°C	$-\Delta G^0$/kJ mol^{-1}	Ref.
$Co(s) + \frac{1}{2}O_2(g) = CoO(s)$	727−1227	163.2−124.3	[8.8]
$2Cu(s) + \frac{1}{2}O_2(g) = Cu_2O(s)$	800−1050	92.0−72.0	[8.4]
$2Cr(s) + \frac{3}{2}O_2(g) = Cr_2O_3(s)$	727−1227	851.6−736.0	[8.8]
$2(1-x)Fe(s) + O_2(g) = 2Fe_{1-x}O(s)$	730−1230	395.7−329.7	[8.8]
$Nb(s) + \frac{1}{2}O_2(g) = NbO(s)$	727−1127	327.0−290.0	[8.9]
$Ni(s) + \frac{1}{2}O_2(g) = NiO(s)$	750−1140	147.4−114.0	[8.4]
$W(s) + \frac{1}{2}O_2(g) = WO_2(s)$	550−950	447.6−366.9	[8.10]
$2Cu(l) + \frac{1}{2}O_2(g) = Cu_2O(s)$	1180−1300	58.1−52.9	[8.11]
$2Ga(l) + \frac{3}{2}O_2(g) = Ga_2O_3(s)$	800−1000	738.6−672.8	[8.12]
$Mn(l) + \frac{1}{2}O_2(g) = MnO(s)$	1300−1550	271.1−246.9	[8.13]
$2Rh(s) + \frac{3}{2}O_2(g) = Rh_2O_3(s)$	627−917	153.2−59.7	[8.15]
$Ru(s) + O_2(g) = RuO_2(s)$	507−767	174.6−122.4	[8.16]
$Zn(l) + \frac{1}{2}O_2(g) = ZnO(s)$	520−895	269.4−229.0	[8.14]
$Pb(l) + \frac{1}{2}O_2(g) = PbO(l)$	887−1098	124.1−108.2	[8.17]
$Sn(l) + \frac{1}{2}O_2(g) = SnO(l)$	500−700	210.1−188.5	[8.18]
$Sn(l) + \frac{1}{2}O_2(g) = SnO(l)$	1027−1152	152.7−141.5	[8.19]
$CoO + Al_2O_3 = CoAl_2O_4$	725−1225	31.5−25.9	[8.5, 8.8]
$CoO + Cr_2O_3 + CoCr_2O_4$	725−1225	56.9−44.8	[8.5, 8.8]
$CoO + Fe_2O_3 = CoFe_2O_4$	900−1425	38.3−45.4	[8.8, 8.20]
$CuO + Al_2O_3 = CuAl_2O_4$	1100−1200	10.1−12.2	[8.21]
$CuO + Cr_2O_3 = CuCr_2O_4$	725−1225	43.6−39.8	[8.8]
$FeO + (\alpha)Al_2O_3 = FeAl_2O_4$	962−1050	24.1−22.6	[8.22]
$FeO + Cr_2O_3 = FeCr_2O_4$	725−1225	41.1−32.9	[8.8]
$FeO + Fe_2O_3 = Fe_3O_4$	825−1425	29.6−36.5	[8.8]
$MgO + Cr_2O_3 = MgCr_2O_4$	725−1225	35.9−32.3	[8.8]
$MgO + Fe_2O_3 = MgFe_2O_4$	825−1425	22.6−21.8	[8.8]
$MnO + Fe_2O_3 = MnFe_2O_4$	791−1100	877.4−779.6	[8.23, 8.25]
$NiO + Cr_2O_3 = NiCr_2O_4$	725−1225	32.5−21.7	[8.8]
$NiO + Fe_2O_3 = NiFe_2O_4$	900−1200	24.7−26.0	[8.8, 8.20]
$ZnO + \alpha\text{-}Al_2O_3 = ZnAl_2O_4$	700−900	38.6−37.3	[8.26]
$ZnO + Cr_2O_3 = ZnCr_2O_4$	700−900	54.5−52.7	[8.26]

8.2 Determination of Chemical Potentials, Thermodynamic Activities or Partial Pressures

The galvanic cells used for carrying out electrochemical determinations of chemical potentials can be divided into two groups. In analogy to the electrodes used in liquid-state electrochemistry, i.e. electrodes of the first and second kind respectively, we shall distinguish between two types of solid-state galvanic cells. If, for example, a galvanic cell containing an oxygen ion conductor is used in order to measure oxygen activities or one containing a silver ion conductor in order to measure silver activities, we shall refer to galvanic cells of the first kind, if however, the galvanic cell containing a silver ion conductor is used in order to measure sulfur activities — and not silver activities — one refers to a galvanic cell of the second kind.

This point will become clearer in the discussion of the following examples.

8.2.1 Galvanic Cells of the First Kind

Apart from the Helmholtz approach, which emphasizes the cell reaction and the corresponding Gibbs reaction energy, we shall also use the approach based on the various electrochemical potentials of the species mediating the equilibrium. To a certain extent, this is in analogy to the Nernst treatment of electrodes in liquid electrolytes.

As a first example we shall discuss, with the aid of the Helmholtz approach, the determination of the chemical potential μ_{Ag} or the thermodynamic activity a_{Ag} of silver in silver sulfide within the temperature range 200—400 °C. The following galvanic cell is suitable:

$$Pt1 \mid Ag \mid AgI \mid Ag_2S \mid Pt2 \tag{8.2.I}$$

$$\longleftarrow \quad \longrightarrow \quad \longleftarrow$$

1 F: e^- Ag^+ e^-

$$
\begin{aligned}
Ag(s) &\qquad Ag^+(\rightarrow) \\
= Ag^+(\rightarrow) &\qquad + e^-(\leftarrow) \\
+ e^-(\leftarrow) &\qquad = Ag(Ag_2S)
\end{aligned}
$$

$$Ag(s) = Ag(Ag_2S) \ldots \Delta G \tag{8.2.1}$$

Equation (8.2.1) states the cell reaction according to the Helmholtz point of view. The flow of one Faraday through the cell corresponds to the conversion of one mole of silver from the metallic state into silver contained in the sulfide phase. The Gibbs reaction energy of this process is equal to the difference of the chemical potentials of silver in silver sulfide and in the standard state:

$$\Delta G = \mu_{Ag}(Ag_2S) - \mu_{Ag}^0 = -EF. \tag{8.2.2}$$

The emf of cell (8.2.I) is a measure of this quantity. The cell (8.2.I) differs from (8.1.II) in the respect that now, in general, Ag_2S is not in equilibrium with liquid sulfur. Thus, the activity of sulfur will generally be smaller than that of liquid sulfur. Under certain circumstances, the emf of the cell may have the value E=0: Ag_2S is then in equilibrium with silver, i.e. the activity of silver is equal to unity and the chemical potential of silver in silver sulfide is equal to that of metallic silver. Normally, however, the emf E will have a non-zero value; the highest value observed will be equal to the emf of the cell for Ag_2S in equilibrium with liquid sulfur. The cell reaction of cell (8.2.I) is equivalent to a solution reaction and the cell may be considered as a concentration cell in analogy to those found in liquid-state electrochemistry, e.g. to a cell containing two amalgam electrodes with differing activities of a metal, the ions of which are present in the electrolyte and thus determine the emf. In Eq. (8.2.2) the chemical potential of silver may be expressed in terms of the activity of silver which is defined according to the following equation:

$$\mu_{Ag} = \mu_{Ag}^0 + RT \ln a_{Ag}. \tag{8.2.3}$$

Rearrangement of Eq. (8.2.3) and using Eq. (8.2.2) yields the thermodynamic activity a_{Ag} of silver in silver sulfide as follows:

$$a_{Ag} = \exp\left(-\frac{EF}{RT}\right). \tag{8.2.4}$$

We shall now proceed using the approach based on electrical and electrochemical potentials. The quantity measurable from the galvanic cell is the emf E which is determined between the two phases at the ends of the cell, i.e. the two platinum leads 1 and 2 of the galvanic cell (8.2.I). This emf may be written as the electrical potential difference between these two phases, platinum 2 and platinum 1:

$$\varphi(Pt2) - \varphi(Pt1) = E. \tag{8.2.5}$$

Because of the same metal platinum the chemical potentials μ_e of the electrons in the two platinum wires are equal

$$\mu_e(Pt1) = \mu_e(Pt2). \tag{8.2.6}$$

Multiplying Eq. (8.2.5) by F and adding $\mu_e(Pt1) - \mu_e(Pt2)$, which is equal to zero, on the left hand side and rearranging yields

$$[\mu_e(Pt1) - F\varphi(Pt1)] - [\mu_e(Pt2) - F\varphi(Pt2)] = EF. \tag{8.2.7}$$

Because of the relation holding for the electrochemical potential η_e of the electrons, i.e.

$$\eta_e = \mu_e - F\varphi \tag{8.2.8}$$

the terms in square brackets of Eq. (8.2.7) can be replaced by the electrochemical potential of the electrons in platinum 1 and platinum 2 respectively

$$\eta_e(Pt1) - \eta_e(Pt2) = EF. \tag{8.2.9}$$

Equation (8.2.9), which emphasizes the fact that we measure the difference of the electrochemical potentials of the electrons in the outer phases of the cell, is the appropriate starting point for the following discussion, since it is relatively simple to make statements concerning electrochemical potentials.

If two phases, which are good electronic conductors, are in contact, but no electrical current flows through them, then the electrochemical potential of the electrons — in the language of semiconductor physics the Fermi potential — is the same in both phases. With respect to cell (8.2.I) this means that the electrochemical potential of the electrons in platinum 1 is equal to that in the silver electrode

$$\eta_e(Pt1) = \eta_e(Ag), \tag{8.2.10}$$

and the electrochemical potential of the electrons in platinum 2 is equal to that in silver sulfide:

$$\eta_e(Pt2) = \eta_e(Ag_2S). \tag{8.2.11}$$

Using Eqs. (8.2.10) and (8.2.11), we can rewrite Eq. (8.2.9) as follows

$$\eta_e(Ag) - \eta_e(Ag_2S) = EF, \tag{8.2.12}$$

i.e. the measured emf E is a measure of the difference in the electrochemical potentials of the electrons in silver and in silver sulfide of cell (8.2.I).

The electrolyte AgI itself is an almost pure ionic conductor; this fact provides some information on the electrochemical potential of the ions in the cell. If no current flows through the cell, no gradient in the electrochemical potential of the silver ions in AgI exists because of the good silver ion conductivity of AgI, since (as we have seen in Sect. 6.1) a particle flux is due to the gradient of the electrochemical potential of the particles transported and their ionic conductivity. A zero value for the current then means that either the partial conductivity or the gradient of the electrochemical potential becomes zero. In the case considered the partial conductivity is relatively large and we may thus conclude that the gradient of the electrochemical potential of the silver ions is virtually zero:

$$\frac{d\eta_{Ag^+}}{dx}\bigg|_{AgI} = 0. \tag{8.2.13}$$

Thus, the electrochemical potential of the silver ions is constant in AgI and, for equilibrium with the electrodes also equal to the value within the silver and the silver sulfide

$$\eta_{Ag^+}(Ag) = \eta_{Ag^+}(Ag_2S). \tag{8.2.14}$$

Addition of Eqs. (8.2.14) and (8.2.12) and rearrangement of the result obtained will lead to the following expression:

$$[\eta_e(Ag) + \eta_{Ag^+}(Ag)] - [\eta_e(Ag_2S) + \eta_{Ag^+}(Ag_2S)] = EF. \tag{8.2.15}$$

The sum of the electrochemical potentials of the electrons and silver ions is equal to the chemical potential of neutral silver; it thus follows from Eq. (8.2.15) that

$$\mu_{Ag}(Ag_2S) - \mu_{Ag}^0 = -EF, \tag{8.2.16}$$

i.e. we have the same result as was obtained using the Helmholtz approach.

Figure 8.2.1 schematically shows how the electrochemical potentials of the electrons and the ions and the chemical potential of silver vary across the galvanic cell (8.2.I); it also shows the variation of the electrical potential across the cell. It is, however, necessary to gain further information, particularly about the disorder in the electrolyte and in silver sulfide before statements concerning the electrical potential can be made. Firstly, it can be expected that Galvani potential jumps will occur at the various interfaces. The electrical potential in the platinum leads, the silver electrode and the silver sulfide respectively will be constant because of the good electronic conductivity in these phases; for this reason, no space charges are to be expected. Taking into account the particular type of disorder in silver iodide we may assume that the chemical potential of the silver ions is constant in this phase. Since the electrochemical potential of the silver ions also undergoes no change therein, it follows that the electrical potential must also be constant in the silver iodide. We must now ask ourselves which of the electrical potential differences occurring at the interfaces changes, if the chemical potential of silver in silver sulfide varies. If equilibrium is attained at a phase boundary, the electrical potential difference — identical with the Galvani potential difference — is

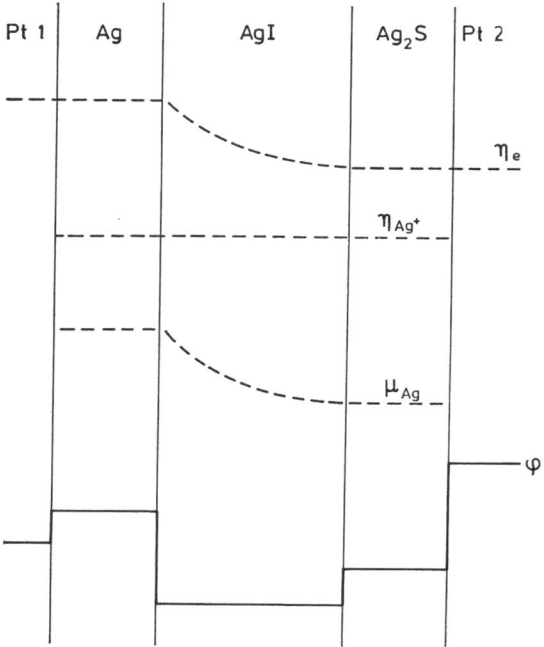

Fig. 8.2.1. Schematic representation of the variations of the electrochemical potentials of electrons and ions, of the chemical potential of silver and of the electrical potential across the galvanic cell (8.2.I)

determined by the difference of the chemical potential of electrons and/or ions along this phase boundary. The chemical potentials of the electrons at the phase boundary Pt1/Ag are invariable even if the chemical potential of silver in Ag$_2$S varies; thus, the Galvani potential difference here remains constant.

At the phase boundary Ag/AgI we may regard the electrochemical equilibrium of the silver ions. Because of the disorder of the silver ions their activity remains unchanged in AgI, so that the Galvani potential difference also remains invariable here. It follows from the special type of disorder in Ag$_2$S that the activity of the silver ions in Ag$_2$S also remains constant, so that the Galvani potential difference at the phase boundary AgI/Ag$_2$S, determined by the silver ions, is also invariable. In the silver sulfide, however, the activity of the electrons varies with the activity of silver (see Chap. 5). Thus, the electrical potential difference (the Galvani potential difference) at the phase boundary Ag$_2$S/Pt2 changes. We thus have here the interesting result that changes in the activity of silver in silver sulfide, are only accompanied by variations in the electrical or Galvani potential difference at the phase boundary Ag$_2$S/Pt2, while all other Galvani potential differences remain constant; apart from this, no electrical potential gradients occur in any of the phases of the cell.

This example was discussed in order to show that it is necessary to know which types of disorder occur in the various phases involved before one can draw conclusions concerning changes in the electrical potential in a galvanic cell. The potential variations shown in Fig. 8.2.1 for cell (8.2.I) are thus only valid for this particular system. The situation may be different for other solid electrolytes and other electrodes; this depends on the type of disorder occurring in the solid phases. It must further be emphasized that no single Galvani potential difference can be measured. Thus, the Galvani potential jumps have no absolute significance.

We shall now discuss further examples of electrochemical measurements of chemical potentials or thermodynamic activities.

For the electrochemical determination of oxygen partial pressures in gases at temperatures of about 800 °C it is possible to use doped zirconium dioxide as an auxiliary solid electrolyte which exhibits virtually pure ionic conduction for oxygen ions. The galvanic cell (8.2.II) is used; we shall once again first apply the Helmholtz approach which leads in a straightforward manner to the required result.

$$\text{Pt, } O_2(g)(p'_{O_2}) \mid ZrO_2(+MgO) \mid O_2(g)(p''_{O_2}), \text{ Pt} \qquad (8.2.II)$$

$$4F: \qquad 4e^- \qquad 2O^{2-} \qquad 4e^-$$

$$
\begin{aligned}
2O^{2-}(\leftarrow) & \quad O_2(g)(p''_{O_2}) \\
= O_2(g)(p'_{O_2}) & \quad +4e^-(\leftarrow) \\
+4e^-(\leftarrow) & \quad = 2O^{2-}(\leftarrow)
\end{aligned}
$$

$$O_2(g)(p''_{O_2}) \qquad = O_2(g)(p'_{O_2}) \ldots \Delta G \qquad (8.2.17)$$

According to Eq. (8.2.17) the cell reaction is equivalent to the transport of gaseous oxygen from one electrode to the other. The experimental arrangement must of course be such that the electrodes are separated in a gas-tight manner. For example, a zirconium dioxide tube coated with two layers of porous platinum on the in- and outside may be used; the inside and the outside of the tube form the two electrodes as shown in Fig. 8.2.2. The Gibbs reaction energy for the transport of oxygen from one electrode to the other may be expressed by the chemical potentials of oxygen or by oxygen partial pressures; it is equal to $-4EF$:

$$\Delta G = \mu'_{O_2} - \mu''_{O_2} = RT \ln \frac{p'_{O_2}}{p''_{O_2}} = -4EF. \qquad (8.2.18)$$

Solving this equation for the oxygen partial pressure p'_{O_2}, we obtain the expression

$$p'_{O_2} = p''_{O_2} \exp\left(-\frac{4EF}{RT}\right). \qquad (8.2.19)$$

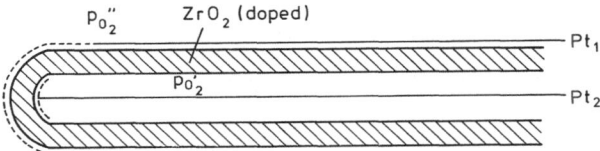

Fig. 8.2.2. Experimental arrangement (schematic) for the electrochemical determination of oxygen partial pressures using doped zirconium dioxide as solid electrolyte

Using a cell of this type, it is principally possible to measure oxygen partial pressures down to 10^{-20} atm at $800\,°C$; at still lower oxygen partial pressures zirconium dioxide becomes an electronic conductor.

Discussing cell (8.2.II) in terms of electrochemical potentials, we may again start with the statement that the difference of the electrochemical potentials of the electrons η_e'' and η_e' at the two electrodes is equal to $-\,EF$:

$$\eta_e'' - \eta_e' = -EF. \tag{8.2.20}$$

Because of the good oxygen ion conductivity in zirconium dioxide the difference of the electrochemical potentials of the oxygen ions in both electrodes is equal to zero:

$$\eta_{O^{2-}}'' - \eta_{O^{2-}}' = 0. \tag{8.2.21}$$

The combination of Eqs. (8.2.20) and (8.2.21) leads to the expression

$$[2\eta_{O^{2-}}' - 4\eta_e'] - [2\eta_{O^{2-}}'' - 4\eta_e''] = -4\,EF \tag{8.2.22}$$

or, since the terms in the brackets of Eq. (8.2.22) are equal to the chemical potentials of oxygen

$$\mu_{O_2}' - \mu_{O_2}'' = -4\,EF, \tag{8.2.23}$$

we obtain the same result as that resulting from the previous approach.

We may use the following galvanic cell to determine the activity of oxygen in liquid copper at $1100\,°C$ [8.27]:

$$\begin{array}{lcccc}
\text{Pt} \mid \text{Ni,} & \mid \text{ZrO}_2 & \mid & \text{Cu(l)} & \mid \text{Pt} \\
\text{NiO} & (+\text{CaO}) & & +\text{O(diss.)} &
\end{array} \tag{8.2.III}$$

$$\begin{array}{lll}
\text{2F:} & \overset{\longleftarrow}{2\,e^-} \quad \overset{\longleftarrow}{\text{O}^{2-}} & \overset{\longleftarrow}{2\,e^-} \\[2mm]
& \text{Ni} & \text{O(diss.)} \\
& +\text{O}^{2-}(\leftarrow) & +2\,e^-(\leftarrow) \\
& = \text{NiO} & = \text{O}^{2-}(\leftarrow) \\
& +2\,e^-(\leftarrow) &
\end{array}$$

Once again doped zirconium dioxide is used as an auxiliary solid electrolyte which exhibits virtually pure ionic conduction. A mixture of nickel and nickel oxide serves as a reference electrode. The cell reaction is given by the following equation

$$\text{Ni} + \text{O[dissolved in Cu(l)]} = \text{NiO} \ldots \Delta G. \tag{8.2.24}$$

ΔG may be written in terms of chemical potentials

$$\begin{aligned}
\Delta G &= \mu_{NiO} - \mu_{Ni} - \mu_O[\text{Cu(l)}] \\
&= \mu_O[\text{Ni, NiO}] - \mu_O[\text{Cu(l)}] \tag{8.2.24a}
\end{aligned}$$

The chemical potential of oxygen corresponding to equilibrium between nickel and nickel oxide minus the chemical potential of oxygen in liquid copper is equal to the Gibbs reaction energy ΔG of this reaction.

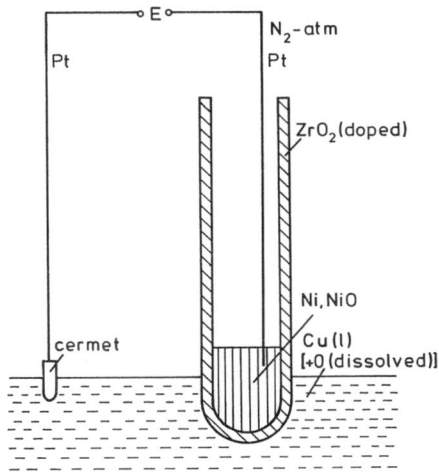

Fig. 8.2.3. Experimental arrangement for the electrochemical measurement of the oxygen activity in liquid copper

Thus, the following expression applies

$$\Delta G = \mu_0[\text{Ni, NiO}] - \mu_0[\text{Cu(l)}] = -2\,\text{EF}. \tag{8.2.25}$$

This relation yields the chemical potential of oxygen in liquid copper. Figure 8.2.3 shows schematically a set-up which is suitable for such measurements; investigations of this type have been carried out by various authors [8.27—8.37] and for different liquid metals as exemplified in Table 8.2.1.

8.2.2 Galvanic Cells of the Second Kind

We shall discuss two examples of the use of galvanic cells of the second kind, a) the measurement of the chemical potential of sulfur in silver sulfide and b) the measurement of the activity of nickel in copper-nickel alloys.

Table 8.2.1. Oxygen-activity measurements in different liquid metals carried out by various authors

Year	Investigator	Metal (liquid)	T/°C	Ref.
1963	Blumenthal, Whitmore	Sn	530—750	[8.28]
1964	Alcock, Belford	Pb	530—750	[8.29]
1964	Goto, Matsushita	Fe, steel	1000—1550	[8.30]
1965	Belford, Alcock	Sn	510—700	[8.31]
1965	Plushkell, Engell	Cu	1150	[8.32]
1965	Fischer, Ackermann	Fe	1110—1430	[8.33]
1965	Kolodney, Minushkin, Sternmetz	Na	327—527	[8.34]
1966	Rickert, Wagner	Cu	1145	[8.35]
1966	Fischer, Ackermann	Ag	1000—1400	[8.36]
1966	Fischer, Ackermann,	Co, Ni, Cu	1460—1700	[8.37]

a) The emf E of the following galvanic cell (8.2.I) discussed above

$$\text{Pt1} \mid \text{Ag} \mid \text{AgI} \mid \text{Ag}_2\text{S} \mid \text{Pt2} \tag{8.2.I}$$

is not only (as stated in Eq. (8.2.2)) a measure of the chemical potential of silver in silver sulfide, but also of the chemical potential of sulfur in Ag_2S, since these two chemical potentials are connected by the Gibbs-Duhem equation. If we are interested in the measurement of the chemical potential of sulfur, we must regard the galvanic cell as being of the second kind, because no sulfur ions are present in the solid electrolyte AgI. Since the deviations from ideal stoichiometry are small in Ag_2S, the Gibbs-Duhem relation applied to this compound appears in the following form:

$$2(\mu_{Ag} - \mu_{Ag}^0) = \mu_S' - \mu_S \tag{8.2.26}$$

where μ_S' is the chemical potential of sulfur in Ag_2S at equilibrium with silver. If we denote the chemical potential of silver in Ag_2S at equilibrium with liquid sulfur by μ_{Ag}^* and taking liquid sulfur as the standard state for sulfur: $\mu_S = \mu_S^0$, it follows from Eq. (8.2.26) that

$$2(\mu_{Ag}^* - \mu_{Ag}^0) = \mu_S' - \mu_S^0. \tag{8.2.27}$$

If E* is the emf E of cell (8.2.I) corresponding to equilibrium with liquid sulfur we can write

$$(\mu_{Ag}^* - \mu_{Ag}^0) = -E^* F; \tag{8.2.28}$$

It then follows from Eqs. (8.2.2), (8.2.26−28) that

$$\mu_S - \mu_S^0 = 2(E - E^*) F, \tag{8.2.29}$$

i.e. the emf E of cell (8.2.I) is also a measure of the chemical potential of sulfur in Ag_2S and, if Ag_2S is in equilibrium with a further phase, e.g. the gas phase, also a measure of the chemical potential of sulfur in the gas phase and thus of the partial pressures of the various types of sulfur molecules or of the activity of sulfur. The activity a_S of sulfur may be obtained using the expression

$$\mu_S = \mu_S^0 + RT \ln a_S \tag{8.2.30}$$

and Eq. (8.2.29):

$$a_S = \exp \frac{2(E - E^*) F}{RT}. \tag{8.2.31}$$

b) The activity of nickel in copper-nickel alloys has been determined by Rapp and Maak [8.38] with the aid of the galvanic cell

$$\text{Pt} \mid \text{Ni, NiO} \mid \text{ZrO}_2(+\text{CaO}) \mid \text{Cu—Ni—alloy, NiO} \mid \text{NiO} \mid \text{Pt} \tag{8.2.IV}$$

which uses doped zirconium dioxide as a virtually pure ionic conductor for oxygen ions. The emf E of the galvanic cell is a measure of the difference of the chemical potential of oxygen at both sides of the electrolyte, i.e. at the two electrodes

$$\mu_O' - \mu_O'' = 2EF, \tag{8.2.32}$$

μ'_O and μ''_O denote the chemical potentials of oxygen on the right- and left-hand side of the solid electrolyte respectively. Since pure nickel oxide is present on both sides of the cell, its chemical potential on both sides must be equal to that in the standard state. Since the chemical potential of NiO may be divided into that of nickel and that of oxygen, the following equations are valid for the two sides of the galvanic cell:

$$\mu'_{Ni} + \mu'_O = \mu^0_{NiO} \tag{8.2.33}$$

$$\mu''_{Ni} + \mu''_O = \mu^0_{NiO} \tag{8.2.34}$$

Combination of Eqs. (8.2.32), (8.2.33) and (8.2.34) gives the following expression:

$$\mu'_{Ni} - \mu''_{Ni} = -2\,EF, \tag{8.2.35}$$

i.e. the emf E is also a measure of the difference of the chemical potentials of nickel on both sides of the galvanic cell (8.2.IV). Since pure nickel is present on the left-hand side, it follows that $\mu''_{Ni} = \mu^0_{Ni}$ with the activity of nickel being equal to unity on this side. Using the relationship between the chemical potential of Ni and its activity according to $\mu'_{Ni} = \mu^0_{Ni} + RT \ln a'_{Ni}$, the following expression for the activity of nickel in the alloy results

$$\ln a'_{Ni}(\text{alloy}) = -2\,EF/RT. \tag{8.2.36}$$

It was possible using this cell to measure the activity of nickel as a function of the composition of Ni—Cu alloys and thus to obtain further thermodynamic data.

8.3 Determination of ΔS and ΔH from the Temperature-Dependence of the Electromotive Force of Galvanic Cells

Just as for galvanic cells with liquid electrolytes, it is possible to determine reaction enthalpies and reaction entropies using cells with solid electrolytes; in order to achieve this it is necessary to measure the temperature-dependence of the emf E. Applying the Gibbs-Helmholtz equation, which is given by

$$\left(\frac{\partial \Delta G}{\partial T}\right)_{p,N_i} = -\Delta S \tag{8.3.1}$$

to Eq. (8.1.1), we obtain the following expression for the entropy of the cell reaction

$$\Delta S = nF \left(\frac{\partial E}{\partial T}\right)_{p,N_i}. \tag{8.3.2}$$

The enthalpy of the cell reaction is related to the corresponding Gibbs energy and the entropy according to the equation

$$\Delta H = \Delta G + T\,\Delta S. \tag{8.3.3}$$

Inserting Eq. (8.1.1) and (8.3.2) into Eq. (8.3.3) in order to eliminate ΔG and ΔS we can write for ΔH

$$\Delta H = \Delta G + T \, \Delta S = -nF \left[E - T \left(\frac{\partial E}{\partial T} \right)_{p, N_i} \right]. \qquad (8.3.4)$$

8.4 References

[8.1]   de Bethune, A. J.: J. Electrochem. Soc. *102*, 288 C (1955)
[8.2]   Reinhold, H.: Z. anorg. allg. Chem. *171*, 181 (1928)
[8.3]   Rickert, H.: Festkörperprobleme VI. Schottky, W. (ed.), p. 85. Braunschweig: Vieweg 1967
[8.4]   Kiukkola, K., Wagner, C.: J. Electrochem. Soc. *104*, 308, 379 (1957)
[8.5]   Schmalzried, H.: Z. Phys. Chem. N.F. *25*, 178 (1960)
[8.6]   Egan, J. J.: J. Phys. Chem. *68*, 978 (1964)
[8.7]   Heus, R. J., Egan, J. J.: Z. Phys. Chem. N.F. *49*, 38 (1966)
[8.8]   Tretyakov, Yu. D., Schmalzried, H.: Ber. Bunsenges. Phys. Chem. *65*, 396 (1965)
[8.9]   Barbi, G. B.: Z. Naturforsch. *23*, 800 (1968)
[8.10]   Barbi, G. B.: J. Phys. Chem. *68*, 1025 (1964)
[8.11]   Osterwald, J.: Z. Phys. Chem. N.F. *49*, 138 (1966)
[8.12]   Katayama, I., Keinori, N., Kozuha, Z.: Nippon Kinzohn Gakhaischi *39*, 188 (1975)
[8.13]   Schwerdtfeger, K.: Trans. Met. Soc. AIME *239*, 1276 (1967)
[8.14]   Wilder, T. C.: Trans. Met. Soc. AIME *245*, 1370 (1969)
[8.15]   Kleykamp, H.: Z. Phys. Chem. N.F. *67*, 277 (1969)
[8.16]   Kleykamp, H.: Z. Phys. Chem. N.F. *66*, 131 (1969)
[8.17]   Charette, G. G., Flengas, S. N.: J. Electrochem. Soc. *115*, 796 (1968)
[8.18]   Belford, T. N., Alcock, C. B.: Trans. Faraday Soc. *61*, 443 (1965)
[8.19]   Kozuha, Z., Siahaan, O. P., Moriyama, J.: Trans. Japan Inst. Met. *9*, 200 (1968)
[8.20]   Schmalzried, H., Tretyakov, Yu. D.: Ber. Bunsenges. Phys. Chem. *70*, 180 (1966)
[8.21]   Jacob, K. T., Alcock, C. B.: J. Am. Ceram. *58*, 192 (1975)
[8.22]   Razukhina, T. N., Levilskii, V. A., Ozhigov, P.: Russ. J. Phys. Chem. *46*, 1687 (1972)
[8.23]   Tretyakov, Yu. D.: Izv. Akad. Nauk SSSR Neorg. Mater. *1*, 1928 (1965)
[8.24]   Roshchupkin, V. I., Lavrentev, V. I.: Izv. Akad. Nauk SSSR Neorg. Mater. *3*, 551 (1967)
[8.25]   Gordeev, I. V., Tretyakov, Yu. D., Khomyakov, K. G.: Russ. J. Inorg. Chem. *9*, 89 (1969)
[8.26]   Jacob, K. T.: Thermochim. Acta *15*, 79 (1976)
[8.27]   Rickert, H., Wagner, H.: Electrochim. Acta *11*, 83 (1966)
[8.28]   Blumenthal, R. N., Whitmore, D. H.: J. Electrochem. Soc. *110*, 92 (1963)
[8.29]   Belford, T. N., Alcock, C. B.: Trans. Faraday Soc. *60*, 822 (1964)
[8.30]   Goto, K., Matsushita, Y.: Tetsu To Hagane *50*, 1818 (1964)
[8.31]   Belford, T. N., Alcock, C. B.: Trans. Faraday Soc. *61*, 443 (1965)
[8.32]   Plushkell, W., Engell, H. J.: Z. Metallk. *56*, 450 (1965)
[8.33]   Fischer, W. A., Ackermann, W.: Arch. Eisenhüttenw. *36*, 643, 695 (1965)
[8.34]   Kolodney, M., Minushkin, B., Sternmetz, H.: Electrochem. Technol. *3*, 214 (1965)
[8.35]   Rickert, H., Wagner, H.: Electrochim. Acta *11*, 83 (1966)
[8.36]   Fischer, W. A., Ackermann, W.: Arch. Eisenhüttenw. *37*, 697 (1966)
[8.37]   Fischer, W. A., Ackermann, W.: Arch. Eisenhüttenw. *37*, 43 (1966)
[8.38]   Rapp, R. A., Maak, F.: Acta Met. *10*, 63 (1962)

9 Technical Applications of Solid Electrolytes — Solid-State Ionics

Recent years have seen the development of important technical applications involving the use of galvanic cells with solid electrolytes. Reviews on some or all of these applications are given by several authors [9.1]. In the following we will discuss a number of examples.

9.1 Sensors

Galvanic cells with solid electrolytes can be used for rapid and direct measurements of partial pressures in gases and of concentrations or activities in liquids and melts.

Weissbart and Ruka [9.2] first applied the galvanic cell (8.2.II) discussed in Sect. 8.2, to the determination of oxygen partial pressures in gases. According to Eq. (8.2.18) the emf of this cell is a measure of the ratio of the oxygen partial pressures on both sides of the electrolyte. The most important part of the arrangement used for measurements is a zirconium dioxide tube closed on one side as shown in Fig. 8.2.2. Cells of this type can be used over a wide range of oxygen partial pressures in gases. These sensors are employed between about 500 and 1000 °C; under these conditions, zirconium dioxide exhibits considerable electron conduction only at oxygen partial pressures lower than ca. 10^{-16} atm, i.e. even oxygen partial pressures of this magnitude can be detected by these sensors. Such small partial pressures occur in practice in systems involving mixtures of H_2 and H_2O or CO and CO_2, or equilibria between metals and metal oxides. The wide range of oxygen partial pressures covered by zirconium dioxide sensors makes them excellent analytical instruments for carrying out measurements in gases. It should also be mentioned that the signal is obtained almost instantaneously, i.e. one has the advantage of very short response times. Measurements are normally carried out with respect to a reference system, i.e. air or a metal/metal oxide is located on one side of the zirconium dioxide in order to determine the oxygen partial pressure at one electrode; the oxygen partial pressure at the other electrode is then determined from the emf according to Eq. (8.2.19). The gas can be passed through the sensor during measurements. Commercial devices for measuring the oxygen content of for example highly purified inert gases are available [9.3]. Further interesting applications lie in the field of medicine for studying the oxygen content of exhaled air, and in industry, for example in the analysis of exhaust gases from furnaces or combustion engines. It is often advisable to locate the zirconium dioxide sensor in the exhaust gas stream near the reaction area.

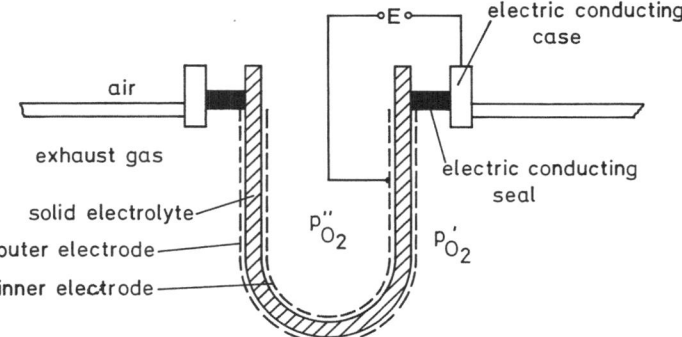

Fig. 9.1.1. Construction of a ZrO_2 probe for measuring oxygen partial pressures (schematic)

Figure 9.1.1 shows such a sensor that can be used for the control and regulation of the combustion process in petrol engines of motor vehicles [9.4]. The feasibility of very exact control of gasoline/air mixtures is of particular interest in view of the possibility of reducing air pollution.

A sensor similar to that described above can be used to measure the concentration or the thermodynamic activity of oxygen in liquid metals. The procedure to be carried out has already been discussed in Sect. 8.2. For example, the sensor may be dipped into the iron melt during steel production [9.5]; the immediate and direct display of the oxygen activity is extremely advantageous. A number of publications dealing with measurements in different liquid metals have appeared recently [9.6]; this method has thus become very important with respect to metallurgical applications. If the oxygen partial pressures or the oxygen activities to be measured are too low, so that electron conduction occurs in zirconium dioxide, doped thorium dioxide can be used, since it allows the determination of much lower oxygen activities. For example $ThO_2(+Y_2O_3)$ has been used to measure the oxygen content of liquid sodium. The partial pressures of gases other than oxygen, for example sulfur, can also be detected using sensors made of solid electrolytes [9.7].

9.2 Fuel Cells and Electrolyzers

The feasibility of developing fuel cells with solid electrolytes was discussed as early as 1935 by Schottky [9.8] and demonstrated by Baur and Preis [9.9] in 1937. However, most of the investigations on the use of solid oxide electrolytes in high-temperature fuel cells running at about $800-1000\,°C$ have been carried out over the past few years.

Figure 9.2.1 shows schematically the mode of operation of a fuel cell containing a zirconium dioxide tube as the solid electrolyte; the voltage of such a cell is about 1 V. The zirconium dioxide, which may be in the form of a tube or a disc, separates two electrode compartments, one containing air or pure oxygen and the other a fuel gas, hydrogen or carbon monoxide. The zirconium dioxide is coated

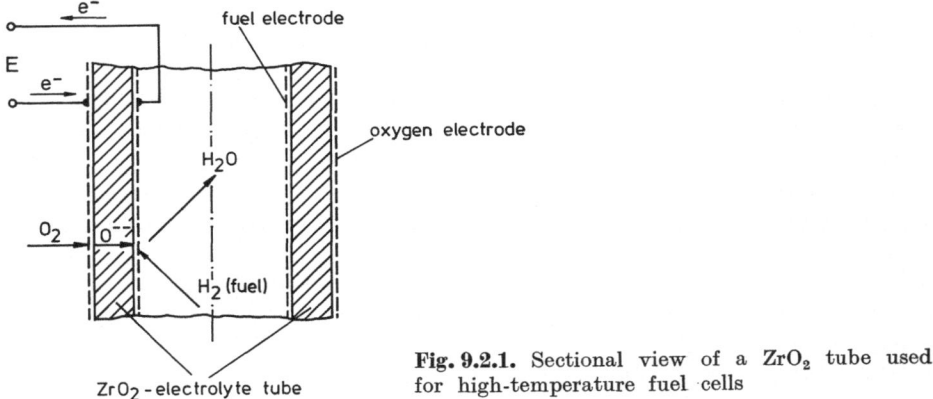

Fig. 9.2.1. Sectional view of a ZrO$_2$ tube used for high-temperature fuel cells

with two porous electrodes; for example nickel on the fuel side, and on the oxygen side, lanthanum-nickel oxide or some other electron-conducting oxide compatible with zirconium dioxide. The zirconium dioxide must be thick enough — about 1 mm — to achieve mechanical stability of the cell. To meet the needs for high power and high voltage a number of cells must be connected together in a gas-tight manner as shown in Fig. 9.2.2. This electrical and mechanical connection is accomplished with the aid of an electron-conducting so-called interconnection material, which must be chemically stable in the oxidizing atmosphere of the oxygen side as well as in the reducing atmosphere of the fuel side of such cells.

Other developments involve the use of zirconium dioxide as a thin-layer electrolyte considerably thinner than 1 mm; the mechanical strength of such a cell must then be achieved in other ways, e.g. by strengthening the electrodes or by using a mechanically stable and porous support.

The process of direct energy conversion taking place in a high-temperature fuel cell may be described as follows: gaseous oxygen takes up electrons at one elec-

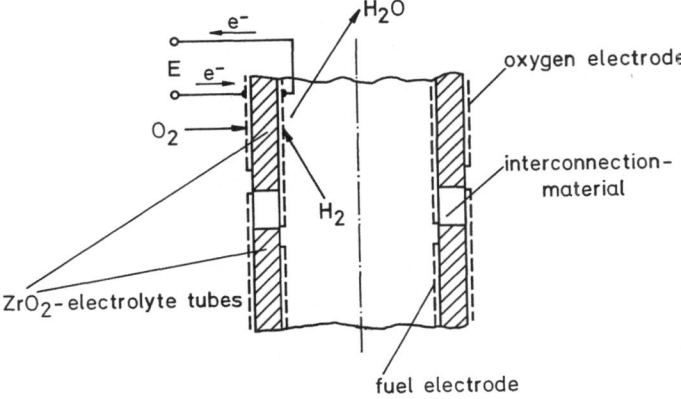

Fig. 9.2.2. Sectional view of a high-temperature fuel cell, which is constructed by adjoining several ZrO$_2$-tubes to a series connection

trode, thus being reduced to oxygen ions which then move through the electro-lyte and combine at the fuel side with H_2 or CO to give H_2O and CO_2 respectively. The electrons given up at this electrode enter the external circuit and flow back to the other electrode. In this way, the desired energy is made available to the user. The high-temperature fuell cell has considerable advantages because little or no polarization occurs at the electrodes and high current densities may be achieved, e.g. 600 mA/cm² with a power density of 400 mW/cm². However, difficulties arise concerning the choice of suitable construction materials because of the high temperatures required. A further disadvantage lies in the fact that the cell must be heated up slowly to the working temperature. Detailed accounts of high-temperature fuel cells, the development of which is being carried on, may be found in the literature [9.10]. Reversal of the current direction in a high-tempera-ture fuel cell, achieved by input of electrical energy, allows the decomposition of steam. The cell may thus be used as an electrolyzer to produce oxygen and hydrogen. The hydrogen generated in this way can be stored or transported through pipelines to remote places where it may in turn be employed as a fuel in a high-temperature fuell cell for the production of electrical energy. This mode of processing is being discussed in connection with large-scale energy storage and with transport of energy.

Much effort is being devoted to studying the decomposition of steam or carbon dioxide in an electrolysis cell with zirconium dioxide as a solid electrolyte in order to recover the oxygen from exhaled breath; this could be useful for example in manned spaceflight. High-temperature electrolysis cells may also serve for purifying gases if it is desired to remove residual traces of oxygen, or for the de-oxidation of liquid metals [9.11].

9.3 Batteries with Solid Electrolytes. The Sodium-Sulfur Battery

The basic arrangement of a sodium-sulfur cell is shown in Fig. 9.3.1; the open circuit voltage of this cell is 2.08 V. Its development started in 1967 with in-vestigations by Weber and Kummer [9.12]. In this case the solid electrolyte, a sodium-ion conductor, mainly consists of β-Al_2O_3 and has the empirical formula $Na_2O \cdot 11\,Al_2O_3$. (For further details on the composition and structure see Sect. 7.1.2.) This material is generally applied in the form of tubes closed at one end and filled with liquid sodium which is the negative electrode. An iron sponge, for absorbing the liquid sodium serves to improve the wetting of the electrolyte and at the same time to ensure safety. A metal wire connected with the sodium main-tains the electrical contact. The positive electrode consists of liquid sodium polysulfide and liquid sulfur embedded in a porous graphite felt. The cell housing generally consists of steel; α-Al_2O_3 is used for insulating purposes and the cell is made leak-proof by means of metal seals. The working temperature of the sodium sulfur cell is about 300 °C. During operation sodium ions enter the electrolyte from the negative electrode move through the β-alumina phase and react at the other electrode with sulfur to give sodium polysulfide whereby the electrons pass through the external circuit and can perform work. The energy density of the sodium sulfur cell is about 200 Wh/kg and thus many times higher than that of

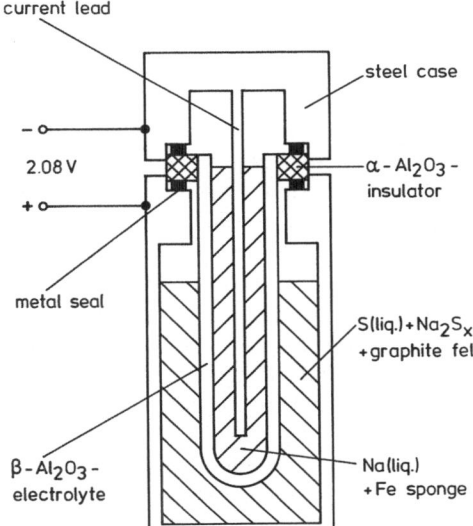

current lead

steel case

− o

2.08 V

+ o

α - Al₂O₃ - insulator

metal seal

S(liq.)+Na₂Sₓ +graphite felt

β - Al₂O₃ - electrolyte

Na(liq.) +Fe sponge

Fig. 9.3.1. Construction of a sodium-sulfur cell (schematic)

common batteries, e.g. of the lead secondary battery. Moreover, the materials constituting the electrolyte and the electrodes are available in large amounts. It is also possible to recharge the cell by reversing the current direction. All these facts explain why there is at present a great interest in the sodium-sulfur battery which consists of many single cells. The electrical network constituted by these cells is equivalent to a circuit of parallel branches, each branch being a series connection of such cells. A great deal of research is being carried out world-wide to develop this battery. Its use for large-scale energy storage and electric traction of electric motor vehicles has also been considered and prototypes of this battery have already been built. Further details on this system may be found in the literature [9.13]. Besides the sodium-sulfur cell, other systems containing solid electrolytes have been developed, some of these meeting the requirement of very long lifetimes. Possible storage times of more than ten years have been reported; such lifetimes cannot be achieved with conventional batteries.

9.4 Chemotronic Building Units

Solid-state galvanic cells are often called chemotronic building units if they are used as components in electrical circuits. In the following we shall describe coulometers and time switches, analog memories and capacitors frequently containing Ag_4RbI_5 as solid electrolyte. This compound has been dealt with in Sect. 7.1.3, where it was pointed out that Ag_4RbI_5 exhibits a remarkably high ion conductivity even at room temperature. Chemotronic building units containing Ag_4RbJ_5 can therefore be applied at and above room temperature. In addition, they have the advantage that it is possible to store them for up to ten years. Such cells are discussed in a review article by Owens and coworkers [9.14].

9.4.1 Coulometers and Time Switches

The following galvanic cell

$$Ag \mid RbAg_4I_5 \mid Au \tag{9.4.I}$$

can serve as a coulometer or time switch. When a current is passing through this cell silver is deposited at the gold electrode whereby the time switch is transformed into the loaded state. Silver may be transported back to the silver electrode during a current flow in the reverse direction. In the course of this stripping process the cell voltage is mainly determined by ohmic losses in the solid electrolyte. These potential drops depend on the magnitude of the current and lie in the millivolt range. If all the silver has been stripped from the gold electrode a sudden rise in the voltage occurs, which is used as a signal. Such electrochemical switches can be constructed to operate over time ranges between seconds and months. The cell described above and a similar one containing Ag_3SI and Ag_3SBr [9.15] can also be employed as coulometers: the amount of charge flowing during the loading process is then determined by the discharge process. Further details are given in [9.16].

9.4.2 Analog Memories

An arrangement working as an analog memory is shown in Fig. 9.4.1. It has been proposed by Takahashi and Yamamoto [9.17]. As can be seen the arrangement is symmetrical: both silver electrodes present are covered by a layer of $RbAg_4I_5$ on the side facing the interior of the device; these electrolyte layers are separated from one another by a silver selenide electrode, which is located in the center of the arrangement and contains a small amount of dissolved silver phosphate. At room temperature this solid selenide shows a remarkably high ion conductivity in addition to good electron conductivity, i.e. it is a mixed conductor. Current flow through one-half of the device leads to silver transport out of the mixed conductor. The chemical potential of silver in this phase is thereby reduced, thus causing an emf to appear between the central electrode and the two outer electrodes. To avoid polarization effects, the emf is measured against the silver electrode through which no current flows. The emf changes continuously with the amount

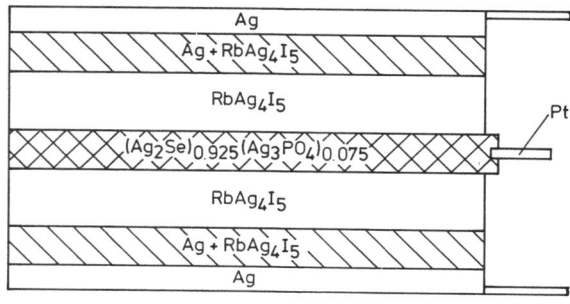

Fig. 9.4.1. Chemotronic device: section through an analog memory (schematic)

of charge transported. Hence, there is a larger content of information compared to the coulometers or time switches described above. Moreover, the reproducibility and accuracy of such devices are appreciably higher.

9.4.3 Capacitors Containing Solid Electrolytes

It is possible to produce capacitors with $RbAg_4I_5$ as electrolyte, one silver electrode, and an electrode consisting of a mixture of graphite and $RbAg_4I_5$. The capacity of these units is of the order of 10 F/cm^3. This is one thousand times larger than that of conventional capacitors, since the capacity of the electrochemical double layer at the graphite electrolyte interface is utilized here directly. This interfacial area is made very large by intimate mixing of the substances concerned. However, the voltage of a single cell is limited to 0.6 V, because at higher voltages the electrolyte begins to decompose. Possible applications of such capacitors are similar to those of the components described above, but in addition the use of these capacitors for energy storage has been discussed because of their very high capacities.

9.5 References

[9.1]   Geller, S.: Solid Electrolytes, Springer-Verlag Berlin, Heidelberg, New York 1977
        Hagenmuller, P., v. Gool, W.: Solid Electrolytes, General Principles, Characterization, Materials, Applications, Academic Press, New York, San Francisco, London 1980
        Subbarao, E. C.: Solid Electrolytes and their Applications, Plenum Press, New York, London 1980
[9.2]   Weissbart, J., Ruka, R.: Rev. Sci. Instrum. 32, 539 (1961)
[9.3]   Heimke, G., Chiari, B., Gugel, E.: Ber. Dtsch. Keram. Ges. 52, 321 (1975)
[9.4]   Dueker, H., Neidhard, H.: Second Automotive Emission Conf., Ann. Arbor, 1973
        Rohr, F. J.: Elektrochemische Energietechnik — Entwicklungsstand und Aussichten. Bundesminister für Forschung und Technologie (ed.), p. 299, Bonn 1981
        Zechnall, R., Baumann, G.: MTZ 34, (1973)
[9.5]   Turkdogan, E. T., Fruehan, R. J.: Amer. Iron Steel Inst. Yearbook, 1, (1968)
        Fischer, W. A., Janke, D.: Arch. Eisenhüttenwes. 39, 89 (1968)
        Fitterer, G. R.: J. Met. 19 (9), 92 (1967)
        Schwerdtfeger, K.: Trans Metall. Soc. AIME 239, 1276 (1967)
        Russel, C. K., Fruehan, Rittiger, J.: J. Met. 23, 44, (1971)
        Matsushita, Y., Goto, K.: Trans. Iron Steel Inst. Japan. 6, 131 (1966)
        Ihida, M., Kawai, Y.: Trans. Iron Steel Inst. Japan 12, 269 (1972)
        Yavoiskii, V. et al.: Steel USSR 1, 513 (1971)
        Oiks, G. N., Asadov, T. M.: Steel USSR 2, 192 (1972)
[9.6]   Blumenthal, R. N., Whitmore, D. H.: J. Electrochem. Soc. 110, 92 (1963)
        Belford, T. N., Alcock, C. B.: Trans. Faraday Soc. 60, 822 (1964)
        Belford, T. N., Alcock, C. B.: Trans. Faraday Soc. 61, 443 (1965)
        Plushhell, W., Engell, H. J.: Z. Metallk. 56, 450 (1965)
        Fischer, W. A., Ackermann, W.: Arch. Eisenhüttenw. 36, 643, 695 (1965)
        Rickert, H., Wagner, H.: Electrochimica Acta 11, 83 (1966)
        Kolodney, M., Minushkin, B., Sternmetz, H.: Electrochem. Technol. 3, 214 (1965)
        Fischer, W. A., Ackermann, W.: Arch. Eisenhüttenw. 37, 43 (1966)
        Fischer, W. A., Ackermann, W.: Arch. Eisenhüttenw. 37, 697 (1966)

Isaacs, H. S.: J. Electrochem. Soc. *119*, 455 (1972)

Ramanarayanan, T. A., Rapp, R. A.: Met. Trans. *3*, 3239 (1972)

[9.7] Dietz, H., Haecker, W., Jahnke, H.: Adv. Electrochem., Electrochem. Eng. *10*, 1 (1977)

[9.8] Schottky, W.: Wiss. Veröff. Siemens *14*, 1 (1935)

[9.9] Baur, E., Preis, H.: Z. Elektrochem. *43*, 727 (1937)

[9.10] Fischer, W. et al.: Chem.-Ing.-Tech. *43*, 1227 (1971)

Fischer, W. et al.: Chem.-Ing.-Tech. *44*, 726 (1972)

University of Oklahoma, Energy Alternatives: A Comprehensive Analysis, US Government Printing Office, Washington, D.C. (1975)

Westinghouse Electric Corp., Final Report, Project Fuel Cell Rep. No. 57 (1970)

Rohr, F. J.: Elektrochemische Energietechnik — Entwicklungsstand und Aussichten, Bundesminister für Forschung und Technologie (ed.), p. 264, Bonn 1981

[9.11] Weissbart, J., Smart, W., Wydeven, T.: Aerospace Medicine *40*, 136 (1960)

[9.12] Weber, N., Kummer, J. T.: Proc. Ann. Power Sources Conf. *21*, 37 (1967)

[9.13] Gelb, G. H., Richardson, N. A.: Proceedings of the Symposium on Batteries, Traction and Propulsion, 1972. Kerr, R. L. (ed.), p. 178, Columbus Section, Electrochemical Society, Battelle Memorial Institute, Columbus, Ohio 1972

Gratch, S. et al.: Proceedings of the 7th Intersociety Energy Conversion Engineering Conference, p. 38, ACS, Washington, D.C. 1972

Miles, L. J., Wynn Jones, I.: Proceedings of the Eight International Power Sources Symposium, Brighton, p. 245, Sussex, England, 1972, International Power Sources Committee, Croydon, Surrey, England

Fischer, W., Haar, W.: Phys. unserer Zeit *9*, 184 (1978)

Fischer, W.: Elektrochemische Energietechnik — Entwicklungsstand und Aussichten. Bundesminister für Forschung und Technologie (ed.), p. 185, Bonn 1981

[9.14] Owens, B. B., Oxley, J. E., Sammells, A. F.: Solid Electrolytes. Geller, S. (ed.), p. 67, Springer, Berlin, Heidelberg, New York 1977

[9.15] Kennedy, J. H.: Phys. Electrolytes *2*, 931, 1972

[9.16] Electronics *40*, 186 (1967)

Kennedy, J. H., Chen, F., Hunter, J.: J. Electrochem. Soc. *120*, 454 (1973), see also [9.15]

[9.17] Takahashi, T., Yamamoto, O.: J. Appl. Electrochem. *3*, 129 (1973)

10 Solid-State Reactions

Defects in solids and adsorbed particles at surfaces, which are generally in equilibrium with one another, are of great importance for the understanding of reactions in and at the surface of solids. Adsorbed atoms, molecules and ions together with simple and complex vacancies in the adsorption layer at higher degrees of coverage play a central role in recent investigations of processes involving interactions between solids and gases [10.1].

Electrochemical concepts known from electrode kinetics in the electrochemistry of liquid electrolytes, are used for the theoretical analysis of reactions at the interface metal/metal compound. In experiments also electrochemical methods using solid electrolytes are becoming more important for the study of solid-state reactions. We shall discuss some examples in the next section.

In this chapter we shall consider in some detail diffusion-controlled solid-state reactions, taking as the first example the formation of tarnishing layers on metals which has been treated quantitatively by C. Wagner [10.2]. A reaction of this type is only possible if defects migrate within the volume of a solid.

To treat such reactions quantitatively we require the flux equations of Chap. 6. In Sect. 10.2.4 we shall then treat a further type of reaction involving formation of double salts and double oxides in solid-state reactions; this process is also made possible by the migration of defects.

10.1 Theory of the Formation of Tarnishing Layers on Metal Surfaces

When metals are oxidized in reactions with for example oxygen, sulfur or a halogen; mainly at higher temperatures, solid-reaction products are often formed (e.g. oxides, sulfides or halides); these grow at the surface of the metal either as a compact protective layer or in a more porous or irregular manner, e.g. as whiskers. The case of the compact plane parallel protective layer is particularly suitable for a quantitative theoretical treatment. We shall thus discuss the formation of a layer of this type, although in practice porous intermediate layers or other irregularities often occur.

Since the metal and the oxidant are separated by the tarnishing layer, either the metal or the nonmetal — in either case generally in the form of ions and electrons or electron defects — must migrate through the tarnishing layer. If we assume that cations (metal ions) and anions (nonmetal ions), electrons or electron defects are mobile particles in the tarnishing layer, the formation of such a compact

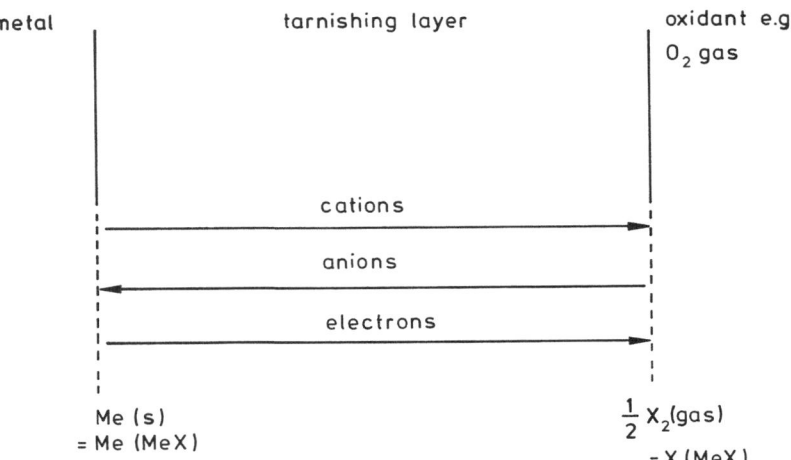

Fig. 10.1.1. Schematic representation of the migration of cations, anions and electrons across a tarnishing layer during the oxidation of a metal

layer is described by the reaction scheme shown in Fig. 10.1.1, which contains three reaction steps:

a) the transition of metal across the phase boundary metal/solid metal compound,
b) the diffusion of cations or anions and electrons or electron defects,
c) the incorporation of the oxidant, e.g. oxygen, at the phase boundary tarnishing layer/oxidant.

The rate law for the growth of the protective layer is determined by the interaction between these reaction steps.

10.1.1 Rate Laws Describing the Formation of Tarnishing Layers

It has been found that various rate laws may apply to the increase in the thickness ΔX of the tarnishing layer as a function of the reaction time t at a constant chemical potential of the oxidant (e.g. constant composition of the gas phase):

a) The linear rate law:

$$\mathrm{d}\,\Delta X/\mathrm{d}t = \text{const.} \tag{10.1.1}$$

or

$$\Delta X = \text{const.} \cdot t \tag{10.1.2}$$

This rate law (see Fig. 10.1.2) may be expected to apply if one or several reactions at phase boundaries are rate-determining.

Examples are:

α) Reaction of iron in a $CO-CO_2$-mixture at 900 °C, which has been studied by Hauffe and Pfeiffer [10.3] and Pettit, Yinger and J. B. Wagner [10.4]. In this case, the reaction at the phase boundary iron (II)oxide/CO_2-CO and in particular the dissociation of adsorbed CO_2 into adsorbed CO and O is rate-determining.

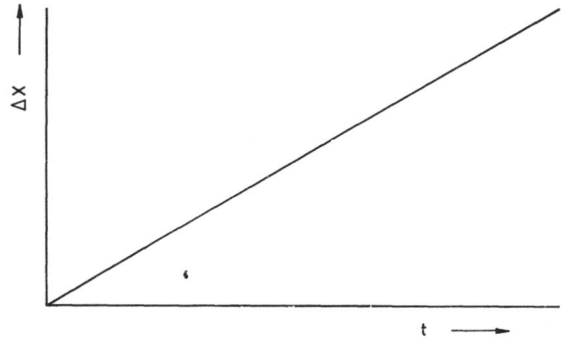

Fig. 10.1.2. Linear rate law for the formation of the tarnishing layer

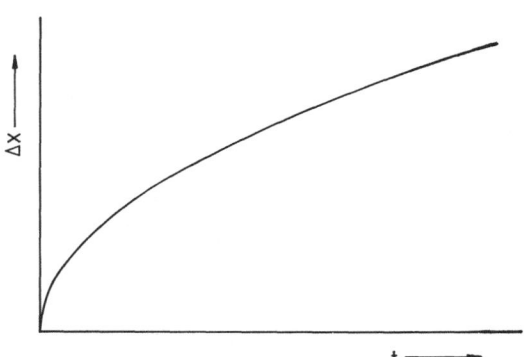

Fig. 10.1.3. Parabolic rate law for the formation of the tarnishing layer

β) The initial reaction of silver with liquid sulfur, for which the reaction at the interface silver/silver sulfide is rate-determining (see Czerski, Mrowec and Werber [10.5] and Rickert [10.6]).

b) The parabolic rate law (Fig. 10.1.3) which was first established by Tammann [10.7] and Pilling and Bedworth [10.8]:

$$\frac{d\,\Delta X}{dt} = \frac{k}{\Delta X} \tag{10.1.3}$$

or in the integrated form (see Fig. 10.1.3)

$$(\Delta X)^2 = 2kT, \tag{10.1.4}$$

k denotes the parabolic tarnishing constant.

The parabolic rate law is expected to hold if diffusion processes in the tarnishing layer are rate-determining. In Sect. 10.1.2 we shall discuss the calculation of k according to the method of C. Wagner [10.9].

c) The logarithmic rate law (Fig. 10.1.4a, b), which was first observed by Tammann and Köster [10.10]:

$$\Delta X = a + b\,\lg t, \tag{10.1.5}$$

where a and b are constants. This rate law is sometimes found in the case of small layer thicknesses. The interpretation is, however, not always simple (compare P. Kofstad [10.11], C. Wagner [10.12]).

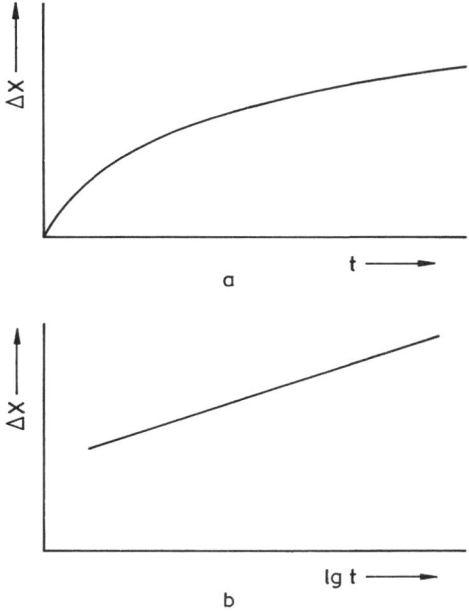

Fig. 10.1.4 a, b. Logarithmic rate law for the formation of the tarnishing layer. **a** Linear plot; **b** semi-logarithmic plot

10.1.2 Calculation of the Parabolic Rate Law Constant Using the Method of C. Wagner [10.9]

The following assumptions are made:

a) The tarnishing layer (e.g. oxide layer) is non-porous and the reaction scheme shown in Fig. 10.1.1 applies.

b) Virtually complete equilibrium at the interfaces between the oxide phase and the neighbouring phases (metal or gas) has been attained.

c) Deviations from the exact stoichiometric composition of the oxide phase are small and local thermodynamic equilibrium has been attained at all points in the oxide phase.

d) The solubility of oxygen in the metal phase is negligibly small.

e) The thickness ΔX of the oxide layer is large in comparison with the thickness of the electrical double layer.

It was shown in Chap. 6 that the following equations give the fluxes of the cations j_1, the anions j_2 and the electrons (or electron defects) j_3 in moles per unit area and unit time for the one-dimensional case:

$$j_1 = -\frac{\sigma_1}{z_1^2 F^2} \frac{\partial \eta_1}{\partial x}, \tag{10.1.6a}$$

$$j_2 = -\frac{\sigma_2}{z_2^2 F^2} \frac{\partial \eta_2}{\partial x}, \tag{10.1.6b}$$

$$j_3 = -\frac{\sigma_3}{F^2} \frac{\partial \eta_3}{\partial x}, \tag{10.1.6c}$$

σ_i denotes the partial conductivity of the particle type i, z_i the charge number of the corresponding particle, η_i the corresponding electrochemical potential and F Faraday's constant.

Equations (10.1.6a—c) may be rewritten using the component diffusion coefficient D_i and the concentration c_i in moles per cm³ of the particle type i:

$$j_1 = -\frac{c_1 D_1}{RT} \frac{\partial \eta_1}{\partial x}, \tag{10.1.7a}$$

$$j_2 = -\frac{c_2 D_2}{RT} \frac{\partial \eta_2}{\partial x}, \tag{10.1.7b}$$

$$j_3 = -\frac{c_3 D_3}{RT} \frac{\partial \eta_3}{\partial x}, \tag{10.1.7c}$$

In general, the problem of calculating parabolic tarnishing constants may be solved with the aid of the flux equations (10.1.6a—c) or (10.1.7a—c). In real cases, however, the mobilities of cations and anions are generally very different. The tarnishing process is then due to migration of either cations (metal ions) and electrons or anions and electrons, depending on the type of ion with the greater mobility. For the following discussion we presume that only metal ions and electrons migrate, i.e. $\sigma_1 \gg \sigma_2$ or $|j_1| \gg |j_2|$; the opposite case may be treated analogously. On the grounds of electrical neutrality, it then follows that the flux of the metal ions is equivalent to that of the electrons:

$$z_1 j_1 = j_3. \tag{10.1.8}$$

Since the following relationship holds for the electrochemical potential η_i

$$\eta_i = \mu_i + z_i F \varphi \tag{10.1.9}$$

(where μ_i is the chemical potential of the particle type i and φ the electrical potential), one can combine Eqs. (10.1.8), (10.1.6a) and (10.1.6c) to obtain:

$$j_1 = -\frac{1}{z_1^2 F^2} \frac{\sigma_1 \sigma_3}{\sigma_1 + \sigma_3} \left(\frac{\partial \mu_1}{\partial x} + z_1 \frac{\partial \mu_3}{\partial x} \right). \tag{10.1.10}$$

Since the local chemical potential μ_{Me} of the metal is related to the chemical potential of the metal ions and to that of the electrons by the equation

$$\mu_{Me} = \mu_1 + z_1 \mu_3 \tag{10.1.11}$$

it follows that

$$j_1 = -\frac{1}{z_1^2 F^2} \frac{\sigma_1 \sigma_3}{\sigma_1 + \sigma_3} \frac{\partial \mu_{Me}}{\partial x}. \tag{10.1.12}$$

Equation (10.1.12) contains only the gradient of the chemical potential of the metal, but not the gradient of the concentration of the metal in the metallic compound. This is of great advantage, since deviations from exact stoichiometry are often very small and are thus very difficult to determine experimentally. The partial conductivities σ_1 and σ_3 are functions of μ_{Me} and are therefore in general not constant within the tarnishing layer.

The values of σ_1 and σ_2 may vary by several orders of magnitude between the metal side of the layer and that in contact with the oxidant. Integration of Eq. (10.1.12) leads to the following expression

$$
\begin{aligned}
j_1 &= -\frac{1}{z_1^2 F^2\,\Delta X} \int\limits_{\mu'_{Me}}^{\mu''_{Me}} \frac{\sigma_1\sigma_3}{\sigma_1+\sigma_3}\,d\mu_{Me} \\
&= \frac{1}{z_1^2 F^2\,\Delta X} \int\limits_{\mu''_{Me}}^{\mu'_{Me}} \frac{\sigma_1\sigma_3}{\sigma_1+\sigma_3}\,d\mu_{Me}
\end{aligned}
\tag{10.1.13}
$$

where μ'_{Me} and μ''_{Me} are the chemical potentials of the metal Me at the interfaces metal/tarnishing layer and tarnishing layer/oxidant respectively.

To establish a connection with the parabolic tarnishing constant introduced via Eqs. (10.1.3) and (10.1.4) respectively, we note that the flux of the metal ions is also given by multiplying the rate of increase in the layer thickness, i.e. $d\,\Delta X/dt$, by the concentration c_1 of the metal in the reacting layer, i.e. we have

$$
j_1 = c_1\,\frac{d\,\Delta X}{dt}.
\tag{10.1.14}
$$

According to Eq. (10.1.3) the term $d\,\Delta X/dt$ can be replaced by the parabolic tarnishing constant k divided by ΔX.

Equation (10.1.14) may now be rewritten to give

$$
j_1 = c_1\,\frac{k}{\Delta X}.
\tag{10.1.15}
$$

We can thus obtain the desired expression for the parabolic tarnishing constant k by comparing Eqs. (10.1.13) and (10.1.15):

$$
k = \frac{1}{z_1^2 F^2 c_1} \int\limits_{\mu''_{Me}}^{\mu'_{Me}} \frac{\sigma_1\sigma_3}{\sigma_1+\sigma_3}\,d\mu_{Me} \quad \text{if} \quad \sigma_1 \gg \sigma_2.
\tag{10.1.16}
$$

For the case $\sigma_2 \gg \sigma_1$, i.e. predominantly anions and electrons migrate through the protective layer, we may apply the same kind of reasoning, which leads to the following expression for k:

$$
k = \frac{1}{z_2^2 F^2 c_2} \int\limits_{\mu'_X}^{\mu''_X} \frac{\sigma_2\sigma_3}{\sigma_2+\sigma_3}\,d\mu_X \quad \text{if} \quad \sigma_2 \gg \sigma_1,
\tag{10.1.17}
$$

μ_X is the chemical potential of the non-metal X, μ'_X that corresponding to the equilibrium between metal and metal compound and μ''_X that corresponding to the equilibrium between metal compound and gas phase.

The partial conductivity of the electrons is often much larger than that of the ions, i.e. $\sigma_3 \gg \sigma_1, \sigma_2$. In this case it follows from Eq. (10.1.16) that for $\sigma_1 \gg \sigma_2$ (migration of metal ions)

$$k = \frac{1}{z_1^2 F^2 c_1} \int_{\mu''_{Me}}^{\mu'_{Me}} \sigma_1 \, d\mu_{Me} \quad \text{if} \quad \sigma_3 \gg \sigma_1 \gg \sigma_2 \qquad (10.1.18)$$

or in terms of diffusion coefficients according to Eqs. (10.1.7 a—c):

$$k = \frac{1}{RT} \int_{\mu''_{Me}}^{\mu'_{Me}} D_1 \, d\mu_{Me}, \quad \text{if} \quad D_3 \gg D_1 \gg D_2. \qquad (10.1.19)$$

The corresponding expressions for electron-conducting tarnishing layers for anion migration ($\sigma_2 \gg \sigma_1$) are:

$$k = \frac{1}{z_2^2 F^2 c_2} \int_{\mu'_X}^{\mu''_X} \sigma_2 \, d\mu_X, \quad \text{if} \quad \sigma_3 \gg \sigma_2 \gg \sigma_1; \qquad (10.1.20)$$

$$k = \frac{1}{RT} \int_{\mu'_X}^{\mu''_X} D_2 \, d\mu_X, \quad \text{if} \quad D_3 \gg D_2 \gg D_1. \qquad (10.1.21)$$

10.2 Examples of the Formation of Tarnishing Layers on Metals and Other Solid-State Reactions

10.2.1 Oxidation of Copper to Cu_2O at 1000 °C [10.13]

The disorder in Cu_2O has been discussed in Chap. 3, where we have seen that, in agreement with the assumption that copper-ion vacancies and electron defects are the predominant defects, the partial conductivity of the electron defects is found experimentally to be proportional to $p_{O_2}^{1/7}$. If the system behaved ideally, a theoretical dependence proportional to $p_{O_2}^{1/8}$ would be expected. Since the concentration of electron defects is equal to that of copper ion vacancies, it is quite reasonable to assume that the conductivity of the copper ions is also proportional to $p_{O_2}^{1/7}$; we may thus write

$$\sigma_1 = \sigma_1(p'_{O_2}) \left(\frac{p_{O_2}}{p'_{O_2}} \right)^{1/7}. \qquad (10.2.1)$$

Because of the higher mobility of electron defects compared to that of copper ion vacancies, the electronic partial conductivity σ_3 is considerably larger than that of the copper ions σ_1. Thus, Eq. (10.1.18) must be applied. According to the Gibbs-Duhem equation we can write:

$$2d\mu_{Cu} = -d\mu_O = -\frac{1}{2}\,d\mu_{O_2} \tag{10.2.2}$$

and, using the expression $\mu_{O_2} = \mu_{O_2}^0 + RT\ln p_{O_2}$:

$$d\mu_{Cu} = -\frac{1}{4}\,RT\,d\ln p_{O_2}. \tag{10.2.3}$$

Substituting Eqs. (10.2.1) and (10.2.3) in Eq. (10.1.18) leads to the expressions

$$k = \frac{RT\sigma_1(p_{O_2}')}{4F^2c_1}\int_{p_{O_2}'}^{p_{O_2}''}\left(\frac{p_{O_2}}{p_{O_2}'}\right)^{1/7}d\ln p_{O_2} \tag{10.2.4}$$

$$= \frac{7}{4}\frac{RT}{F^2c_1}\,\sigma_1(p_{O_2}')\left[\left(\frac{p_{O_2}''}{p_{O_2}'}\right)^{1/7} - 1\right]. \tag{10.2.5}$$

The parabolic tarnishing constant should be a linear function of the seventh root of the external oxygen partial pressure. This is in good agreement with the experimental results [10.13].

10.2.2 Oxidation of Zinc at 400°C

As shown in Chap. 3, virtually equal concentrations of quasi-free electrons and zinc ions occupying interstitial sites may be assumed to be present in ZnO. The partial conductivities of the electrons and of the zinc ions are proportional to $p_{O_2}^{-1/4}$, i.e.

$$\sigma_1 = \sigma_1(p_{O_2}')\left(\frac{p_{O_2}}{p_{O_2}'}\right)^{-1/4}. \tag{10.2.6}$$

Apart from this, the same considerations presented above for the oxidation of copper to Cu_2O apply. Thus, for the oxidation of Zn to ZnO we obtain the following equation:

$$k = const.\left[1 - \left(\frac{p_{O_2}'}{p_{O_2}''}\right)^{1/4}\right] \tag{10.2.7}$$

or, if the external oxygen partial pressure is large enough

$$k = const., \quad if \quad p_{O_2}' \ll p_{O_2}'', \tag{10.2.7a}$$

i.e. in this case k is independent of the external oxygen partial pressure p_{O_2}'' [10.13].

10.2.3 Reaction of Silver with Liquid Sulfur at 400 °C

When silver reacts with liquid sulfur, a porous Ag_2S layer is generally formed adjacent to silver while a compact layer is formed in contact with sulfur. However, with the aid of a special experimental arrangement shown schematically in Fig. 10.2.1 it is possible to suppress the formation of the porous layer, and Ag_2S grows one-dimensionally between silver and sulfur. The formation of a porous layer is primarily due to the fact that, because of the migration of silver ions and electrons in Ag_2S, silver is used up at the interface Ag/Ag_2S; in this case, if no relative movements between the Ag_2S tarnishing layer and silver metal occur, cavities are necessarily generated. The experimental arrangement shown in Fig. 10.2.1 permits relative movements between Ag_2S and silver, since Ag_2S grows into a glass tube which is constantly pressed against the silver. Because of the high mobility of silver ions and electrons, Ag_2S grows very rapidly into the glass tube so that the thickness of the layer may reach several centimetres in the course of one day. At 400 °C equilibrium has virtually been attained at the phase boundaries, while at 200 and 300 °C strong deviations from equilibrium occur at the phase boundary Ag/Ag_2S; investigations of this phase boundary reaction will be discussed in the next chapter. At 400 °C the parabolic tarnishing constant may be calculated using the partial conductivity of the silver ions (since $\sigma_e > \sigma_{Ag^+} > \sigma_{S^{--}}$) and the difference of the chemical potentials of silver along the tarnishing layer.

Silver sulfide is a particularly simple example, since the partial conductivity σ_1 of the silver ions in Ag_2S is virtually constant and independent of the chemical potential of the silver. The value of σ_1 at 400 °C is 5.5 Ω^{-1} cm^{-1}. Thus, the parabolic tarnishing constant k is given by the equation

$$k = \frac{\sigma_1}{F^2 c_1} (\mu'_{Ag} - \mu''_{Ag}). \tag{10.2.8}$$

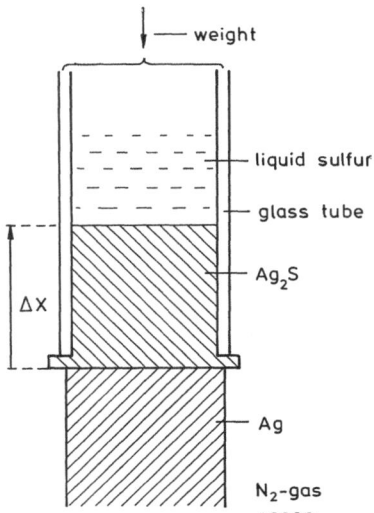

Fig. 10.2.1. Experimental arrangement (schematic) for studying the reaction of silver with liquid sulfur

The term $(\mu'_{Ag} - \mu''_{Ag})$ may be obtained from the emf E of cell (7.2.I) in Sect. 7.2, the value of which is 0.22 V. Thus, the calculated value of k is $2.5 \times 10^{-4} \times cm^2 \, s^{-1}$, in good agreement with the experimental value [10.14].

10.2.4 Formation of Double Salts and Double Oxides via Solid-State Reactions

As an example, we shall discuss the formation of Ag_2HgI_4 from AgI and HgI_2 at 65 °C [10.15] according to the equation

$$2\,AgI + HgI_2 = Ag_2HgI_4. \tag{10.2.9}$$

The reaction product is formed between the reactants and therefore separation of these from one another occurs; this is schematically shown in Fig. 10.2.2. If equilibrium has been attained at the interfaces, the diffusion of Ag^+- and Hg^{2+}- ions in opposite directions, which obeys a parabolic rate law, is rate-determining for the formation of Ag_2HgI_4.

The relatively large iodide ions are virtually not involved in the transport; if this were not the case, the reaction mechanism shown in Fig. 10.2.2 would have to be correspondingly modified. The rate constant of the parabolic rate law can be calculated if the partial conductivities of the ions diffusing in Ag_2HgI_4 or their component diffusion coefficients are known as a function of the chemical potential μ_{AgI} (or μ_{HgI_2}) and if the Gibbs reaction energy for the reaction considered is known. (For details of the calculations see [10.15]).

The formation of double oxides, e.g. spinels, from the simple oxides may be treated analogously. Here, it is necessary to distinguish between systems where, as in the previous example, metal ions diffuse in opposite directions and those in which oxygen ions also take part in the transport processes. In some cases, material transport in the gas phase may be significant. A further example of the type of reaction discussed here is the formation of double sulfides such as Ag_3SbS_3 and $AgSbS_2$ from Sb_2S_3 and Ag_2S [10.16].

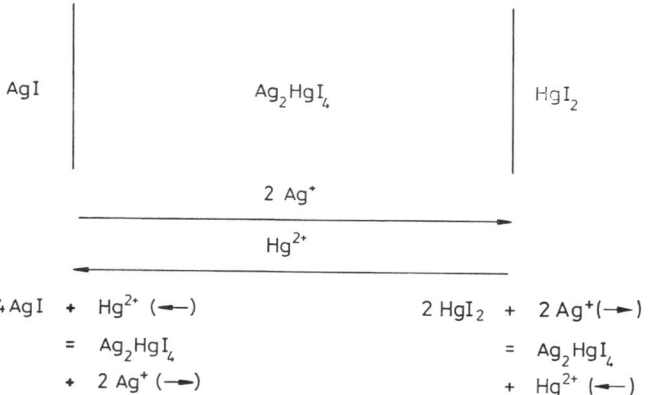

Fig. 10.2.2. Formation of the double salt Ag_2HgI_4 due to a solid-state reaction

10.3 References

[10.1]   Wagner, C.: Adv. Catalysis *21*, 323 (1970)
[10.2]   Wagner, C.: Z. phys. Chem. *B21*, 25 (1933)
[10.3]   Hauffe, K., Pfeiffer, H.: Z. Elektrochem. *56*, 390 (1952)
[10.4]   Pettit, F. S., Yinger, R., Wagner, J. B.: Acta Met. *8*, 617 (1960)
[10.5]   Czerski, L., Mrowec, S., Werber, T.: Arch. Hutnictwa *3*, 49 (1958)
[10.6]   Rickert, H.: Z. phys. Chem. N.F. *23*, 355 (1960)
[10.7]   Tammann, G.: Z. anorg. Chem. *111*, 78 (1920)
[10.8]   Pilling, N. B., Bedworth, R. E.: J. Inst. Met. *29*, 529 (1923)
[10.9]   Wagner, C.: Z. phys. Chem. *B21*, 25 (1933)
[10.10] Tammann, G., Köster, W.: Z. anorg. Chem. *123*, 196 (1922)
[10.11] Kofstad, P.: High-Temperature Oxidation of Metals. New York: Wiley 1966
[10.12] Wagner, C.: Werkst. Korrosion *21*, 886 (1971)
[10.13] Wagner, C., Grünewald, K.: Z. phys. Chem. *B40*, 455 (1938)
[10.14] Rickert, H.: Z. phys. Chem. N.F. *23*, 355 (1960)
[10.15] Ketelaar, J. A. A.: Z. phys. Chem. *B26*, 327 (1934)
         Koch, E., Wagner, C.: Z. phys. Chem. *B34*, 317 (1936)
         Wagner, C.: Z. phys. Chem. *B34*, 309 (1936)
[10.16] Rickert, H., Wagner, C.: Z. Elektrochem. *64*, 793 (1960), *66*, 502 (1962)

11 Galvanic Cells with Solid Electrolytes for Kinetic Investigations

Besides their use for thermodynamic measurements, galvanic cells with solid electrolytes have become increasingly important for kinetic studies in solids and at surfaces and phase boundaries of solids. As has been discussed in Chap. 8, thermodynamic measurements using such cells allow to determine Gibbs reaction energies ΔG or to obtain activities or chemical potentials in mixtures and chemical compounds. With the aid of polarization measurements on suitable galvanic cells, it is possible (see Sect. 6.3) to obtain information about partial conductivities in solids; this already represents an application of galvanic cells in kinetic studies. Other kinetic measurements with solid-state galvanic cells involve diffusion processes or diffusion-controlled reactions, reactions taking place at the phase boundary solid/gas, or solid/solid respectively. Kinetic investigations of this type are made possible because galvanic cells containing solid auxiliary electrolytes, which exhibit virtually pure ionic conductivity, not only provide thermodynamic information via the emf but also allow direct determination of reaction rates via measurements of the current flow within a suitably constructed cell. The combination of rate measurements (via electric currents) with measurements of thermodynamic quantities (via emf's) often permits the analysis of kinetic processes. This will be discussed in the following for six typical examples:

1. Diffusion of oxygen in metals,
2. formation of nickel sulfide on nickel,
3. transfer of silver across the phase boundary silver/silver sulfide,
4. electrochemical studies of evaporation and condensation,
5. the electrochemical Knudsen cell,
6. chemical diffusion in metal oxides and sulfides.

11.1 Electrochemical Measurements of Oxygen Diffusion in Metals at Higher Temperatures Using Zirconium Dioxide as Solid Electrolyte

11.1.1 The Principle of Electrochemical Measurements of Oxygen Diffusion in Metals

Electrochemical investigations on the oxygen diffusion in metals have first been carried out by the author of this book and his coworkers [11.1—11.4] and in the meantime by others [11.6—11.19]. The basic element for such measurements is a solid-state galvanic cell of the type

$$p'_{O_2}, \text{Pt1} \mid \text{ZrO}_2(+10 \text{ mol\% } \text{Y}_2\text{O}_3) \mid \text{Me}(+\text{O dissolved}), \text{Pt2}. \qquad (11.1.\text{I})$$

This cell contains doped ZrO_2 as a solid auxiliary electrolyte with virtually pure ionic conductivity. One side is an oxygen electrode, which may be a porous Pt-electrode in contact with air, in order to establish a reference oxygen activity. The other electrode consists of a metal sample containing dissolved oxygen. Cell (11.1.I) exhibits two important features, a thermodynamic one and a kinetic one:
The thermodynamic one lies in the fact that the emf of this cell is a measure of the difference between the chemical potential μ''_{O_2} and μ'_{O_2} of oxygen on both sides of the cell:

$$\mu''_{O_2} - \mu'_{O_2} = 4 \text{ EF}. \qquad (11.1.1)$$

Since one can write $1/2\mu_{O_2} = \mu_O = \mu_O^0 + \text{RT} \ln a_O$, the emf is also a measure of the ratio of the oxygen activities a'_O and a''_O on both sides of the solid electrolyte:

$$E = \frac{\text{RT}}{2\text{F}} \ln \frac{a''_O}{a'_O}. \qquad (11.1.2)$$

It is thus possible

a) to determine ratios of oxygen activities via the emf of the cell at open circuit and
b) to stipulate certain ratios of oxygen activities or the oxygen activity on one side of the electrolyte, if the activity at the other side is fixed, by applying a definite emf to the cell.

Care must be taken to assure that no errors occur due to polarization effects within the electrolyte and at the reference electrode.

The kinetic feature of the galvanic cell is described by the relationship:

$$\vec{i} = -2\text{F}\vec{j}_{O^{2-}} \qquad (11.1.3)$$

Equation (11.1.3) states that the current flowing through the cell is a measure of the rate, with which oxygen is transported from one side of the electrolyte to the other due to the migration of oxygen ions through the electrolyte and the movement of electrons through the external electron-conducting leads.

The flux of oxygen ions $\vec{j}_{O^{2-}}$ gives the rate of removing or adding oxygen from or to the metal electrode. Thus, oxygen diffuses out of or into the metal electrode on the right-hand side of the galvanic cell. This diffusion current is equivalent to the electric current, the latter being the experimentally determinable quantity. The evaluation depends on the choice of the boundary and the initial conditions. In the following we will discuss both the potentiostatic and the galvanostatic method.

11.1.2 The Potentiostatic Method

Discussing the potentiostatic method, we consider as an example the investigation on the diffusion of oxygen in silver [11.1]. The following galvanic cell has been used for measuring this diffusion:

$$
\begin{array}{c}
\xrightarrow{\quad\quad x \quad\quad} \\
\text{Fe, FeO} \mid \text{ZrO}_2(+\text{Y}_2\text{O}_3) \mid \text{Ag}(+\text{O diss.}) \\
\xleftarrow{\;2\,e^-\;} \xleftarrow{\;\;\text{O}^{2-}\;\;} \xleftarrow{\quad\;\text{O}\;\quad} \\
\xleftarrow{\quad\; 2\,e^- \;\quad}
\end{array}
\qquad (11.1.\text{II})
$$

The left-hand electrode of this cell consists of an Fe/FeO electrode by means of which a reference oxygen partial pressure of ca. 10^{-19} atm is established at 800 °C. The electrode on the other side is metallic silver containing dissolved oxygen. Under currentless conditions, the emf of the cell is a measure of the activity of the oxygen dissolved in the silver and thus of the initial concentration of oxygen in it. The experiment begins at time $t = t_0$ by applying a definite potential difference to the cell. Due to this potential difference the oxygen activity at the phase boundary ZrO_2/Ag will be controlled. If, e.g. a potential difference of ca. 100 mV is potentiostatically applied to cell (11.1.II) with the negative pole at the silver, the resulting oxygen activity at the phase boundary ZrO_2/Ag is smaller than the activity corresponding to equilibrium between iron and iron oxide. Thus, the oxygen concentration at the phase boundary considered is virtually zero. The resulting gradient of the oxygen concentration forces the oxygen to diffuse out of the metal phase to the phase boundary ZrO_2/Ag. From here oxygen ions are transported through the electrolyte because of the potential difference applied whereas the electrons move along the external circuit to the silver electrode. Thus, the electron current is a measure of the diffusion current and may be determined experimentally. From the current, which is equivalent to the diffusion of oxygen out of the metal, the diffusion coefficient of oxygen may be calculated.

The evaluation depends on the geometry of the cell; we shall later discuss a) linear and b) cylindrical geometries.

It is not possible, to impose oxygen partial pressures lower than 10^{-23} atm at 800 °C, since doped zirconia becomes electron-conducting in this pressure range [11.5]. It generally suffices to establish an oxygen concentration at the phase boundary zirconium dioxide/metal, which is smaller by a factor 10^2 to 10^3 than the initial concentration of oxygen in the metal sample. If oxygen is dissolved in the form of atoms (which may eventually have taken up electrons), a change in the oxygen concentration by a factor ten corresponds — according to Eq. (11.1.2) — to a variation of 126 mV in the emf at 800 °C, assuming ideal behaviour. The condition that the other electrode is nonpolarizable is approximately satisfied for the Fe/FeO electrode, which takes up the oxygen entering from the zirconia, since FeO is formed relatively rapidly.

11.1.2.1 The Linear Geometry

The silver sample may have the form of a cylinder, one front area of it being in close contact with a zirconia pellet. To prevent radial diffusion out of the silver the silver cylinder must be surrounded by a material impermeable to oxygen, e.g. by Al_2O_3 (Fig. 11.1.1a). Thus, the diffusion of oxygen takes place only along the axis of the cylinder, which will be taken as the x-axis, the origin of which being located at the phase boundary $ZrO_2(Y_2O_3)/Ag$. Fick's diffusion coefficient of oxygen, sometimes merely called diffusion coefficient, can be determined by solving the diffusion equation (Fick's second law)

$$\frac{\partial c}{\partial t} = D \frac{\partial^2 c}{\partial x^2} \tag{11.1.4}$$

whereby the following initial and boundary conditions are applied:

$$c = c_0 \quad \text{for} \quad 0 \leqq x < \infty \quad \text{and} \quad t = 0,$$
$$c = 0 \quad \text{for} \quad x = 0 \quad \quad \text{and} \quad t > 0.$$

At time $t = 0$ the silver rod contains dissolved oxygen at concentration $c = c_0$. The oxygen concentration at the phase boundary $Ag/ZrO_2(+Y_2O_3)$ is kept at approximately zero for $t > 0$ due to potentiostatic performance. A diffusion-limited current flows through the cell; this phenomenon has been well-known for a long time in the subject of electrode kinetics in liquid-state electrochemistry. Within certain limits this diffusion-limited current must be independent of the potential difference applied to the cell; this has been established experimentally. In order to fulfill the boundary conditions chosen (semi-infinite linear geometry), the silver rod must be sufficiently long so that during the experiments the initial concentration c_0 remains constant at large distances from the origin. The solution of the diffusion Eq. (11.1.4) yields the concentration c of oxygen as a function

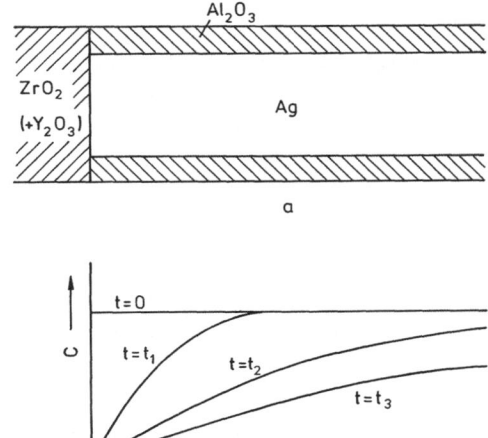

Fig. 11.1.1. a Set-up of the silver sample for the determination of the oxygen diffusion using a linear geometry. b Variations of the oxygen concentration in the direction of the cylinder axis x at different times t_0, t_1, t_2 etc.

of time t and of the distance x from the phase boundary zirconia/Ag with regard to the chosen initial and boundary conditions (see for example Crank [11.20])

$$c = c_0 \, erf \left(\frac{x}{2 \sqrt{Dt}} \right),$$ (11.1.5)

erf (z) is an abbreviated notation for the so-called error function which is defined by:

$$erf \, (z) = \frac{2}{\sqrt{\pi}} \int_0^z e^{-\xi^2} \, d\xi.$$ (11.1.6)

The variation of concentration c as a function of the distance x is sketched in Fig. 11.1.1b for three different times t. Applying Fick's first law, it follows from Eq. (11.1.5) that the diffusion flux j is connected with the concentration gradient of oxygen at the point x = 0 according to

$$|j| = D \left. \frac{\partial c}{\partial x} \right|_{x=0} = \frac{Dc_0}{\sqrt{\pi \, Dt}}.$$ (11.1.7)

Then the following expression results for the electrical current density i

$$i = |2Fj| = 2F \, \frac{c_0 \sqrt{D}}{\sqrt{\pi t}}.$$ (11.1.8)

According to Eq. (11.1.8) the electrical current density is proportional to $1/\sqrt{t}$, i.e. a plot of i versus $1/\sqrt{t}$ gives a straight line with the slope $2Fc_0 \sqrt{D/\pi}$. From this the value of the diffusion coefficient can be calculated, if the initial concentration c_0 is known.

Figure 11.1.2 shows the experimental set-up used. Measurements are performed with a potentiostat imposing constant potential differences on the cell. In Fig. 11.1.3 the current density is shown as a function of the reciprocal square root of time $1/\sqrt{t}$ resulting from the diffusion of oxygen out of silver at 800 °C at two different initial concentrations c_0 (using the linear geometry), namely 11.5×10^{-6} and 3.2×10^{-6} Mol O/cm³ Ag, corresponding to a previous saturation at 800 and 600 °C, respectively. From the measurements we can obtain values for the dif-

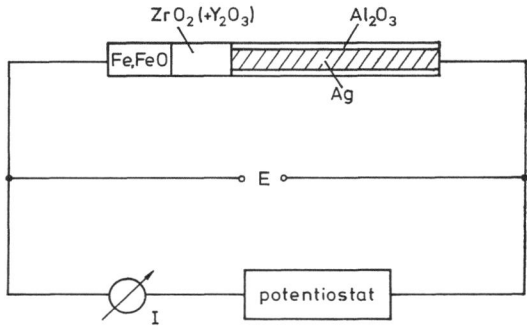

Fig. 11.1.2. Experimental arrangement (schematic) for the potentiostatic determination of oxygen diffusion in silver

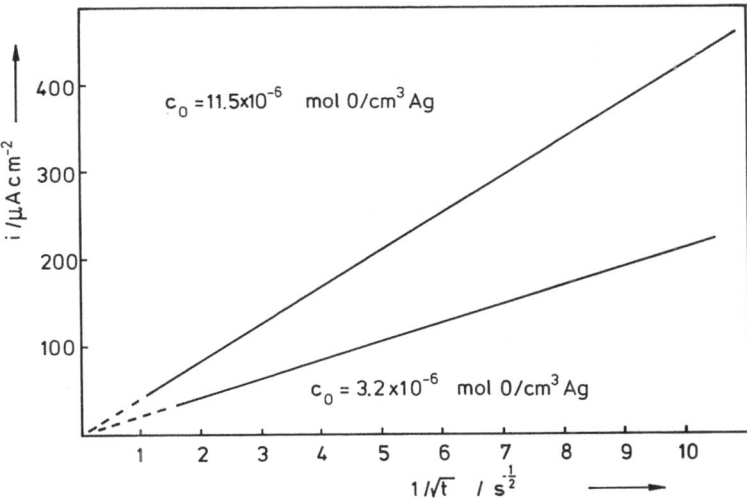

Fig. 11.1.3. Current density i as a function of $t^{-1/2}$ for the diffusion of oxygen in silver at 800 °C with two different initial concentrations c_0 using a linear geometry

fusion coefficient D of oxygen in silver at 800 °C: 1.9×10^{-5} cm^2 s^{-1} and 1.8×10^{-5} cm^2 s^{-1} respectively, in good agreement with values in the literature [11.21] determined with other methods.

11.1.2.2 The Cylindrical Geometry

The cylindrical geometry offers several advantages over the linear one, as will be discussed below. As shown in Fig. 11.1.4 the silver can be fused into a tube of zirconia; in this case oxygen diffuses radially out of the silver in the direction of increasing values of r, whereby the origin of this coordinate has been chosen to be on the cylinder axis. Apart from this the principle of investigation in the case of cylindrical geometry remains the same. Care must be taken that the ratio of the length of the tube to its diameter is large so that diffusion of oxygen through the ZrO$_2$-bottom area of the cylinder is not significant. The counter-electrode can for example be porous platinum surrounded by air at atmospheric pressure. The following boundary conditions must be taken into account for solving the diffusion equation (11.1.4)

$$c = c_0 \quad \text{for} \quad 0 < r < a \quad \text{and} \quad t = 0,$$
$$c = 0 \quad \text{for} \quad r = a \qquad \text{and} \quad t > 0$$

where a denotes the radius of the cylinder and c_0 the initial concentration of oxygen in silver. The plot current versus time may be evaluated in three different ways:

a) For small values of t oxygen has only diffused out of a thin surface layer of the silver cylinder. Hence, the situation is virtually the same as in the case of linear geometry so that the current-time curve may be evaluated correspondingly. Thus, if the current density i is plotted versus $1/\sqrt{t}$, one obtains the product $c_0 D$.

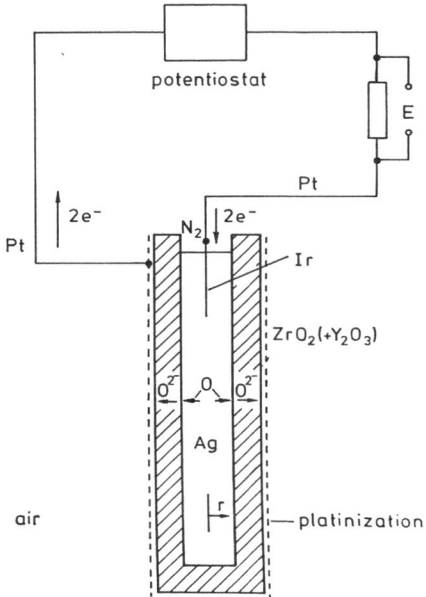

Fig. 11.1.4. Experimental arrangement (schematic) for the electrochemical measurement of oxygen diffusion in silver using a cylindrical geometry

b) For larger values of t the solution of the diffusion equation is as follows (see Crank [11.20]):

$$\frac{m_t}{m_\infty} = 1 - \frac{4}{\xi_1^2} \exp\left(-\frac{\xi_1^2 \, Dt}{a^2}\right). \tag{11.1.9}$$

Here, $m_t = \int_0^t J \, dt = \frac{1}{2F} \int_0^t I \, dt$ denotes the amount of substance, which has diffused out of the cylinder within time t; $m_\infty = \int_0^\infty J \, dt = $ const. is the amount of oxygen initially dissolved in the cylinder and $\xi_1 = 2.405$ the first root of the zero-order Bessel function.

If m_∞, the amount initially dissolved, is known and if m_t, the amount diffused out of the cylinder in time t, is measured, then the diffusion coefficient D may be calculated with Eq. (11.1.9). One can calculate the diffusion coefficient D without knowing the solubility. For this one must differentiate Eq. (11.1.9) with respect to t, whereby the following expression for the electric current I results:

$$I = A \exp\left(-\frac{t}{\tau}\right) \tag{11.1.10}$$

where

$$A = \frac{8 D m_\infty F}{a^2}, \tag{11.1.11}$$

$$\tau = \frac{a^2}{\xi_1^2 D}. \tag{11.1.12}$$

The time constant τ can be obtained from the slope of a logarithmic plot of current versus time; using Eq. (11.1.12) one can calculate the diffusion coefficient D without using the solubility. This is a particular advantage of the cylindrical geometry. By combining the results based upon linear geometry, which, as was stated above, prevails at the beginning of the experiments performed with a cylindrical set-up, with those relying on a cylindrical geometry, it is also possible to determine the initial concentration c_0.

c) The third type of evaluation involves the determination of the current-time integral which is equal to the area under the corresponding curve. This integral is a direct measure of the amount of oxygen initially dissolved in the metal sample, i.e. of m_∞ and thus also of the initial concentration c_0.

Figure 11.1.5 shows a semi-logarithmic plot of current I versus time t for two cylindrical samples at different temperatures (radii of the samples a = 2.55 mm and a = 3.5 mm respectively). The following values for the diffusion coefficient of oxygen as a function of temperature were obtained from these measurements:

$$D(760\,^\circ C) = 1.5 \times 10^{-5}\ cm^2/s, \qquad D(810\,^\circ C) = 2.0 \times 10^{-5}\ cm^2/s,$$

$$D(850\,^\circ C) = 2.3 \times 10^{-5}\ cm^2/s, \qquad D(885\,^\circ C) = 2.9 \times 10^{-5}\ cm^2/s,$$

$$D(900\,^\circ C) = 2.9 \times 10^{-5}\ cm^2/s.$$

A comparison of these values with those obtained using other experimental techniques demonstrates the reliability of the electrochemical method. Deviations of $\pm 20\%$ were observed in studies on different well-sealed samples while the accuracy of measurements on the same sample is characterized by errors of $\pm 5\%$; thus, the method itself is an exact one and allows small irregularities in the preparation of the sample to be distinguished.

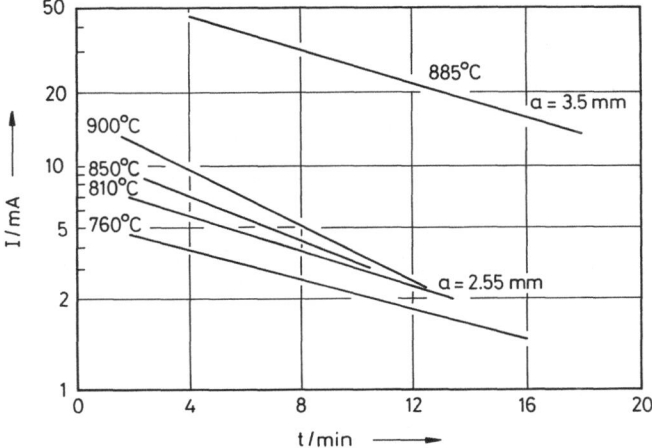

Fig. 11.1.5. Semi-logarithmic plot of current I as a function of time t for oxygen diffusion measurements in silver using a cylindrical geometry. Radii of samples: a = 2.55 mm and 3.5 mm; lengths of cylinders l = 5.95 and 7.0 cm respectively

It has been assumed in the evaluation of measurements that the stoichiometry of the zirconia is practically independent of oxygen pressure. In principle, a small variation in the zirconium-oxygen ratio is to be expected when the oxygen pressure is decreased. This means that when the potential is applied, additional current flows that alters the metal — to — oxygen ratio of the solid electrolyte. Some clarification of this point is necessary before measurements can be carried out on metals having a very small solubility of oxygen and, if necessary, a correction for this additional current can then be made.

Besides the potentiostatic measuring procedure it is also possible to apply a galvanostatic method, as will be discussed below.

11.1.3 The Galvanostatic Method

According to this method, a constant current is applied to the cell at a certain time $t = 0$, causing a constant diffusion current to occur. The emf of the cell is measured as a function of time. We shall first consider the linear geometry; this also serves as a good approximation of the cylindrical geometry for short times where diffusion only takes place within a thin layer of the metal near the zirconium dioxide/metal phase boundary. Choosing the phase boundary metal/zirconia to be the origin of a coordinate system with increasing values of the x-coordinate in the direction of the metal sample, the following boundary conditions are valid for the galvanostatic case:

$$c = c^0 \quad \text{for} \quad 0 \leq x < a \quad \text{and} \quad t = 0,$$
$$c = c^0 \quad \text{for} \quad x \to \infty \qquad \text{and} \quad t > 0, \tag{11.1.13}$$

$$\left(\frac{\partial c(x, t)}{\partial t} \right)_{x=0} = -\frac{i_0}{zFD} = k = \text{const.} \tag{11.1.14}$$

where i_0 denotes the constant electrical current density applied to the cell. Using the initial and boundary conditions, the following expression for the concentration $c(x, t)$ of oxygen as a function of x and t results [11.22]:

$$c(x, t) = c^0 - \frac{2kD^{1/2}t^{1/2}}{\pi^{1/2}} \exp\left(-\frac{x^2}{4Dt} \right) + kx \, \text{erfc} \left(\frac{x}{2D^{1/2}t^{1/2}} \right) \tag{11.1.15}$$

where erf $c(z)$ denotes the complement of the error function erf (z), and is defined as follows:

$$\text{erfc}(z) = 1 - \text{erf}(z) \tag{11.1.16}$$

The dependence of the emf E of cell (11.1.II) on the chemical potential of oxygen on both sides of the solid electrolyte is given according to the following equation:

$$\mu_{O_2}(\text{Me}) - \mu_{O_2}(\text{Fe, FeO}) = 4EF \tag{11.1.17}$$

The initial concentration of dissolved oxygen may be denoted as c^0, and the corresponding value of the emf as E^0. Then, the following relation holds:

$$\mu_{O_2}(\text{Me}, c = c^0) - \mu_{O_2}(\text{Fe, FeO}) = 4E^0F. \tag{11.1.18}$$

Using the definition of the chemical potential of oxygen according to

$$\mu_O = \mu_O^0 + RT \ln a_O \tag{11.1.19}$$

where a_O denotes the activity of oxygen, the following expression results with the aid of Eqs. (11.1.17—19)

$$\frac{RT}{2F} \ln \frac{a_O}{a_O^0} = E - E^0 \tag{11.1.20}$$

or

$$E = E^0 + \frac{RT}{2F} \ln \frac{a_O}{a_O^0}. \tag{11.1.21}$$

For low oxygen concentrations the following relationships are approximately valid [11.23]:

$$\frac{\sqrt{p_{O_2}}}{\sqrt{p_{O_2}^0}} \approx \frac{a_O}{a_O^0} \approx \frac{c}{c^0}. \tag{11.1.22}$$

Equation (11.1.21) may therefore be rewritten as follows

$$E = E^0 + \frac{RT}{2F} \ln \frac{c}{c^0}. \tag{11.1.23}$$

Inserting the oxygen concentration at $x = 0$ — given by Eq. (11.1.15) — into Eq. (11.1.13), the following relation holds for the potential difference of cell (11.1.II).

$$E = E^0 + \frac{RT}{2F} \ln \frac{1}{c^0} \left(c^0 - \frac{2kD^{1/2}t^{1/2}}{\pi^{1/2}} \right)$$

$$= E^0 + \frac{RT}{2F} \ln \frac{\tau^{1/2} - t^{1/2}}{\tau^{1/2}} \tag{11.1.24}$$

where

$$\tau^{1/2} = \frac{z\pi^{1/2}FD^{1/2}c^0}{2i_0}, \tag{11.1.25}$$

τ is the transition time. According to Eq. (11.1.24), the potential difference theoretically becomes infinite, if the denominator is zero; in this case, the time τ corresponds to the concentration $c(0, \tau) = 0$.

From measurements of the transition time an expression containing the product $D^{1/2}c^0$ may be obtained with the aid of Eq. (11.1.25).

The diffusion coefficient of oxygen in various metals has been determined by several authors, which applied the potentiostatic and/or the galvanostatic method. The results obtained from such measurements are listed in Table 11.1.1.

Table 11.1.1. Determination of the diffusion coefficient of oxygen in various solid and liquid metals. (The data and other information are listed according to the year of publication.)

Year	Investigator	Metal	Solid: s Liquid: l	Technique	Geometry	Temperature range/°C	Diffusion coefficient/ cm² s⁻¹	Ref.
1965	Rickert	Ag	s	Potentiostatic	Linear	800	1.8×10^{-5}	[11.1]
1966	Rickert, Steiner	Ag	s	Potentiostatic	Linear and cylindrical	760—900	$1.5 \times 10^{-5} - 2.9 \times 10^{-5}$	[11.2]
1967	Masson, Whiteway	Ag	l	Potentiometric	Linear	970—1200	$8.2 \times 10^{-5} - 1.7 \times 10^{-4}$	[11.7]
1968	Rickert, El Miligy	Ag Cu	l l	Potentiostatic and galvanostatic	Cylindrical	990—1220 1100—1250	$1.4 \times 10^{-4} - 2.2 \times 10^{-4}$ $6.2 \times 10^{-5} - 1.0 \times 10^{-4}$	[11.3] [11.4]
1969	Pastorek, Rapp	Cu	s	Potentiostatic and galvanostatic	Linear	800—1030 950—1030	$9.3 \times 10^{-6} - 3.5 \times 10^{-5}$ $2.8 \times 10^{-5} - 4.1 \times 10^{-5}$	[11.6]
1972	Ramanarayanan, Rapp	Sn Ag Ni	l s s	Potentiostatic	Cylindrical	750—950 750—950 1393	$4.5 \times 10^{-5} - 7.4 \times 10^{-5}$ $1.6 \times 10^{-5} - 4.1 \times 10^{-5}$ 1.3×10^{-6}	[11.8]
1970	Sano, Honma, Matsushita	Ag	l	Potentiometric	Linear	1000—1200	$9.6 \times 10^{-5} - 1.5 \times 10^{-4}$	[11.12]
1971	Sano, Honma, Matsushita	Pb	l	Potentiometric	Linear	800—1100	$1.0 \times 10^{-5} - 1.7 \times 10^{-5}$	[11.13]
1972	Szwarc, Oberg, Rapp	Pb	l	Potentiostatic		740—1080	$6.6 \times 10^{-5} - 1.4 \times 10^{-4}$	[11.9]
1972	Oberg, Friedman, Szwarc, Boorstein, Rapp	Fe	l	Potentiostatic		1620	1.5×10^{-4}	[11.10]
1972	Ramanarayanan, Rao, Tare	Cu	s	Potentiostatic	Linear	800—1000	$2.3 \times 10^{-6} - 7.6 \times 10^{-6}$	[11.11]

Year	Author	Metal		Method	Model	Temperature	D	Ref.
1973	Otsuka, Kozuka	Ag	l	Potentiometric	Linear	1000—1150	$9.0 \times 10^{-5} - 1.3 \times 10^{-4}$	[11.14]
1974	Ramanarayanan, Worrell	Cu	s	Galvanostatic		800—1000	$3.9 \times 10^{-6} - 1.5 \times 10^{-5}$	[11.15]
1975	Otsuka, Kozuka	Pb	l	Combined potentiostatic, potentiometric	Linear	900—1100	$2.0 \times 10^{-4} - 2.7 \times 10^{-4}$	[11.16]
1976	Otsuka, Kozuka	Cu	l	Combined potentiostatic, potentiometric	Linear	1125—1300	$7.8 \times 10^{-5} - 1.3 \times 10^{-4}$	[11.17]
1976	Kawahami, Goto	Fe	l	Galvanostatic		1550	1.9×10^{-4}	[11.18]
		In	l				$1.7 \times 10^{-6} - 4.5 \times 10^{-6}$	
		Ga	l				$4.3 \times 10^{-5} - 8.3 \times 10^{-5}$	
1981	Heshtmatpour	Sb	l	Potentiostatic	Cylindrical	750—950	$1.4 \times 10^{-6} - 2.9 \times 10^{-6}$	[11.19]
		Bi	l				$8.6 \times 10^{-6} - 1.4 \times 10^{-5}$	

11.2 Electrochemical Investigations on the Formation of Nickel Sulfide in the Solid-State at Higher Temperatures
[11.24]

We will consider the formation of nickel sulfide on a nickel surface to demonstrate that it is possible to investigate diffusion-controlled solid-state reactions electrochemically using a solid-state galvanic cell.

According to Dravnieks [11.25], the reaction of nickel with liquid sulfur obeys a parabolic rate law. The reaction layer mainly consists of NiS, with a very thin layer of NiS_2 between NiS and sulfur. The same result was obtained by Czerski, Mrowec and Werber [11.26] using sulfur vapour at a pressure of one atmosphere in the temperature range between 480 and 640 °C. In these cases, the formation of NiS is obviously diffusion-controlled. The chemical potential of sulfur at the phase boundary NiS/NiS_2 is determined by the equilibrium between these two phases. It is, however, very important for a theoretical analysis of the reaction mechanism, to vary the chemical potential of sulfur at the phase boundary nickel sulfide/sulfur, thus obtaining the dependence of the parabolic tarnishing constant k on the chemical potential of sulfur μ_S at this site (see also Chap. 8). This may be achieved by either varying the sulfur partial pressure or by using different H_2-H_2S mixtures. However, the electrochemical method described below is more advantageous; here, the rate of formation of NiS at the nickel surface is studied as a function of the chemical potential μ_S of sulfur at the phase boundary NiS/S [11.24].

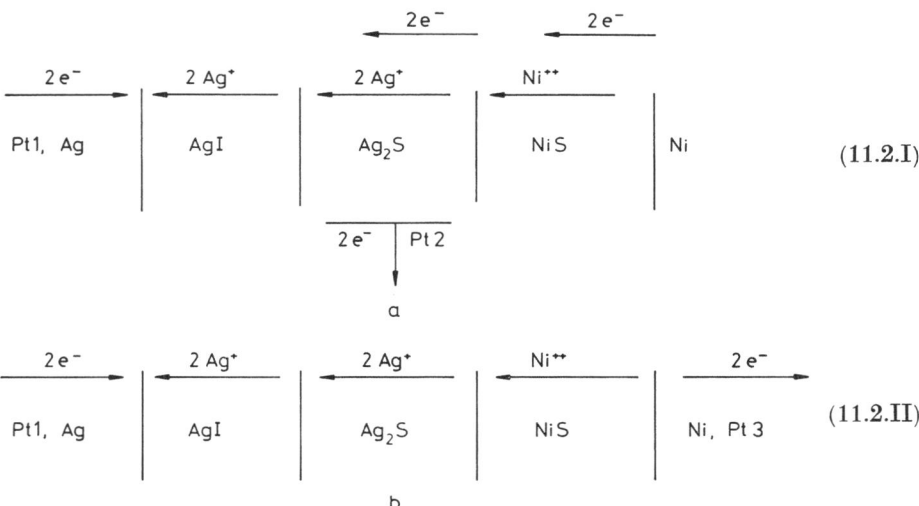

Fig. 11.2.1a, b. Solid-state galvanic cell for the investigation of the formation of solid NiS on the metallic nickel surface. **a** For the migration of nickel ions and electrons in NiS. **b** For the migration of nickel ions in NiS without simultaneous migration of electrons

This method exhibits several important advantages; thus, NiS grows one-dimensionally to form a compact layer at the front area of a nickel cylinder. Thus, the formation of a porous sulfide layer next to the metal will be suppressed, which is normally observed during the sulfidization of nickel.

The experimental set-up suitable for such investigations contains a solid-state galvanic cell sketched in Fig. 11.2.1; AgI serves as a solid electrolyte with virtually pure silver-ion conductivity. The negative pole is on the left-hand side at $Pt\,1$; the positive pole is formed either by $Pt\,2$ (case a) or $Pt\,3$ (case b). The electrical current passing through this cell is a measure of the rate, with which silver is removed from Ag_2S. The silver ions move through AgI and the electrons through leads $Pt\,2$ or $Pt\,3$ respectively. If the chemical potential of silver in Ag_2S, $\mu_{Ag}(Ag_2S)$ is smaller than that of pure silver μ_{Ag}^0 (this case corresponds to $E > 0$), the removal of silver from Ag_2S is equivalent to a delivery of sulfur by the Ag_2S, since at $300\,°C$ the latter can only exist with respect to the composition between $Ag_{2.0024}S$ and $Ag_{2.0000}S$. If, in addition, evaporation of sulfur can be excluded, as is the case with small values of the chemical potential of sulfur [11.27], the removal of silver from Ag_2S is also equivalent to the formation of NiS according to the equation

$$Ni + Ag_2S = NiS + 2\,Ag. \tag{11.2.1}$$

It is of course necessary to check that NiS is in fact the main product of the reaction between Ni and Ag_2S [11.24]. It must also be ensured that the presence of Ag_2S does not affect the reaction. The reaction mechanism proposed is sketched in Fig. 11.2.1.

Case a): nickel ions and electrons migrate through NiS; an equivalent current of silver ions and electrons flows out of the Ag_2S through AgI or lead $Pt\,2$ respectively and may thus be measured electrically.

Case b): only nickel ions migrate through NiS without simultaneous migration of electrons.

Cases a) and b) have led to the same experimental results; this means that the flow of electrons has no influence on the reaction rate so that electron equilibrium is virtually always attained.

Important experimental quantities are the current I flowing through the cell shown in Fig. 11.2.1 which is a measure of the reaction rate, and the emf of this cell, which is equivalent to the following cell (11.2.III) concerning E

$$Pt \mid Ag \mid AgI \mid Ag_2S \mid Pt. \tag{11.2.III}$$

The emf of (11.2.III), as has been discussed in Chap. 8, is primarily a measure of the chemical potential of silver in Ag_2S $\mu_{Ag}(Ag_2S)$, according to the equation

$$\mu_{Ag}(Ag_2S) - \mu_{Ag}^0 = -EF \tag{11.2.2}$$

where μ_{Ag}^0 is the chemical potential of pure silver and F Faraday's constant. The emf is also a measure of the chemical potential of sulfur in Ag_2S, $\mu_S(Ag_2S)$, as has been shown in Chap. 8.

$$\mu_S(Ag_2S) - \mu_S^0 = 2(E - E^*)\,F, \tag{11.2.3}$$

E* is the emf of the solid-state galvanic cell (11.2.III) for the equilibrium between Ag$_2$S and liquid sulfur. Since no ternary phases occur between Ag$_2$S and NiS [11.24], the chemical potential of the sulfur μ_S(Ag$_2$S) for equilibrium with Ag$_2$S is equal to that at the phase boundary NiS/Ag$_2$S, if thermodynamic equilibrium has been attained there and if no NiS$_2$ has been formed. Because of the high mobilities of silver ions and electrons, the potential drop in Ag$_2$S can be neglected. In contrast to previous methods for monitoring the progress of a solid-state reaction, in particular of tarnishing processes (e.g. measurement of the increase in the thickness of the layer by cooling the sample and preparing a polished section), the electrochemical method offers remarkable experimental feasibilities. Thus, via the limiting case of galvanostatic measurements, it is possible to stipulate a definite rate of reaction (11.2.1) by applying a constant current. The resulting chemical potential of sulfur in Ag$_2$S can then be measured. The other limiting case of potentiostatic performance is more related to the normal type of investigations.

11.2.1 Potentiostatic Measurements

A constant potential difference E is applied to the cell sketched in Fig. 11.2.1; the current I is then measured as a function of time. If the growth of the NiS-layer obeys a parabolic rate law, i.e. if thermodynamic equilibrium has been attained at the phase boundaries, Tammann [11.28] and Pilling and Bedworth [11.29] have pointed out that

$$dX/dt = k/X \qquad (11.2.4)$$

where X denotes the thickness of the NiS layer and t the time. The parabolic tarnishing constant k must be regarded as a function of temperature and of the emf E of cell (11.2.I) or (11.2.II). Since the current density i measured and the flux of nickel ions are equivalent to one another, the following relationship holds for the rate of growth of the NiS layer

$$dX/dt = jV_m = \frac{iV_m}{2F} \qquad (11.2.5)$$

where V_m is the molar volume of NiS and j the flux of nickel ions in moles per unit area and unit time. Thus, according to Eq. (11.2.5), measurement of the current density i allows a straightforward determination of the quantity dX/dt, i.e. the rate of growth of the NiS layer; it is only necessary to multiply the current density i by a known factor of constant magnitude. Previous methods of studying solid-state reactions, in particular scale formations, yield quantity X as a function of time t by measuring the thicknesses of the layers or by weighing the samples. The electrochemical method involving measurements of currents opens up new theoretical and practical possibilities, since it permits direct determination of dX/dt. Integration of Eq. (11.2.4) leads to

$$X^2 = 2kt. \qquad (11.2.6)$$

By combining Eqs. (11.2.4—6) the following expression for i results:

$$i = \frac{F}{V_m} \sqrt{\frac{2k}{t}}, \tag{11.2.7}$$

i.e. a plot of i versus $1/\sqrt{t}$ gives a straight line passing through the origin, the slope of which is $FV_m^{-1}\sqrt{2k}$, from which k may be determined. A second method of determining k is as follows: X is calculated as a function of time using the following relation obtained by integration of Eq. (11.2.5).

$$X = \frac{V_m}{2F} \int_0^t i \, dt. \tag{11.2.8}$$

Measurements are evaluated in the usual way, e.g. plotting X^2 versus t and determining k according to Eq. (11.2.6) from the slope of the resulting straight line.

11.2.2 Galvanostatic Measurements

In galvanostatic measurements a constant current, and thus also a constant current density, is applied; E is measured as a function of t.

We must then distinguish between the following two cases:

a) k is not a function of E. Up to a certain thickness of the layer corresponding to a definite period of time, displacement of silver would govern the rate of the removal of silver due to the current. Thus, silver would be liberated in Ag_2S and E would become equal to zero. For large values of t, the amcunt of silver removed from Ag_2S exceeds that formed during the displacement reaction (11.2.1). Thus, the silver originally liberated in Ag_2S will also be transported away. If this process is finished, the emf suddenly increases at time t_k and is determined by the evaporation of sulfur out of the Ag_2S.

At time t_k the amount of silver $\frac{it_k}{F}$ transported by the current is exactly equal to that generated by the displacement of silver due to the formation of NiS, i.e. $2X/V_m$:

$$\frac{it_k}{F} = \frac{2X}{V_m}. \tag{11.2.9}$$

It follows from Eqs. (11.2.9) and (11.2.6) that

$$\frac{it_k}{F} = 2 \frac{\sqrt{2kt_k}}{V_m}, \tag{11.2.10}$$

k may be determined with the aid of Eq. (11.2.10).

b) k depends on E. In this case the potential difference E is at first zero, as long as the layer thickness is small and the reaction rate and thus the displacement of silver is high. At larger values of t, the potential difference E increases; the value of k, however, also increases and is adjusted to the thickness of the layer. Hence, the following is valid:

$$\frac{dX}{dt} = \frac{k}{X} = \frac{iV_m}{2F} = const..$$ (11.2.11)

In spite of a parabolic rate law valid for E = const., the rate of growth of the NiS layer must be constant as long as i is constant apart, of course, from the initial stages.

Since Faraday's law applies, the following holds for the layer thickness

$$X = \frac{iV_m}{2F} t.$$ (11.2.12)

It follows from Eqs. (11.2.11) and (11.2.12) that

$$\frac{i^2 V_m^2}{4F^2} t = k.$$ (11.2.13)

According to Eqs. (11.2.12) and (11.2.13) X and k increase linearly with time t. Following Eq. (11.2.13) one obtains a series of values for the quantity $\left(\frac{iV_m}{2F}\right)^2 t$, which corresponds to the series of measured values of E. Thus, a plot of

$$\log\left[\left(\frac{iV_m}{2F}\right)^2 t\right]$$

versus E shows how log k depends on E. The interpretation of such a dependence is important for the theoretical discussion.

From Eq. (11.2.13) it can be concluded that equal values of $i^2 t$ correspond to equal values of k and thus also of E, i.e. for different current densities i_1 and i_2 the following equation must apply for times t_1 and t_2 belonging to equal potential differences

$$i_1^2 t_1 = i_2^2 t_2.$$ (11.2.14)

By using Eq. (11.2.14) it is for example possible to check if a parabolic rate law holds.

Experimental investigations have established that at 400 °C the parabolic tarnishing constant k increases by a factor 10, if the emf E increases by 67 mV. Thus, k is proportional to the activity of sulfur. This result is in agreement with the disorder model accepted for NiS; it is assumed that nickel ion vacancies are present, which are responsible for the diffusion of nickel ions, while the electrons exhibit intrinsic disorder, which is virtually independent of the chemical potential of nickel in NiS. From non-stationary experiments (decay curves) it is also possible to determine diffusion coefficients for nickel ion vacancies and to calculate the nickel deficit in NiS at a certain chemical potential of sulfur, e.g. for the equilibrium NiS/NiS_2; this will however not be discussed in detail here.

11.3 Electrochemical Investigations of the Transfer of Silver, Silver Ions and Electrons Across the Phase Boundary Solid Silver/Solid Silver Sulfide

In this section we use the solid-state galvanic cell

$$\text{Pt} \mid \text{Ag} \mid \text{AgI} \mid \text{Ag}_2\text{S} \mid \text{Pt} \qquad\qquad (11.3.\text{I})$$

to investigate the transfer of silver across the phase boundary $\text{Ag(s)}/\text{Ag}_2\text{S(s)}$. These experiments [11.30] aim at the determination of quantitative relationships between the chemical potential difference of silver, the electrochemical potential differences of silver ions and electrons, and the fluxes of neutral silver, of silver ions and of electrons across the phase boundary $\text{Ag(s)}/\text{Ag}_2\text{S(s)}$. The first two fluxes correspond to the rate of dissolution of silver at the phase boundary considered.

It is immediately obvious, that the problem stated is closely connected to the anodic dissolution of metals in electrolyte solutions; there is, however, one essential difference: an anodic current flowing through the phase boundary metal/aqueous solution is, on condition that other electrode reactions (e.g. redoxreactions) are not allowed to occur, a direct measure of the rate of dissolution of the metal. Silver sulfide, however, predominantly exhibits electronic conduction. The phase boundary under study is thus one where exchange of electrons is allowed to take place. Nevertheless, using the solid-state galvanic cell (11.3.I), i.e. applying an electrochemical method, it is possible to make definite statements even in this case.

The following properties of cell (11.3.I) are of main interest in our case:
At first, the emf of cell (11.3.I) is, according to Eq. (8.2.2), a measure of the chemical potential μ_{Ag} of silver in Ag_2S.

$$\mu_{\text{Ag}} - \mu_{\text{Ag}}^0 = -\text{EF}. \qquad\qquad (11.3.1)$$

Secondly, the current flowing through cell (11.3.I) — the right-hand side of the latter being connected with the positive pole of an external current source — is a measure of the rate of removal of silver from Ag_2S whereby silver ions migrate through AgI and electrons through the external circuit.

11.3.1 Simplified Experimental Arrangement

Figure 11.3.1 shows a simplified set-up for the investigation of polarization effects at the phase boundary $\text{Ag}/\text{Ag}_2\text{S}$.

With the aid of this set-up it is possible to measure the sum of the fluxes of silver ions and electrons, and thus of neutral silver through the phase boundary $\text{Ag}/\text{Ag}_2\text{S}$; in addition, we can determine the chemical potential difference of silver across this interface. The extension of this set-up will be discussed below.

In the experimental arrangement according to Fig. 11.3.1, a silver sulfide pellet with a thickness of some tenths of a millimetre lies on a silver iodide cylinder

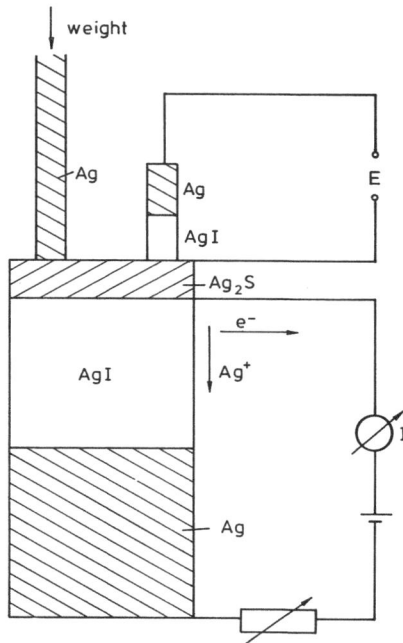

Fig. 11.3.1. Experimental arrangement for the electrochemical measurement of the transfer of silver across the phase boundary Ag/Ag_2S as a function of the chemical potential difference of silver across the interface

which, in turn, is placed on a silver cylinder. Platinum leads are connected with Ag_2S and Ag so that a current circuit is formed. The current can be adjusted by a resistor and measured with an amperemeter. Under current flow, silver is removed from Ag_2S; the chemical potential of silver in Ag_2S would continuously decrease whereby elementary sulfur would be caused to evaporize out of Ag_2S, if silver were not added to it from the other side. This is achieved by a silver rod placed upon the silver sulfide pellet. In the stationary state, the same amount of silver is transported from silver to Ag_2S as is withdrawn from the latter through AgI and platinum in the form of silver ions and electrons respectively. The steady state is distinguished by the fact that the chemical potential of silver in Ag_2S does not vary with time. This may be checked by using the Ag/AgI probe shown in the figure which, together with Ag_2S, forms a galvanic cell of type (11.3.I) the emf of which is, according to Eq. (11.3.1), a measure of the difference between the chemical potential of silver in silver sulfide and that of pure silver. Current and emf measurements are thus separated from one another so that measurements of the potential difference cannot be disturbed by possible polarization effects. The probe corresponds to a Luggin capillary, which has been used in liquid-state electrochemistry for many years.

11.3.2 Extended Experimental Arrangement

By the use of suitable additional probes it is possible to measure not only the chemical potential difference of neutral silver but also the electrochemical potential differences of silver ions and electrons and, furthermore, to allow silver ions and

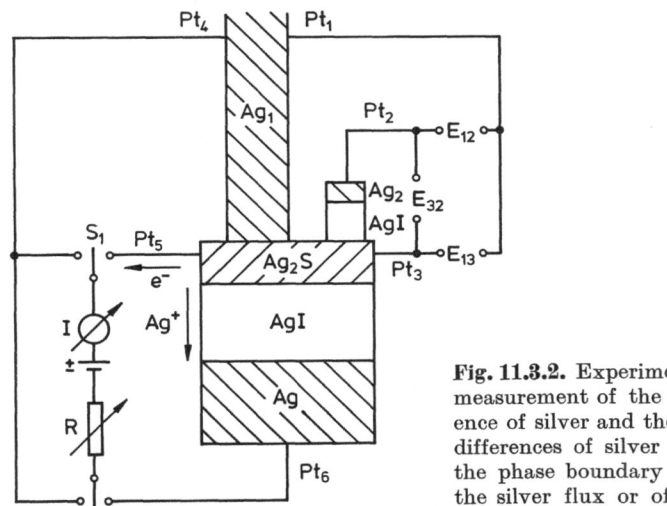

Fig. 11.3.2. Experimental arrangement for the measurement of the chemical potential difference of silver and the electrochemical potential differences of silver ions and electrons across the phase boundary Ag/Ag$_2$S as a function of the silver flux or of the current densities of silver ions or electrons

electrons to flow separately through the phase boundary. This is achieved with the set-up shown in Fig. 11.3.2. In order to measure the rate of transfer of silver ions alone an electric current is passed through the following cell [11.31]

$$\text{Pt} \mid \text{Ag} \mid \text{Ag}_2\text{S} \mid \text{AgI} \mid \text{Ag} \mid \text{Pt}. \qquad (11.3.\text{II})$$

When the positive pole of a current source is connected to the left-hand side of the cell, AgI, being a pure silver ion conductor, provides that under steady state conditions only Ag$^+$-ions are transported to the Ag$_2$S. Then, in order to obtain a silver ion flux, a current has to pass a sequence of different phases, as shown in cell (11.3.II). For this purpose, the left contact of switch S$_1$ and the right contact of switch S$_2$ were closed.

Conversely, the rate of transfer of electrons alone can be obtained using the cell

$$\text{Pt} \mid \text{Ag} \mid \text{Ag}_2\text{S} \mid \text{Pt} \qquad (11.3.\text{III})$$

with the negative pole of a current source connected to the left-hand side, where due to the presence of the phase boundary Ag$_2$S/Pt only electrons can move through the Ag$_2$S under steady state conditions, since only these particles are exchanged at the interface considered. In this case, the right contact of S$_1$ and the left contact of S$_2$ must be closed. In each case, apart from the chemical potential difference E$_{32}$F of silver, the electrochemical potential differences of both silver ions and electrons can be determined by measuring the electrical potential differences E$_{12}$ and E$_{13}$. Using the symbol φ for the electrical potential of the phase as indicated, we obtain

$$\text{E}_{12} = \varphi(\text{Pt}_1) - \varphi(\text{Pt}_2) = \varphi(\text{Ag}_1) - \varphi(\text{Ag}_2). \qquad (11.3.2)$$

Multiplying Eq. (11.3.2) by F and adding the term $\mu_{\text{Ag}^+}(\text{Ag}_1) - \mu_{\text{Ag}^+}(\text{Ag}_2)$, which is equal to zero, on the right-hand side of Eq. (11.3.2) and rearranging gives

$$\text{E}_{12}\text{F} = [\mu_{\text{Ag}^+}(\text{Ag}_1) + \text{F}\varphi(\text{Ag}_1)] - [\mu_{\text{Ag}^+}(\text{Ag}_2) + \text{F}\varphi(\text{Ag}_2)]. \qquad (11.3.2\text{a})$$

Introducing the electrochemical potential of the silver ions $\eta_{Ag^+} = \mu_{Ag^+} + F\varphi$, Eq. (11.3.2a) reads

$$E_{12}F = [\eta_{Ag^+}(Ag_1) - \eta_{Ag^+}(Ag_2)]. \tag{11.3.3}$$

Furthermore, since no current flows through the phase boundaries Ag_2S/AgI and AgI/Ag_2 the electrochemical potential of silver ions in these phases are the same; then, $\eta_{Ag^+}(Ag_2S) = \eta_{Ag^+}(Ag_2)$. It then follows from Eq. (11.3.3) that

$$E_{12}F = \eta_{Ag^+}(Ag_1) - \eta_{Ag^+}(Ag_2S). \tag{11.3.4}$$

For E_{13} one can write

$$E_{13} = \varphi(Pt_1) - \varphi(Pt_3). \tag{11.3.5a}$$

Multiplying Eq. (11.3.5a) by F and adding $\mu_{e^-}(Pt_3) - \mu_{e^-}(Pt_1)$, which is equal to zero, on the right-hand side of Eq. (11.3.5a) and rearranging leads to Eq. (11.3.5b)

$$E_{13}F = [\mu_{e^-}(Pt_3) - F\varphi(Pt_3)] - [\mu_{e^-}(Pt_1) - F\varphi(Pt_1)] \tag{11.3.5b}$$

Introducing the electrochemical potential of electrons $\eta_{e^-} = \mu_{e^-} - F\varphi$ it follows from Eq. (11.3.5b) that

$$E_{13}F = \eta_{e^-}(Pt_3) - \eta_{e^-}(Pt_1) = -[\eta_{e^-}(Pt_1) - \eta_{e^-}(Pt_3)]. \tag{11.3.6}$$

Due to the good conductivity of silver and platinum $\eta_{e^-}(Pt_1) = \eta_{e^-}(Ag_1)$. Since no current passes through the phase boundary Ag_2S/Pt_3 the electrochemical potentials of the electrons in the phases Ag_2S and Pt_3 are equal and $\eta_{e^-}(Ag_2S) = \eta_{e^-}(Pt_3)$. Thus, from Eq. (11.3.6) it follows that

$$-E_{13}F = \eta_{e^-}(Ag_1) - \eta_{e^-}(Ag_2S). \tag{11.3.7}$$

From Eqs. (11.3.4) and (11.3.7) it follows that the potential differences E_{12} and E_{13} multiplied by F are equal to the differences of the electrochemical potentials of Ag^+ and e^- between the two sides of the phase boundary Ag/Ag_2S. These electrochemical potential differences are used in the general formal theory as the driving forces for the transfer of Ag^+ and e^-.

11.3.3 Results and Discussion of Polarization Measurements at the Phase Boundary $Ag(s)/Ag_2S(s)$

The following results were obtained from these measurements using the galvanostatic and potentiostatic method, respectively [11.32].

a) If only electrons pass through the phase boundary $Ag(s)/Ag_2S(s)$ no significant emf values E_{12}, E_{13} and E_{32} are obtained for current densities up to 0.75 A/cm^2.

b) If only silver ions are transferred across the interface, E_{12} is nearly equal to E_{32}. Values between 0.1 and 0.2 V result, while E_{13} is lower than 0.02 V.

c) If equivalent amounts of silver ions and electrons are transferred, the emf values observed are equal to those obtained for pure silver ion transfer.

Figure 11.3.3 shows stationary current-density potential curves corresponding to case c), obtained at 200, 300 and 400 °C. The silver rod Ag_1 was pressed against the Ag_2S pellet with a load pressure of 7 kp/cm^2. Load-pressure variations between

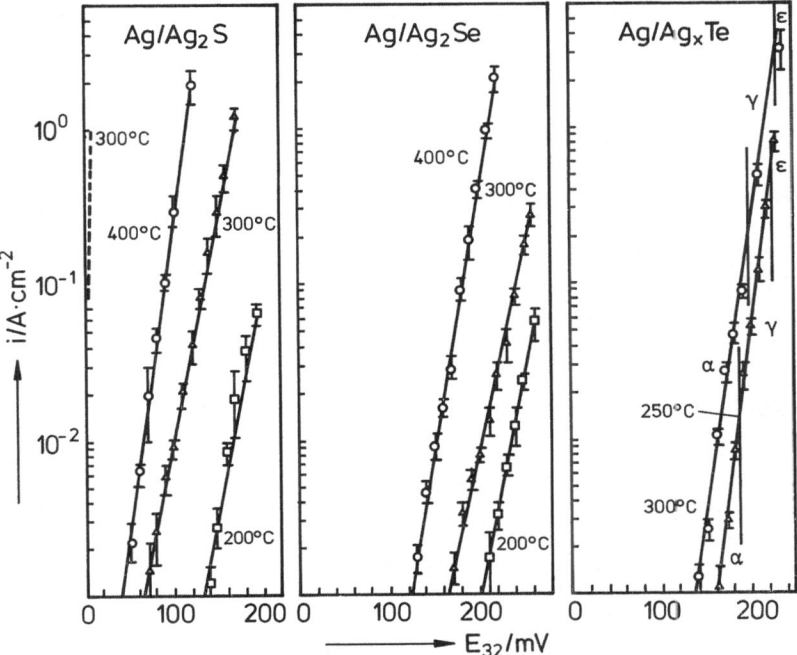

Fig. 11.3.3. Stationary current-density potential curves at 200, 300 and 400 °C. Pressure 7 kp cm^{-2}

1 and 10 kp/cm² show only a slight influence on the measuring points. At very high load pressures (more than 50 kp/cm²), the polarization effects of the dissolution of silver in Ag$_2$S nearly vanishes. A similar behaviour was found using in situ generated silver whiskers. The dotted line in Fig. 11.3.3 is an example of a high load pressure measurement (200 kp/cm²) at 300 °C.

The results of the high-pressure measurements where no polarization effects, occur indicate that the transfer of silver through the phase boundary Ag/Ag$_2$S is a very fast reaction step, if efficient microcontacts are present at the phase boundary. At high pressures, silver sulfide exhibits plastic deliquescence so that silver and silver sulfide are in close contact at all sites, i.e. we have a situation similar to that existing in liquids. It may be assumed that this very good contact partly disappears if there is no plastic deformation, expecially if half crystal positions on the silver surface loose their contacts with silver sulfide. Other mechanisms are then necessary to establish again the contact with the reacting sites. One of these may involve evaporation of S$_2$-molecules from silver sulfide and their diffusion to the silver over microdistances whereby reaction with the silver may take place, thus leading to new microcontacts for dissolution. This picture is consistent with the fact that for low pressures the current-potential curves have slopes corresponding to an increase of current by one order of magnitude, if the rise in voltage is about 30 mV. This shows that the stationary rates of dissolution of silver at low load pressures are proportional to a$_{S_2}$, which indicates a participation of S_2-molecules in the process. Further arguments which support this mechanism may be found in [11.32].

Using an analogous experimental set-up similar results were found for the transfer of silver across the phase boundaries Ag/Ag_2Se and Ag/Ag_2Te respectively [11.32]. The results for these systems are also given in Fig. 11.3.3.

The phase boundary reaction Ag/Ag_2Se has also been investigated by Mizusaki, Fueki and Mukaibo [11.33]. Similar studies on the phase boundary Cu/Cu_2S have been carried out by Donner and Rickert [11.34]. The rate of the heterogeneous reaction occurring at the interface $Ag_2S/S(l)$ has been determined by Sasaki, Mizusaki and Fueki [11.35]. The rate of the heterogeneous reaction at the phase boundary $Ag_2Se/Se(l)$ has also been determined [11.33].

11.4 Electrochemical Studies of Evaporation and Condensation

In this section we will take as a first example the evaporation of sulfur molecules from the silver sulfide surface in vacuo in the temperature range $200-400\,°C$. Subsequently, the reverse reaction, the incorporation of sulfur into Ag_2S from the gas phase, is discussed. The solid-state galvanic cell

$$Pt \mid Ag \mid AgI \mid Ag_2S \mid Pt \qquad (11.4.I)$$

in which silver iodide is a pure ionic conductor, has been taken as the basic system for these studies. This cell has both thermodynamic and kinetic properties which are important for the investigations described in this section.

The thermodynamic property is, as discussed in Chapter 8, that in open circuit, and also under current flow conditions when polarization effects are negligible, the emf of this cell is related to the chemical potential μ_{Ag} of silver in the Ag_2S sample according to

$$\mu_{Ag} - \mu_{Ag}^0 = -EF \qquad (11.4.1)$$

where μ_{Ag}^0 is the chemical potential of silver in the standard state, which is pure metallic silver.

With this cell it is also possible (see. Chap. 8) to measure the chemical potential of sulfur μ_S.

Let the emf of cell (11.4.I) for Ag_2S, being at equilibrium with sulfur, be $E*$. Then

$$\mu_S - \mu_S^0 = 2(E - E*)\,F. \qquad (11.4.2)$$

In this way the measurement of the emf yields the difference in the chemical potential of sulfur in silver sulfide and in the standard state μ_S^0, which is chosen as the chemical potential of liquid sulfur. The activity of sulfur, a_S, in Ag_2S is obtained by

$$\mu_S = \mu_S^0 + RT \ln a_S \qquad (11.4.3)$$

and from Eq. (11.4.2)

$$a_S = \exp\left(\frac{2(E - E*)\,F}{RT}\right). \qquad (11.4.4)$$

When the Ag$_2$S in cell (11.4.I) is equilibrated with the gas phase, the emf of this cell is not only a measure of the activity of sulfur within the Ag$_2$S. Due to the equilibrium between the Ag$_2$S and the surrounding gas the chemical potential of sulfur is the same in both phases. Thus the emf is also a measure of the activity of sulfur or of the partial pressure of sulfur molecules in the gaseous phase.

In kinetic applications, the electric current passing through cell (11.4.I) is a measure of the rate of addition to or removal of silver from the silver sulfide. In the steady state, i.e. when the silver-to-sulfur ratio in the sulfide remains constant, and when no other side reaction occurs, the silver lost from the silver sulfide corresponds to the sulfur evolved by the Ag$_2$S phase. The reverse occurs when sulfur is absorbed by the Ag$_2$S phase.

In the simplest case, cell (11.4.I) can be placed in vacuum in the temperature range 200—400 °C and, on passing a current through the cell, silver can be removed from the sulfide phase. In the steady state an equivalent amount of sulfur evaporates from the silver sulfide surface into the vacuum; thus, the current is a measure of the rate of evaporation [11.36]. This experiment is carried out with a set-up schematically shown in Fig. 11.4.1. This arrangement includes (11.4.I) cells. Current is passed through one cell (lower part of Fig. 11.4.1) and the magnitude of this current controls the rate of sulfur evaporation. Voltage measurements on the second cell (top section in Fig. 11.4.1) yield the chemical potential of silver and hence of sulfur in Ag$_2$S. If the negative pole of the current-carrying circuit (lower part of Fig. 11.4.1) is connected with the silver and the positive pole with the platinum sheet, which is in contact with the Ag$_2$S tablet, the current results in the withdrawal of silver from Ag$_2$S, the silver ions migrating through AgI and electrons through the platinum sheet. This is indicated by the arrows in

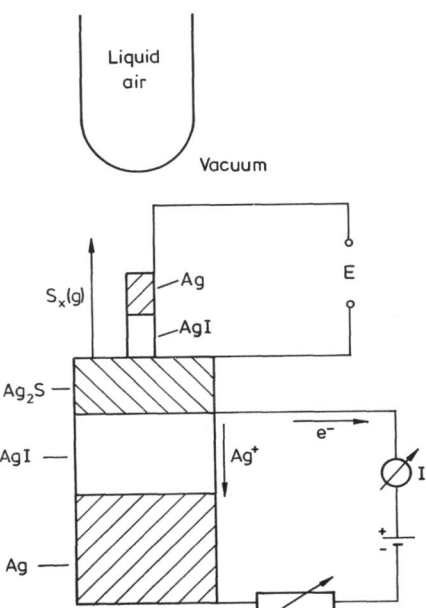

Fig. 11.4.1. Experimental arrangement for the electrochemical measurement of the rate of evaporation of sulfur from solid silver sulfide into vacuum as a function of the chemical potential of sulfur

Fig. 11.4.1. The current I can be controlled by a series of resistances and measured with an ammeter. Two processes now occur simultaneously. Firstly, silver is removed from Ag_2S through the passage of current. This alone causes a decrease of the silver potential in Ag_2S and a corresponding increase of the sulfur potential. Secondly, sulfur evaporates with an increasing rate from the Ag_2S surface as the sulfur potential rises. This evaporated material condenses as a gold-coloured film on the cold finger, which is placed above the sulfide. This evaporation process alone would cause the sulfur potential of the Ag_2S to decrease; thus, under the prevailing conditions, a steady state is set up. In the stationary state, which is automatically achieved, an amount of sulfur being evaporated is equivalent to the amount of silver withdrawn from the sulfide due to the passage of the current through the cell. The current in the steady state thus permits a direct measurement of the evaporation rate of the sulfur, i.e.

$$j = I/FA. \tag{11.4.5}$$

Here j is the evaporation rate of sulfur in equivalents per unit area and unit time, I the electrical current passing through cell (11.4.I), F Faraday's constant, and A the free surface are of the Ag_2S sample. The presence of the steady state is indicated by the fact that the silver and sulfur potentials in Ag_2S reach constant values. These potentials can be measured by means of the Ag/AgI probe (upper part of Fig. 11.4.1) which forms a cell of type (11.4.I) whose emf is related to the chemical potential of silver and that of sulfur through Eqs. (11.4.1) and (11.4.2). In this way current and potential measurements are carried out separately and errors in the potentials due to polarization effects are reduced to a minimum. The achievement of equilibrium in the Ag_2S tablets is very rapid as compared to the establishment of the steady state. Because of the large diffusion coefficient of the silver ions (10^{-5} cm^2 s^{-1}), the high electronic conductivity and the narrow range of stoichiometry, which gives a chemical diffusion coefficient of 10^{-1} cm^2 s^{-1} [11.37], potential differences in a 1 mm thick Ag_2S tablet are eliminated in less than one second. The achievement of a steady state depends on the magnitude of the current which is drawn, as well as the thickness of the tablet. The time taken is about one minute when large currents are drawn, but requires several hours for the smallest currents used.

It should be particularly noted that in this procedure the rate of evaporation of sulfur can be obtained at constant silver and constant sulfur potentials in the steady state and at certain different potentials. If the silver is not removed from Ag_2S the silver-to-sulfur ratio changes as sulfur vaporization occurs until silver is finally liberated. Then, silver sulfide is in equilibrium with pure silver, i.e. $\mu_{Ag} = \mu_{Ag}^0$. This is a typical example of thermal decomposition. The evaporation of sulfur which occurs under these conditions is for one sulfur potential only and no conclusion on the potential dependence of this reaction without further studies can be drawn. Moreover, a measurable rate of decomposition would only occur at high temperatures. The experimental results obtained at 300 °C by use of the electrochemical technique are shown in Fig. 11.4.2 [11.40]. Here, the current density, with reference to the exposed Ag_2S area, is plotted as a function of the emf of cell (11.4.I). Sulfur vapour contains a number of molecular species at thermodynamic equilibrium ranging from S_2 to S_8 [11.38]. From the kinetic

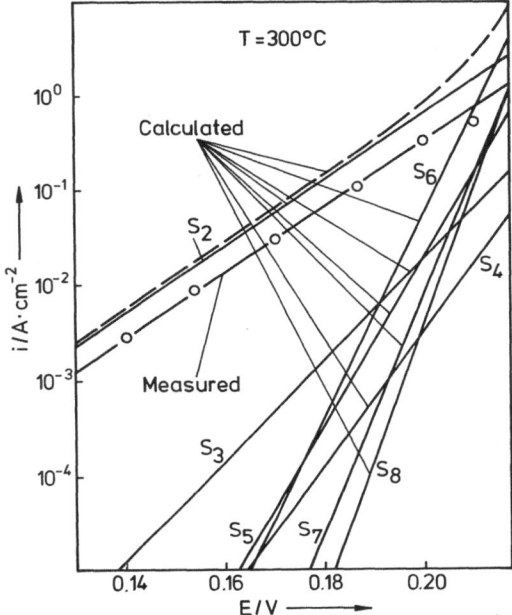

Fig. 11.4.2. ⊃, Experimental results; — — —, maximum theoretical rate for the evaporation of sulfur into vacuum from an Ag_2S surface represented as current density i as a function of the emf of cell (11.4.I). *Straight lines* represent the maximum partial current densities of the various sulfur species

theory of gases it is possible to calculate the maximum rate of evaporation of a single species as a function of the chemical potential of sulfur in Ag_2S. This rate is equal to the rate at which molecules strike the surface of an Ag_2S sample in equilibrium with a gas phase. The maximum possible rates of evaporation equal to the collision rates of sulfur molecules from an equilibrium vapour to the surface of an Ag_2S sample, are shown for each species as straight lines in Fig. 11.4.2. These have been converted into electrical units and shown as current densities. The equilibrium vapour pressures of the different S_x-molecules and thus the maximum possible rates of evaporation can be determined by using an electrochemical Knudsen cell described in the next section of this chapter. It can be seen that the measured rates are about a factor 1/2 smaller than the sum of these maximum possible rates which is denoted by a dashed line. According to early studies [11.36, 11.39] the measured rates were about a factor 10^{-2} smaller than the maximum possible rates but by improvement of the experimental set-up [11.40] larger rates could be obtained. These high rates of evaporation indicate that reactions in the adsorption layer are fast and that the adsorbed molecules are nearly in equilibrium with the solid Ag_2S. As a further example the evaporation of iodine from copper(I) iodide [11.41] will be discussed.

Electrochemical Study of the Evaporation of Iodine from Copper(I) Iodide

In this study the same method as that employed in the study of the evaporation of sulfur from silver sulfide was used. A cell with copper(I) bromide as solid electrolyte was applied. The galvanic cell was

$$Pt \mid Cu \mid CuBr \mid CuI \mid Pt. \tag{11.4.II}$$

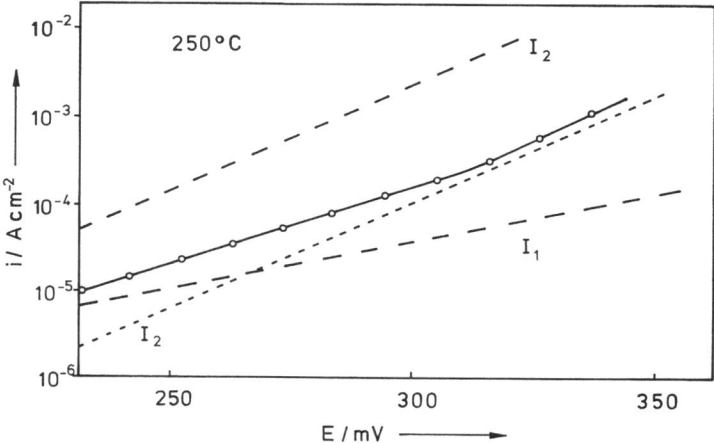

Fig. 11.4.3. Experimentally measured and maximum theoretical rates for iodine evaporation in electrical units as a function of the emf of cell (11.4.II) The *circles* denote the measured results, the *dotted line* the experimental rate of evaporation of I_2 molecules and the *dashed curves* the maximum theoretical rates for iodine atoms and iodine molecules

The electrical conduction of copper(I) bromide is almost entirely due to copper(I) ions [11.41]. Copper(I) iodide has a sufficiently large electronic conductivity under the conditions of the experiment to permit rapid equilibration. The results of the evaporation of iodine from CuI at 250 °C are illustrated in Fig. 11.4.3. According to these results the assumption is made that iodine atoms evaporate at the maximum evaporation rate whereas the rate of evaporation of iodine molecules, is by a factor of about ten less than the maximum rate. This indicates that the formation of I_2-molecules in the adsorption layer is the rate-determining step in their evaporation.

Electrochemical Study of the Incorporation of Sulfur Molecules into Solid Ag_2S from the Vapour Phase [11.40]

The set-up which was used for this investigation is shown in Fig. 11.4.4. Three cells of type (11.4.I) are assembled as shown in Fig. 11.4.4 so that the surfaces of the Ag_2S tablets form part of the inner walls of the evacuated vessel which is connected directly with the high vacuum system. The three cells, which are identical, have three separate functions. Cell *3* through which a predetermined current is passed is used to transport sulfur into the evacuated vessel. This supply of sulfur compensates for the loss of sulfur from the vessel due to evaporation into the vacuum system and the incorporation of sulfur into Ag_2S in cell *2*. In the steady state, a constant sulfur pressure inside the vessel is attained. The sulfur vapour in the vessel reaches equilibrium with cell *1* which is maintained under open-circuit conditions. According to Eq. (11.4.2), the emf of this cell is a measure of the chemical potential of sulfur in Ag_2S. Once equilibrium with the vapour phase has been established, as is soon achieved under steady-state conditions, this

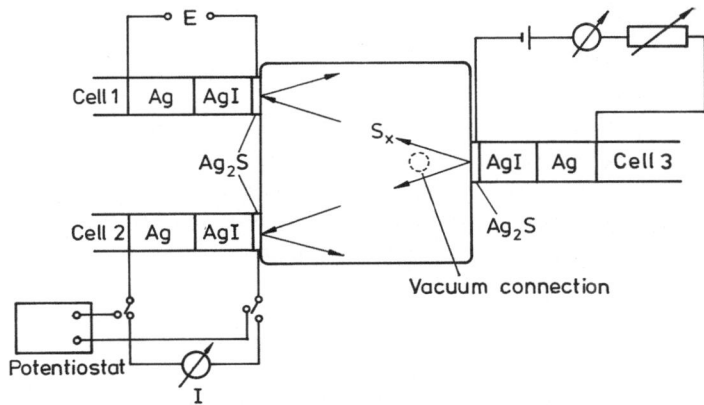

Fig. 11.4.4. Principle of the set-up for the electrochemical measurement of the rate of addition of sulfur to solid Ag_2S from the gas phase

cell is also a measure of the sulfur potential in the vapour phase and hence of the pressure of each individual sulfur species. This follows from the mass action law

$$p_{S_x} = a_S^x p_{S_x}^0 = p_{S_x}^0 \exp \left[x(\mu_S - \mu_S^0)/RT \right] \tag{11.4.6}$$

and from Eq. (11.4.4)

$$p_{S_x} = p_{S_x}^0 \exp \left[2x(E - E^*) \, F/RT \right] \tag{11.4.7}$$

where $p_{S_x}^0$ is the partial pressure of the S_x-molecules in the saturated sulfur vapour. Cell 2 is used to measure the rate of incorporation of sulfur into the Ag_2S tablet of this cell. For this purpose, the chemical potential of sulfur in Ag_2S is held at a low value, the silver being connected with the Ag_2S through the shunt. The resistance of the shunt was so low that the flowing current gave rise to a potential drop of a few millivolt which could be used for the measurement of the current. This current is thus a measure of the rate of transfer of silver to Ag_2S. Under steady-state conditions, this rate is equivalent to the rate of incorporation of sulfur into Ag_2S. The results reveal that from about two collisions on the surface one sulfur molecule is added under conditions (ca. 400 °C) where S_2 is the predominant species in the vapour.

Other electrochemical measurements of the rates of reactions, which occur at the phase boundary solid/gas, have been carried out using solid electrolytes. For example, the reaction between Ag_2S and H_2 leading to the formation of H_2S has been studied by Kobayashi and Wagner [11.42], Roy and Schmalzried [11.43], and Bechthold and Schwabe [11.44].

11.5 An Electrochemical Knudsen Cell for the Investigation of the Thermodynamics of Gases

In 1909, Knudsen [11.45] carried out kinetic measurements on equilibrium vapour pressures of mercury at very low pressures. A normal Knudsen cell sketched in Fig. 11.5.1 consists of a small vessel containing a condensed phase and a vapour phase in equilibrium with one another; outside the Knudsen cell is a vacuum, and the vessel itself contains a small hole connecting the interior of the cell with the vacuum. The vapour effuses out of this small hole from the Knudsen cell into the vacuum. The hole must be small compared to the surface area of the condensed phase so that equilibrium between vapour and condensed phase is always attained inside the vessel. It follows from gas kinetic considerations that j, the rate of effusion in moles per unit area and unit time is given by

$$j = \frac{p}{(2\pi MRT)^{1/2}} \tag{11.5.1}$$

where p is the vapour pressure inside the Knudsen cell, R the universal gas constant, T the thermodynamic temperature and M the mass per mole gas particles which effuse out of the cell. If different types of gas particles are present, Eq. (11.5.1) applies to each type and the total rate of effusion is given by the sum of the various effusion rates. It is thus evident that vapour pressures can easily be calculated only if the composition of the vapour is known, i.e. one must know whether the vapour consists of atoms, molecules or of mixtures of both. Further information is therefore necessary if the vapour phase has a more complicated composition, this being obtained by combining a normal Knudsen cell with a solid-state galvanic cell whereby an electrochemical Knudsen cell [11.46] is

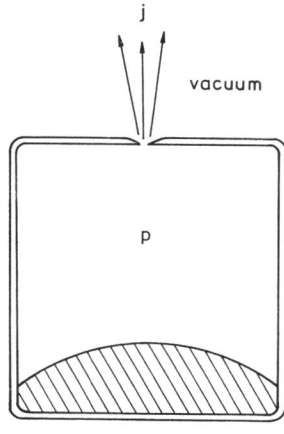

$$j = \frac{p}{\sqrt{2\pi MRT}}$$

Fig. 11.5.1. Schematic diagram of a Knudsen cell

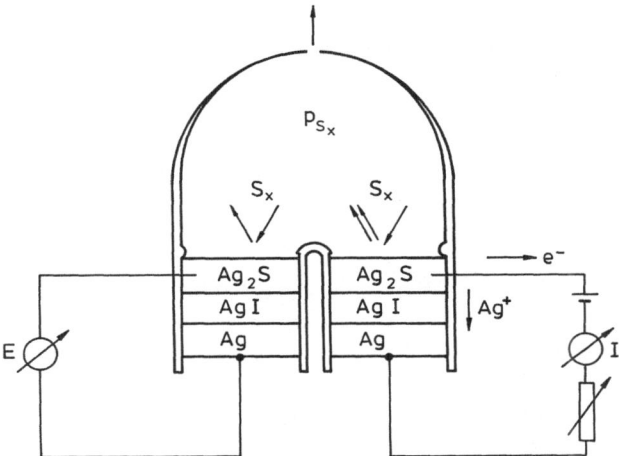

Fig. 11.5.2. Electrochemical Knudsen cell (schematic diagram)

formed. Further improvement is achieved by combination of an electrochemical Knudsen cell with a mass spectrometer [11.47].

In the following the principle of the electrochemical Knudsen cell will be demonstrated in the discussion of the investigations of the thermodynamics of sulfur vapour. An electrochemical Knudsen cell is shown schematically in Fig. 11.5.2; one of its important components is formed by the following solid-state galvanic cells:

$$\text{Pt} \mid \text{Ag} \mid \text{AgI} \mid \text{Ag}_2\text{S} \mid \text{Pt}. \tag{11.5.I}$$

These cells contain silver iodide, a virtually pure silver ion conductor. The Ag_2S pellets of cells (11.5.I) form parts of the inner surface of the Knudsen cell. Through one of these cells a definite current flows, thus causing sulfur to evaporize from the Ag_2S pellet and from the Knudsen cell; the second cell measures potential differences under open circuit conditions. Hence, measurements of currents in one cell and of emf's in the other simultaneously provide information on the rate of evaporation of sulfur from the cell and on the activity of sulfur or on the partial pressures of sulfur molecules within the Knudsen cell. Measuring these two quantities with the aid of two independent galvanic cells avoids failures in the potential measurements due to possible polarization effects or deviations from equilibrium. The particular significance of combining a Knudsen cell with solid-state galvanic cells lies in the fact that measurements can be performed not only at a certain chemical potential, and thus at a certain vapour pressure of sulfur, but, by changing the current through cell (11.5.I) and thereby the rate of effusion, these quantities may be varied. This technique thus provides new possibilities of determining the equilibrium partial pressure of the various types of sulfur molecules existing in the vapour; for each rate of effusion we can determine the chemical potential of the sulfur from emf measurements. In this way the dependence of the rate of effusion on the activity may be established. Emf measurements using cell (11.5.I) permit the determination of the chemical potential of silver in silver sulfide;

however, as has been shown in Chap. 8, data on the chemical potential of sulfur in Ag$_2$S are also obtained. Referring to the arguments given there, the following expression applies:

$$\mu_S - \mu_S^0 = 2(E - E^*)\, F \tag{11.5.2}$$

where μ_S is the chemical potential of sulfur in silver sulfide, μ_S^0 the chemical potential of sulfur under standard conditions, chosen to be liquid sulfur, E the emf of the cell, E* the emf for equilibrium of silver sulfide with liquid sulfur and F Faraday's constant.

If, under open circuit conditions equilibrium is attained between silver sulfide within the galvanic cell and sulfur molecules in the Knudsen cell (under stationary conditions this is always the case after a certain period of time), then the thermodynamic quantities of sulfur in silver sulfide are equal to those of sulfur in the vapour, i.e. inside the Knudsen cell. Using the definition of the activity a_S of sulfur atoms $\mu_S = \mu_S^0 + RT \ln a_S$, it follows from Eq. (11.5.2) that

$$a_S = \exp\left[2(E - E^*)\, F/RT\right]. \tag{11.5.3}$$

The equilibrium partial pressures p_{S_x} of the different types of sulfur molecules in the gas phase (x = number of atoms of the molecule) result from the relation $\mu_{S_x} = x\mu_S$ which is valid in the case of thermodynamic equilibrium

$$p_{S_x} = a_S^x p_{S_x}^0, \tag{11.5.4}$$

i.e. the partial pressure of S_x-sulfur molecules is obtained by multiplying the partial pressure $p_{S_x}^0$ in the saturated vapour by the activity of the sulfur atoms to the x-th power. It follows from Eqs. (11.5.3) and (11.5.4) that

$$p_{S_x} = p_{S_x}^0 \exp\left[2x(E - E^*)\, F/RT\right], \tag{11.5.5}$$

i.e. the partial pressures of the various types of sulfur molecules generally show an exponential dependence on the emf E of cell (11.5.I), the exponent being proportional to the number x, i.e. to the number of atoms in the type of molecules considered. Thus, each partial pressure is characterized by its emf dependence.

Under stationary conditions the electric current flowing through the second solid galvanic cell of type (11.5.I), with the positive pole of the current source attached to the Ag$_2$S pellet, is a measure of the rate of effusion of sulfur out of the Knudsen cell. Primarily, the current is a measure of the rate of removal of silver from silver sulfide out of which silver ions and electrons flow via AgI and platinum lead respectively (indicated in Fig. 11.5.2 by arrows); however, under stationary conditions this rate of removal of silver is equivalent to the rate of evaporation of sulfur from the pellet, and also from the Knudsen cell. Denoting the electrical current by the symbol I, the area of the hole in the Knudsen cell by the symbol A and the current density (with respect to this hole) by the symbol i, we obtain

$$i = \frac{I}{A} = \sum i_{S_x} = 2F \sum x j_{S_x} \tag{11.5.6}$$

where i_{S_x} is the partial current density corresponding to the S_x-type of sulfur molecules and j_{S_x} the rate of effusion of S_x-molecules per unit area out of the hole.

According to Eq. (11.5.1), we can write the following expression for the effusion fluxes:

$$j_{S_x} = \frac{p_{S_x}}{(2\pi M_{S_x} RT)^{1/2}}.$$ (11.5.7)

Characteristic of this method is the fact that the rate of effusion and thus the partial pressures within the Knudsen cell may be varied by means of the electrical current flowing through one galvanic cell, the corresponding emf being measured with the other cell. According to Eqs. (11.5.5—7) the following relationship holds for the dependence of the current densities i_{S_x} on the emf E:

$$\ln i_{S_x} = \ln 2xFj_{S_x}^0 + 2x(E - E^*) F/RT.$$ (11.5.8)

Thus, a plot of $\ln i_{S_x}$ versus the emf gives a straight line with a slope equal to $2xF/RT$. Using an electrochemical Knudsen cell alone, these partial current densities can, however, only be measured if a single type of molecules is present or predominant since the current measured is, according to Eq. (11.5.6), equal to the sum of all partial current densities.

In order to obtain further information the electrochemical Knudsen cell was combined with a mass spectrometer. The sulfur vapour effusing out of the Knudsen cell was ionized by electron impact; the intensities of the various ions were then measured in the usual way. The intensity $I_{S_x}^+$ of the S_x^+-ions, which are generated in the mass spectrometer is composed of two parts. The first part corresponds to S_x-molecules primarily present in the Knudsen cell, the other is due to the formation of S_x-fragments from higher sulfur polymers during ionisation. The intensity $I_{S_x}^+$ of primary ions generated by electron impact is related to the partial pressure p_{S_x} of the S_x-molecules in the Knudsen cell according to

$$I_{S_x}^+ = \frac{p_{S_x} A_{S_x}}{T}.$$ (11.5.9)

A_{S_x} denotes the sensitivity of the mass spectrometer with respect to the S_x-molecules.

This method thus provides a possibility of distinguishing between primary and fragment ions. Plotting $\ln I_{S_x}^+$ versus E yields, according to Eqs. (11.5.9) and (11.5.5), a straight line for the primary ions of each type of S_x-molecules with the characteristic slope $2xF/RT$. In contrast the intensity of S_x-fragment ions exhibits an emf dependence like that found for $S_{x'}$-molecules from which they are formed $(x' > x)$. Apart from the possibility of distinguishing between primary and fragment ions, the combination of an electrochemical Knudsen cell with a mass spectrometer also allows the mass spectrometer to be calibrated. If a certain type of S_x-molecules predominantes in a potential range, the total current density i_{tot} is nearly equal to i_{S_x}, and from Eqs. (11.5.6), (11.5.7), (11.5.4), and (11.5.9) we can determine the sensitivity A_{S_x} for this type of molecules. Then, the electric current is a direct measure of the pressure of this type of molecules, and we can thus determine the sensitivity of the mass spectrometer. If two types of molecules are present, whereby the sensitivity for one type is known, it is possible to determine the sensitivity for the other. If several types of molecules occur, the sensitivities of the other types of ions may be calculated by means of subtractions and by fitting the sum of the partial current densities to the total current density.

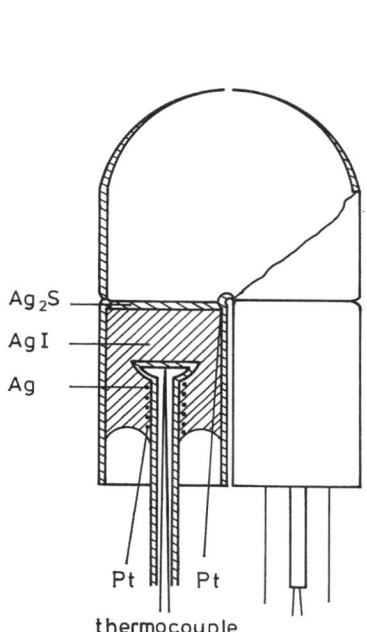

Fig. 11.5.3. Set-up of the electrochemical Knudsen cell

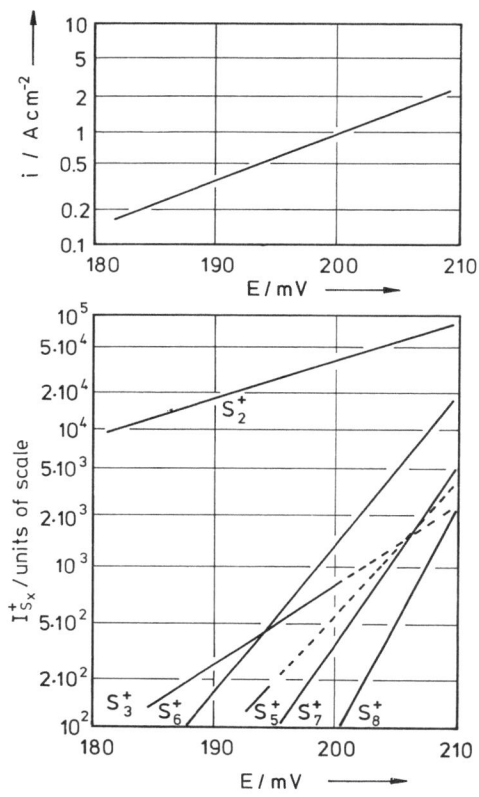

Fig. 11.5.4. Ion intensities of S_x-molecules and total current density i as a function of the emf E at 300 °C

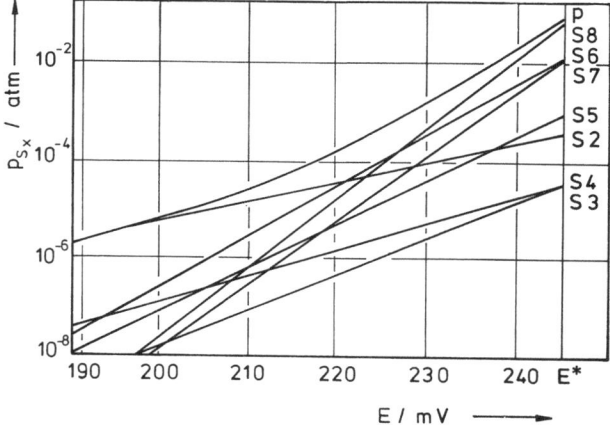

Fig. 11.5.5. Partial pressures p_{S_x} of S_x-molecules as a function of the emf E at 300 °C

For those types of ions, which make up no significant contribution to the total current density and thus to the total pressure over the full range studied, the sensitivities can only be estimated.

The experimental set-up of the electrochemical Knudsen cell is shown in Fig. 11.5.3. Since both sleeves containing the solid-state galvanic cells must be closed in a gas-tight manner, silver iodide was molten for a short period of time and then (without cooling), being in a solidified state, placed into an oven within the mass spectrometer and adjusted there.

Measurements were carried out between 200 and 400 °C. Figures 11.5.4 and 11.5.5 exemplarily show the results obtained from measurements carried out at 300 °C. In Fig. 11.5.4, the ion intensities of the different S_x-molecules and the total current density are plotted versus the emf. Figure 11.5.5 contains the partial pressures calculated from these values.

Extrapolating the straight lines to the saturation point corresponding to $E = E^*$, we obtain the saturation partial pressures $p_{S_x}^0$. Addition of these partial pressures gives the total pressures at various temperatures which are listed in Table 11.5.1; these values are in good agreement with the total pressures measured directly by West and Menzies [11.48]. However, in order to avoid larger inaccuracies in the total pressure, which are to be expected by simply adding partial pressures, the relative partial pressures were combined with the total pressure measured by West and Menzies. The absolute partial pressures obtained in this way are compiled in Table 11.5.2. Table 11.5.3 contains some thermodynamic quantities of the sulfur molecules calculated from these data.

Table 11.5.1. Total pressures of saturated sulfur vapour in the temperature range from 200 to 400 °C

$T/°C$	p_{tot}^0/atm
200	3.03×10^{-3}
250	1.96×10^{-2}
300	8.09×10^{-2}
350	2.43×10^{-1}
400	6.11×10^{-1}

Table 11.5.2. Total pressures p_{tot}^0 according to West and Menzies [11.48] and partial pressures $p_{S_x}^0$ of saturated sulfur vapour in the temperature range from 200 to 400 °C

$T/°C$	200	250	300	350	400
$p_{S_2}^0$	1.40×10^{-6}	2.60×10^{-5}	2.68×10^{-4}	1.90×10^{-3}	9.40×10^{-3}
$p_{S_3}^0$	1.70×10^{-7}	3.38×10^{-6}	3.66×10^{-5}	2.68×10^{-4}	1.34×10^{-3}
$p_{S_4}^0$	2.65×10^{-7}	3.04×10^{-6}	3.25×10^{-5}	2.15×10^{-4}	1.04×10^{-3}
$p_{S_5}^0$	1.56×10^{-5}	1.72×10^{-4}	9.64×10^{-4}	4.20×10^{-3}	1.43×10^{-2}
$p_{S_6}^0$	5.50×10^{-4}	3.60×10^{-3}	1.60×10^{-2}	5.25×10^{-2}	1.37×10^{-1}
$p_{S_7}^0$	3.28×10^{-4}	2.63×10^{-3}	1.27×10^{-2}	4.55×10^{-2}	1.26×10^{-1}
$p_{S_8}^0$	1.89×10^{-3}	1.02×10^{-2}	3.64×10^{-2}	9.70×10^{-2}	2.14×10^{-1}
p_{tot}^0	2.79×10^{-3}	1.66×10^{-2}	6.62×10^{-2}	2.00×10^{-1}	4.98×10^{-1}

Table 11.5.3. Thermodynamic data of sulfur molecules: Standard reaction enthalpy ΔH_T^0, standard reaction entropy ΔS_T^0, standard entropy of S_x-molecules $S_T^0(S_x)$; standard enthalpy of the evaporation of S_x-molecules $\Delta_v H_{0K}^0(S_x)$

Molecule	Equilibrium reaction	T/K	ΔH_T^0 / kJ mol^{-1}	ΔS_T^0 / JK^{-1} mol^{-1}	$S_T^0(S_x)^{[a]}$ / JK^{-1} mol^{-1}	$\Delta_v H_{0K}^0(S_x)$ / kJ mol^{-1}
S_2	$2S(l) \rightleftharpoons S_2(g)$	460—670	117.4 ± 1.5	32.6	59.6 (565 K)	129.12 ± 1.46
S_3	$2S_3(g) \rightleftharpoons 3S_2(g)$	566—669	111.2 ± 8.4	(37.6 ± 3.0)	(71.8 ± 1.5) (618 K)	139.3 ± 5.4
S_4	$S_4(g) \rightleftharpoons 2S_2(g)$	615	117.9 ± 8.4	(36.7 ± 2.5)	(83.9 ± 2.5) (615 K)	140.2 ± 9.2
S_5	$2S_5(g) \rightleftharpoons 5S_2(g)$	565—620	399.2 ± 16.7	(111.4 ± 8.0)	(94.3 ± 4.0) (592 K)	121.8 ± 10.5
S_6	$\frac{3}{4}S_8(g) \rightleftharpoons S_6(g)$	435—625	26.2 ± 1.4	(7.6 ± 0.8)	(101.6 ± 3.3) (512 K)	103.8 ± 11.3
S_7	$\frac{7}{8}S_8(g) \rightleftharpoons S_7(g)$	435—625	24.1 ± 1.3	7.2 ± 1.0	116.9 ± 3.8 (512 K)	117.6 ± 11.3
S_8	$S_8(g) \rightleftharpoons 4S_2(g)$	460—625	406.6 ± 17.6	110.2 ± 4.2	126.8 ± 4.2 (530 K)	108.8 ± 11.3

[a] The values for S_5, S_7, S_8 were calculated from experimental data and from literature values (see Sect. 11.5.3). Entropy values of S_2, S_3, S_4 and S_5 were calculated on the basis of statistical thermodynamics.

Table 11.5.4. Saturation partial pressures $p_{Se_x}^0$ of selenium molecules

T/°C	200	250	300	350	400	450
$p_{Se_2}^0$	3.52×10^{-8}	9.43×10^{-7}	1.27×10^{-5}	1.07×10^{-4}	6.51×10^{-4}	3.02×10^{-3}
$p_{Se_3}^0$	4.20×10^{-11}	2.36×10^{-9}	5.56×10^{-8}	7.36×10^{-7}	6.63×10^{-6}	4.43×10^{-3}
$p_{Se_4}^0$	9.69×10^{-11}	6.14×10^{-9}	1.45×10^{-7}	1.90×10^{-6}	1.70×10^{-5}	1.09×10^{-4}
$p_{Se_5}^0$	3.03×10^{-7}	5.24×10^{-6}	4.07×10^{-5}	2.02×10^{-4}	7.97×10^{-4}	2.53×10^{-3}
$p_{Se_6}^0$	1.48×10^{-6}	2.11×10^{-5}	1.33×10^{-4}	5.57×10^{-4}	1.86×10^{-3}	4.89×10^{-3}
$p_{Se_7}^0$	3.98×10^{-7}	6.58×10^{-6}	4.34×10^{-5}	1.88×10^{-4}	6.49×10^{-4}	1.80×10^{-3}
$p_{Se_8}^0$	3.57×10^{-8}	7.25×10^{-7}	5.18×10^{-6}	2.20×10^{-5}	8.01×10^{-5}	2.36×10^{-4}
p_{tot}^0	2.25×10^{-6}	3.46×10^{-5}	2.35×10^{-4}	1.08×10^{-3}	4.06×10^{-3}	1.27×10^{-2}

Similar investigations on selenium vapours have revealed the existence of molecules like Se_2, Se_3, Se_4, Se_5, Se_6, Se_7 and Se_8. The saturation partial pressures of the various Se_x-molecules are listed in Table 11.5.4. Thermodynamic quantities have also been calculated [11.50]. The use of electrochemical Knudsen cells in studies of other systems, particularly oxide systems, seems to be possible.

11.6 Electrochemical Measurements of the Chemical Diffusion Coefficient of Wüstite and Silver Sulfide

The electrochemical method for determining the chemical diffusion coefficient described in Sect. 6.1.10 will be applied to the model substances FeO and Ag_2S [11.51]. The basic element of the set-up used for such measurements on $Fe_{1-\delta}O$ is the solid-state galvanic cell [11.52]

$$\text{Pt } 1, p_{O_2} \mid ZrO_2(+10 \text{ Mol}\% Y_2O_3) \mid Fe_{1-\delta}O \mid N_2, \text{Pt } 2 \qquad (11.6.I)$$

containing doped ZrO_2 as a solid auxiliary electrolyte with virtually pure oxygen ion conductivity. On one side of the cell is a mounted Pt-electrode surrounded by air; the other electrode consists of the compound to be studied, i.e. wüstite. The principle of the measurements is that, starting from a suitable initial state, the potential difference E or the current I respectively of cell (11.6.I) are varied systematically; the variable I or E respectively is measured as a function of time and analyzed with respect to the diffusion coefficient of wüstite. At the beginning of the measurements, a certain chemical potential of oxygen was first set up throughout the sample, due to the application of a certain potential difference to the cell. The emf E of the cell, which is a measure of the oxygen concentration in $Fe_{1-\delta}O$ at the phase boundary $Fe_{1-\delta}O/ZrO_2$, was then suddenly increased by a value of $\Delta E = 5 \cdots 20 \text{ mV}$; the resulting current was then measured as a function of time. The current is a measure of the rate at which oxygen is exchanged via the phase boundary ZrO_2/FeO, and thus of the diffusion of iron from the interior of the compound to this phase boundary or vice versa. In this process the stoichiometry of $Fe_{1-\delta}O$ changes as a function of time until a new δ-value is attained.

Applying the following initial and boundary conditions for the concentration c of the metal

$$c(x = 0, t) = c' \quad \text{for} \quad t > 0,$$

$$c(x, t = 0) = c'' \quad \text{for} \quad 0 < x < l, \qquad (11.6.1)$$

$$\left(\frac{\partial c}{\partial x}\right)_{x=1} = 0 \quad \text{for} \quad t > 0,$$

we obtain as the solution of Fick's second law (6.1.5) the following expression for the time dependence of the electrical current I on condition that $t \ll l^2/\tilde{D}$ [11.53]:

$$I = 2A\, e(c' - c'')\left(\frac{\tilde{D}}{\pi t}\right)^{1/2} \quad \left(\text{if } t \ll \frac{l^2}{\tilde{D}}\right), \qquad (11.6.2)$$

A is the cross sectional area of the metal oxide sample, l its length and e the elementary charge. Denoting the total amount of charge flown during the relaxation process by Q which, apart from direct measurements, may also be obtained from coulometric titrations [11.54], Eq. (11.6.2) can be rewritten to

$$I = \frac{Q}{l}\left(\frac{\tilde{D}}{\pi t}\right)^{1/2} \quad \left(\text{if } t \ll \frac{l^2}{\tilde{D}}\right). \qquad (11.6.3)$$

Here, we have made use of the fact that Q equals $2e \cdot V \cdot (c' - c'')$, where V is the volume of the sample. According to Eqs. (11.6.2) and (11.6.3) the current is proportional to the reciprocal of the square root of time t for small values of t. If $t > l^2/4\tilde{D}$, the following expression is valid [11.55]:

$$\lg I = \lg \frac{4A\, e\tilde{D}(c' - c'')}{l} - \frac{1.071\tilde{D}}{l^2}t \quad \left(t > \frac{l^2}{4\tilde{D}}\right), \qquad (11.6.4)$$

i.e. if $t > l^2/4\tilde{D}$ the current decreases exponentially. From the slopes of the straight lines resulting from a $\lg I$ versus t plot, the chemical diffusion coefficient can be determined, even without knowledge of the concentration difference $c' - c''$. The intercept with the ordinate axis may also be used to calculate $\tilde{D}$; however, in this case, $c' - c''$ must be known. The thermodynamic factor $d \ln a_{Me}/d \ln c_{Me}$, which is essential for a comparison of the chemical with the component or tracer diffusion coefficient, can be obtained as a function of stoichiometry by performing a coulometric titration of cell (11.6.I). A plot of E versus δ yields $dE/d\delta$, the latter quantity being connected with the thermodynamic factor according to:

$$\frac{d \ln a_{Me}}{d \ln c_{Me}} = \frac{2F}{RT}\frac{dE}{d\delta}. \qquad (11.6.5)$$

The experimental arrangement is sketched in Fig. 11.6.1. The course of a typical relaxation process at T = 1000°C and δ = 0.106 is shown in Figs. 11.6.2—4. The agreement with theory is good for small and large values of t. The following values were obtained for the chemical diffusion coefficient $\tilde{D}$ at T = 1000°C and the deviation δ = 0.106 from ideal stoichiometry: $\tilde{D}$ = 3.4 × 10^{-6} cm² s⁻¹ for small values of t and 3.1 × 10^{-6} cm² s⁻¹ for large values. From the intercept with the ordinate axis one obtains: $\tilde{D}$ = 3.2 × 10^{-6} cm² s⁻¹. The value of $\tilde{D}$ calculated on the basis of tracer diffusion measurements [11.56], using the correlation factor f = 0.78 and the thermodynamic factor d ln a_{Fe}/ d ln c_{Fe} = 32, is 3.5 × 10^{-6} cm² s⁻¹.

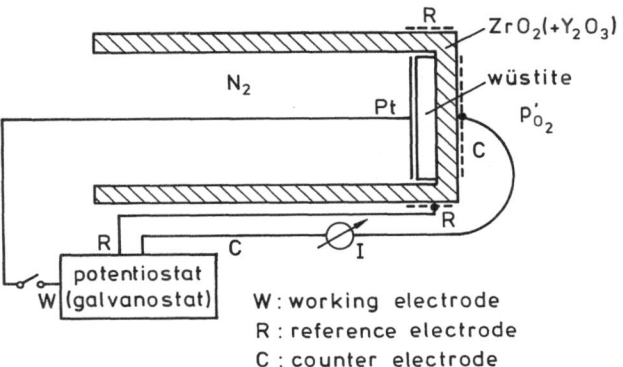

Fig. 11.6.1. Experimental arrangement (schematic) for the measurement of the chemical diffusion coefficient of wüstite

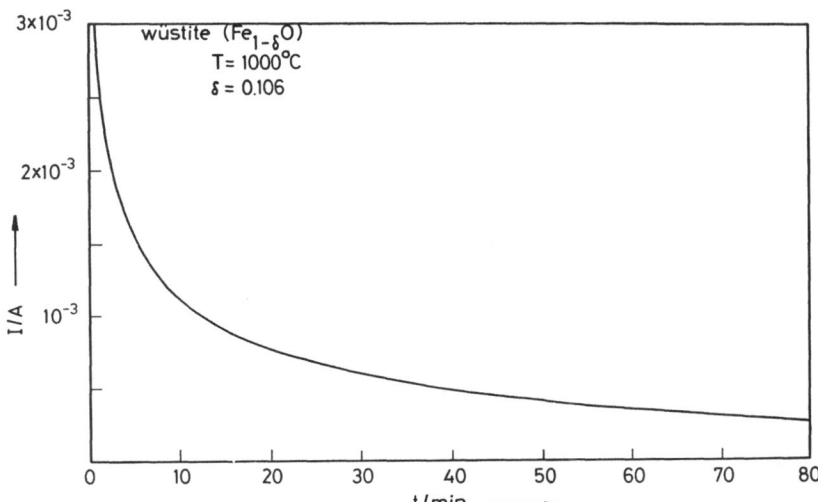

Fig. 11.6.2. Dependence of current I on time t after a change of the potential difference ΔE = 20 mV using cell (11.6.I) at T = 1000°C and δ = 0.106 for the determination of the chemical diffusion coefficient of FeO

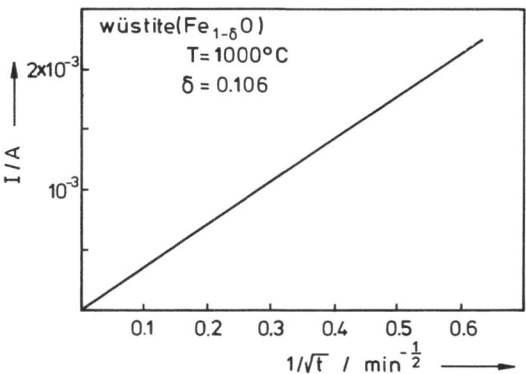

Fig. 11.6.3.
Relaxation behaviour at small times t using the experimental data from Fig. 11.6.2

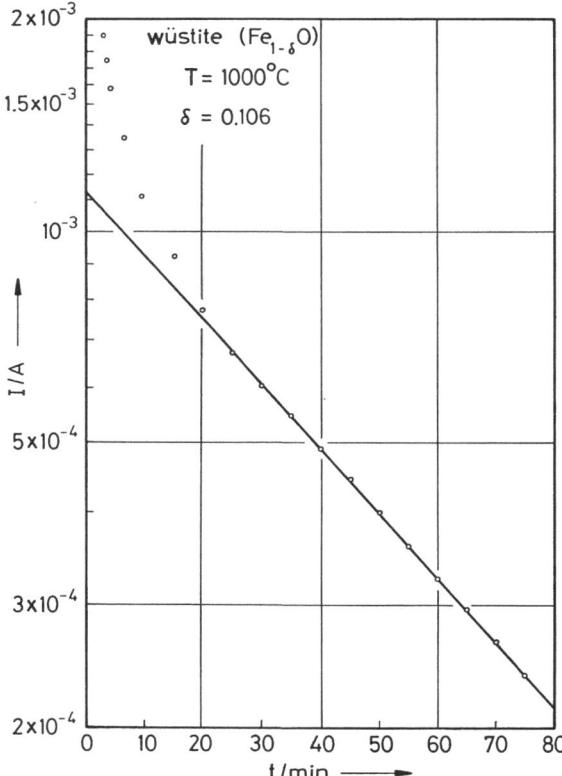

Fig. 11.6.4. Logarithmic plot of the experimental values of current I (from Fig. 11.6.2) as a function of time t for the calculation of $\tilde{D}$ from the slope and from the intercept on the ordinate axis respectively of the resulting straight line for $t > l^2/4\tilde{D}$

The basic element of the arrangement used for measuring the chemical diffusion coefficient of Ag_2S was the galvanic cell

$$Pt \mid Ag \mid AgI \mid Ag_{2+\delta}S \mid Pt \qquad (11.6.II)$$

If polarization effects can be neglected, i.e. under open circuit conditions, the emf E is a measure of the chemical potential of silver in Ag_2S

$$-EF = \mu_{Ag} - \mu_{Ag}^0 \qquad (11.6.6)$$

where μ_{Ag}^0 is the chemical potential of pure silver. A current flowing through the cell is a measure of the rate at which silver is added to or removed from $Ag_{2+\delta}S$. These two properties are essential for relaxation measurements.

The following galvanic cell was used for measuring the chemical diffusion coefficient of $Ag_{2+\delta}S$, the latter phase being equilibrated with liquid sulfur on one side:

$$Pt \mid Ag \mid AgI \mid Ag_{2+\delta}S \mid S(l). \qquad (11.6.III)$$
$$\mid$$
$$Pt$$

At the beginning of measurements a constant current or a constant potential difference was applied; as a result, a stationary concentration gradient of silver was built up between the phase boundaries $Ag_{2+\delta}S/AgI$ and $Ag_{2+\delta}S/S(l)$. The voltage applied in these experiments differed from the value of the emf in the case of equilibrium with sulfur only by $10-20$ mV. The external circuit was then disconnected. From the time-dependence of the emf of the cell, the variation of the chemical potential of silver and thus the change of the Ag-concentration (the relation between these quantities follows from coulometric titration curves) were measured as a function of time.

The theoretical relationship giving the concentration as a function of time may be obtained by the solution of Fick's second law using the initial and boundary conditions

$$c = c'' + (c' - c'') \frac{x}{l} \quad \text{for} \quad t = 0$$

$$c = c' \qquad\qquad\quad \text{for} \quad x = l, t > 0 \qquad (11.6.7)$$

$$\left(\frac{\partial c}{\partial x}\right)_{x=0} = 0 \qquad\quad \text{for} \quad t > 0$$

The solution is as follows [11.57]:

$$c = c'' - \frac{2(c'' - c')}{l}\left(\frac{\tilde{D}t}{\pi}\right)^{1/2} \quad \left(t \ll \frac{4l^2}{\pi^2\tilde{D}}\right). \qquad (11.6.8)$$

The experimental arrangement is shown in Fig. 11.6.5. In order to avoid polarization effects, a separate Ag/AgI probe was used for the emf measurements. Prior to the actual experiments, silver sulfide was grown one-dimensionally into a glass tube. Figure 11.6.6 shows the relaxation of the emf with time. The concentration $c(x = 0)$ obtained from this relaxation with the aid of titration

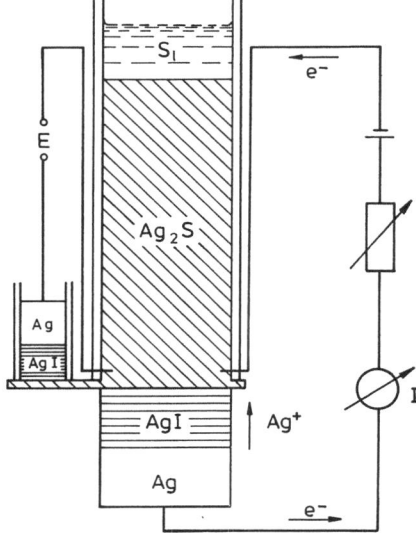

Fig. 11.6.5. Experimental arrangement for the determination of the chemical diffusion coefficient of silver sulfide

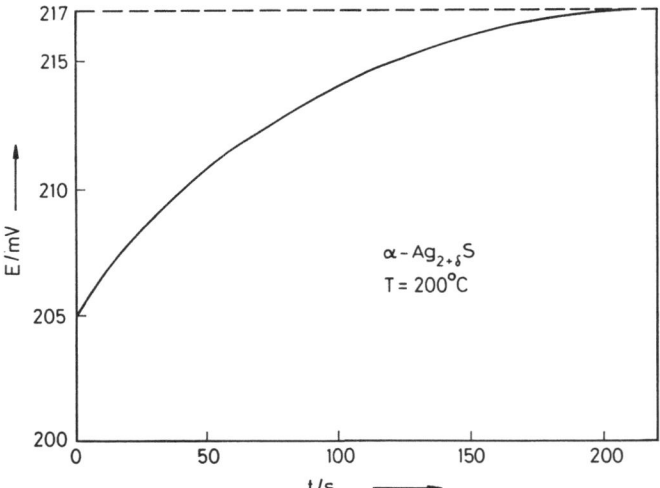

Fig. 11.6.6. Relaxation of the emf E of cell (11.6.III) as a function of time

curves is plotted in Fig. 11.6.7 as a function of $\sqrt{t}$. According to Eq. (11.6.8) the result is a straight line the slope of which allows the calculation of $\tilde{D}$.

At 200 °C the value of $\tilde{D}$ obtained from the experiment equals 0.47 cm² s⁻¹. The value calculated from partial conductivities [11.58] and $\dfrac{d \ln a}{d \ln c}$ is 0.39 cm² s⁻¹.

The agreement between these two values can be considered as satisfactory, particularly since the thermodynamic factor is of the order of magnitude of 10⁴.

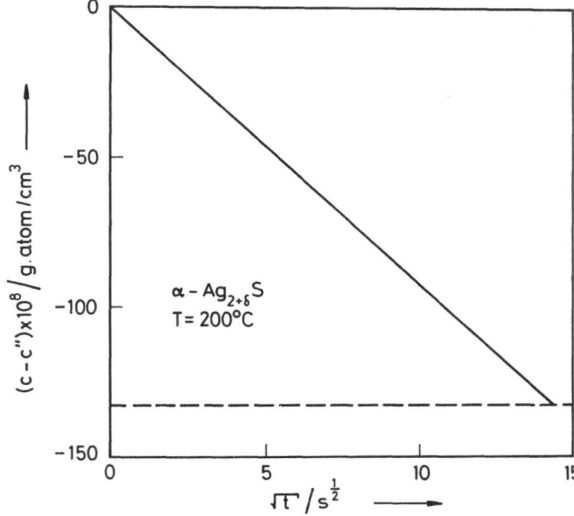

Fig. 11.6.7. Concentration $c(x = 0)$ as a function of $\sqrt{t}$, calculated from Fig. 11.6.6

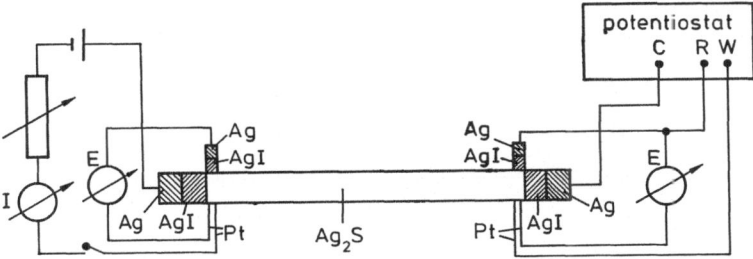

Fig. 11.6.8. Experimental arrangement (schematic) used for measuring the chemical diffusion coefficient in $\alpha\text{-}Ag_{2+\delta}S$ as a function of stoichiometry [11.37]

The chemical diffusion coefficients $\tilde{D}$ in $Ag_{2+\delta}S$ and $Ag_{2+\delta}Se$ over the total range of stoichiometric composition [11.37] may be determined by use of an improved experimental set-up (Fig. 11.6.8). As can be seen it contains four galvanic cells of type (11.6.II), a pair of such cells being located at each end of the $Ag_{2+\delta}S$ (or $Ag_{2+\delta}Se$) sample. These samples were ca. 10—15 cm long cylindrical rods.

The cells at the right-hand side of the sample serve for stipulating the chemical potential of the silver in silver sulfide at this site. For each measurement the deviation from ideal stoichiometry can thus be fixed to a definite value within the range of possible stoichiometric compositions. One cell at the left-hand side is connected with an external current circuit. A certain current flowing through this circuit thus determines the rate at which silver is either added to or withdrawn from this site of the sample. Under steady-state conditions the corresponding removal or addition of silver from or to the sample at the right-hand side attains the same rate. Thus, a concentration or — to be more general — an

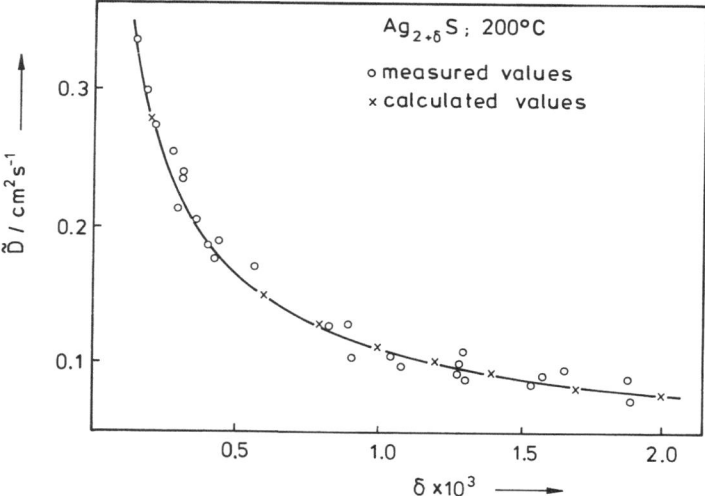

Fig. 11.6.9. Chemical diffusion coefficient of silver in α-$Ag_{2+\delta}S$ as a function of the deviation from ideal stoichiometry at 200 °C. $\circ$, measured values obtained by using an electrochemical method; $\times$, values calculated from partial conductivities and values of the thermodynamic factor

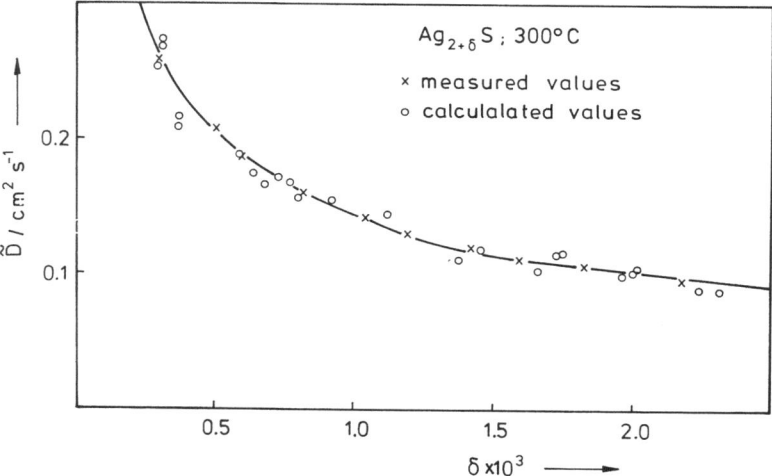

Fig. 11.6.10. Chemical diffusion coefficient of silver in α-$Ag_{2+\delta}S$ as a function of the deviation from ideal stoichiometry at 300 °C. $\circ$, measured values obtained by using an electrochemical method; $\times$, values calculated from partial conductivities and values of the thermodynamic factor

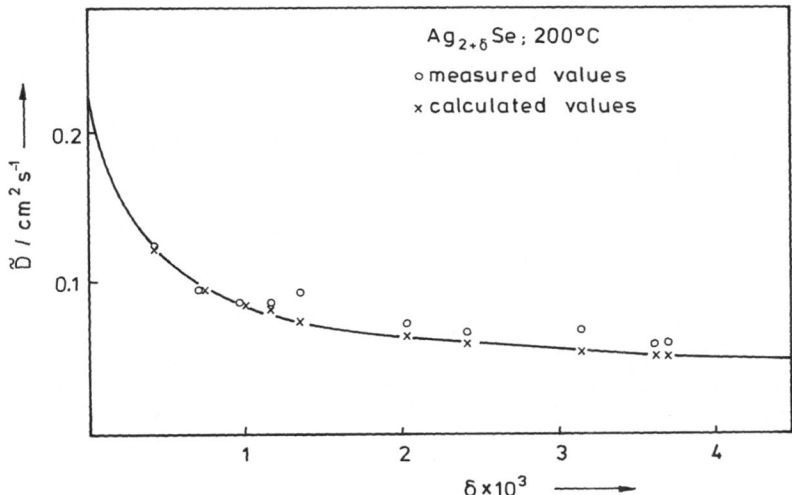

Fig. 11.6.11. Chemical diffusion coefficient of silver in α-Ag$_{2+δ}$Se as a function of the deviation from ideal stoichiometry at 200 °C. ○, measured values obtained by using an electrochemical method; ×, values calculated from partial conductivities and values of the thermodynamic factor

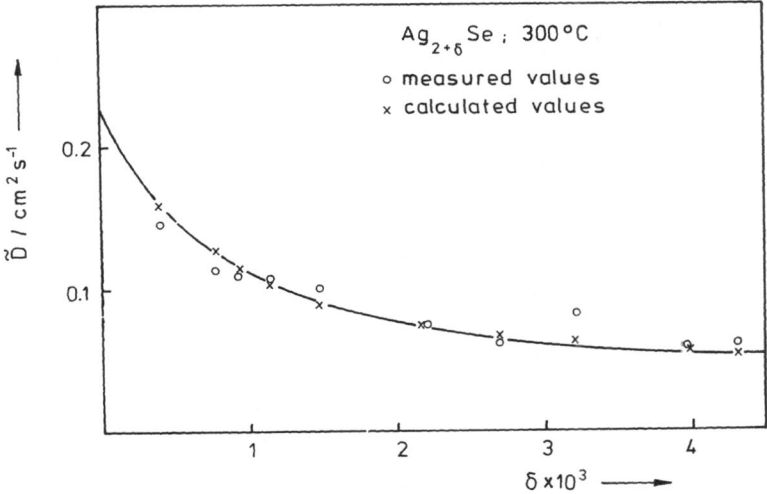

Fig. 11.6.12. Chemical diffusion coefficient of silver in α-Ag$_{2+δ}$Se as a function of the deviation from ideal stoichiometry at 300 °C. ○, measured values obtained by using an electrochemical method; ×, values calculated from partial conductivities and values of the thermodynamic factor

activity gradient of silver is established in the stationary state. Since the variation of δ along the sample, which is associated with this gradient, is small compared to the total range of stoichiometry, the concentration gradient may be considered to be constant, i.e. the variation of the silver concentration along the sample is approximately linear. Hence, the initial and boundary conditions, which must be known for evaluating the results obtained from measurements, are similar to those holding for the case discussed above.

The chemical diffusion coefficients $\tilde{D}$ of $Ag_{2+\delta}S$ and $Ag_{2+\delta}Se$ respectively could be determined at 200 and 300 °C as a function of the deviation from ideal stoichiometry δ by performing emf relaxation measurements [11.37].

Table 11.6.1. Chemical diffusion coefficient $\tilde{D}$ in Ag_2S at 200 °C as a function of the deviation δ from ideal stoichiometry calculated from the value of the partial conductivity $\sigma_{Ag^+} = 3.80\ \Omega^{-1}\ cm^{-1}$ (corresponding to the component diffusion coefficient $D_{Ag^+} = 2.76 \times 10^{-5}\ cm^2\ s^{-1}$) and from the values of the thermodynamic factor, which are also included in the table

$\delta \times 10^3$	$\dfrac{d\ln a_{Ag}}{d\ln c_{Ag}}$	$\tilde{D}/cm^2\ s^{-1}$
0.12	132	0.36
0.2	101	0.28
0.4	67	0.18
0.6	53	0.14
0.8	47	0.13
1.0	41	0.11
1.2	37	0.10
1.4	35	0.09
1.7	31	0.08
2	30	0.08

Table 11.6.2. Chemical diffusion coefficient $\tilde{D}$ in Ag_2S at 300 °C as a function of the deviation δ from ideal stoichiometry calculated from the value of the partial conductivity $\sigma_{Ag^+} = 4.84\ \Omega^{-1}\ cm^{-1}$ (corresponding to the component diffusion coefficient $D_{Ag^+} = 4.19 \times 10^{-5}\ cm^2\ s^{-1}$) and from the values of the thermodynamic factor, which are also included in the table

$\delta \times 10^3$	$\dfrac{d\ln a_{Ag}}{d\ln c_{Ag}}$	$\tilde{D}/cm^2\ s^{-1}$
0.285	61	0.25
0.47	49	0.20
0.60	44	0.18
0.80	38	0.15
1.0	33	0.14
1.2	30	0.12
1.4	28	0.11
1.6	26	0.11
1.8	25	0.10
2.2	22	0.09
2.5	19	0.08

In Figs. 11.6.9—12 the diffusion coefficients are plotted versus δ. By means of coulometric titration curves of $Ag_{2+\delta}S$ and $Ag_{2+\delta}Se$ respectively the values of δ corresponding to the diffusion coefficients could be identified. It is also possible to calculate chemical diffusion coefficients both from conductivities [11.58] and the thermodynamic factor, the latter being determined from titration curves (see Chap. 5). The values of $\tilde{D}$ resulting from such calculations are also represented in Figs. 11.6.9—12 and in Tables 11.6.1—4. As can be seen from Figs. 11.6.9—12 there is satisfactory agreement between the calculated values of $\tilde{D}$ and those obtained from direct measurements.

Table 11.6.3. Chemical diffusion coefficient $\tilde{D}$ in Ag_2Se at 200 °C as a function of the deviation δ from ideal stoichiometry calculated from the value of the partial conductivity $\sigma_{Ag^+} = 3.08 \, \Omega^{-1} \, cm^{-1}$ (corresponding to the component diffusion coefficient $D_{Ag^+} = 2.37 \times 10^{-5} \, cm^2 \, s^{-1}$) and from the values of the thermodynamic factor, which are also included in the table

$\delta \times 10^3$	$\dfrac{d \ln a_{Ag}}{d \ln c_{Ag}}$	$\tilde{D}/cm^2 \, s^{-1}$
0.428	51	0.12
0.713	41	0.09
0.975	37	0.08
1.182	34	0.08
1.363	32	0.07
2.044	27	0.06
2.433	25	0.06
3.148	23	0.05
3.608	21	0.05
3.699	21	0.05

Table 11.6.4. Chemical diffusion coefficient $\tilde{D}$ in Ag_2Se at 300 °C as a function of the deviation δ from ideal stoichiometry calculated from the value of the partial conductivity $\sigma_{Ag^+} = 2.97 \, \Omega^{-1} \, cm^{-1}$ (corresponding to the component diffusion coefficient $D_{Ag^+} = 3.75 \times 10^{-5} \, cm^2 \, s^{-1}$) and from the values of the thermodynamic factor, which are also included in the table

$\delta \times 10^3$	$\dfrac{d \ln a_{Ag}}{d \ln c_{Ag}}$	$\tilde{D}/cm^2 \, s^{-1}$
0.415	40	0.15
0.784	34	0.12
0.935	30	0.11
0.158	27	0.10
1.492	24	0.09
2.213	19	0.07
2.703	17	0.06
3.227	16	0.06
3.988	15	0.05
4.321	14	0.05

11.7 References

[11.1] Rickert, H., Steiner, R.: Naturwissenschaften *15*, 451 (1965)
[11.2] Rickert, H., Steiner, R.: Z. phys. Chem. N.F. *49*, 9 (1966)
[11.3] Rickert, H., El Miligy, A. A.: Z. Metallk. *59*, 635 (1968)
[11.4] Rickert, H., El Miligy, A. A.: Reactivity of Solids. Mitchell, J. W. (ed.), Vol. 17. New York: Wiley-Interscience, 1969
[11.5] Burke, L. D., Rickert, H., Steiner, R.: Z. phys. Chem. N.F. *74*, 146 (1971) Patterson, J. W., Bogren, E. C., Rapp, R. A.: J. Electrochem. Soc. *114*, 752 (1967)
[11.6] Pastorek, R. L., Rapp, R. A.: Trans. Met. Soc. AIME *245*, 1711 (1969)
[11.7] Masson, C. R., Whiteway, S. G.: Can. Met. Quart. *6*, 199 (1967)
[11.8] Ramanarayanan, T. A., Rapp, R. A.: Met. Trans. *3*, 3239 (1972)
[11.9] Szwark, R., Oberg, K. E., Rapp, R. A.: High-Temp. Sci. *4*, 347 (1972)
[11.10] Oberg, K. E. et al.: J. Iron Steel Inst. *210*, 359 (1972)
[11.11] Ramana Rao, A. V., Tare, V. B.: Z. Metallk. *63*, 70 (1972)
[11.12] Sano, N., Honma, S., Matsushita, Y.: Met. Trans. *1*, 301 (1970)
[11.13] Sano, N., Honma, S., Matsushita, Y.: Met. Trans. *2*, 1494 (1971)
[11.14] Otsuka, S., Kozuka, Z.: J. Japan Inst. Met. *37*, 364 (1973)
[11.15] Ramanarayanan, T. A., Worell, W. L.: Met. Trans. *5*, 1773 (1974)
[11.16] Otsuka, S., Kozuka, Z.: Met. Trans. *6 B*, 389 (1975)
[11.17] Otsuka, S., Kozuka, Z.: Met. Trans. *7 B*, 147 (1976)
[11.18] Kawakami, M., Goto, K. S.: Trans. Iron Steel Inst. Jpn. *16*, 204 (1976)
[11.19] Heshmatpour, B., to be published in: J. Electroanal. Chem., Interf. Electrochem. (1981)
[11.20] Crank, J.: The Mathematics of Diffusion. London, Oxford: Clarendon Press 1957
[11.21] Eichenauer, W., Müller, G.: Z. Metallk. *53*, 321 (1962)
[11.22] Sand, H. J.: Phil. Mag. *1*, 45 (1901); Z. phys. Chem. *35*, 641 (1900)
[11.23] Rickert, H., Wagner, H.: Electrochim. Acta *11*, 83 (1966)
[11.24] Mrowec, S., Rickert, H.: Z. phys. Chem. N.F. *36*, 329 (1963)
[11.25] Dravnieks, A.: J. Electrochem. Soc. *102*, 435 (1955)
[11.26] Czerski, L., Mrowec, S., Werber, T.: J. Electrochem. Soc. *109*, 273 (1962)
[11.27] Rickert, H.: Z. Electrochem. Ber. Bunsenges. phys. Chem. *65*, 463 (1961)
[11.28] Tammann, G.: Z. anorg. allg. Chem. *111*, 78 (1920)
[11.29] Pilling, N. B., Bedworth, R. E.: J. Inst. Met. *29*, 599 (1923)
[11.30] Rickert, H., O'Brian, C. D.: Z. phys. Chem. N.F. *31*, 71 (1962); Rickert, H., Wagner, C.: Z. phys. Chem. N.F. *31*, 32 (1962)
[11.31] Hebb, M. H.: J. Chem. Phys. *20*, 185 (1952)
[11.32] Contreras, L., Rickert, H.: Ber. Bunsenges. phys. Chem. *82*, 292 (1978)
[11.33] Mizusaki, J., Fueki, K., Mukaibo, T.: Bull. Chem. Soc. Japan *47*, 2851 (1974)
[11.34] Donner, D., Rickert, H.: Z. phys. Chem. N.F. *60*, 11 (1968)
[11.35] Sasaki, J., Mizusaki, J., Fueki, K.: Bull. Chem. Soc. Japan *51*, 1027 (1978)
[11.36] Rickert, H.: Z. Electrochem. *65*, 463 (1961)
[11.37] Hartmann, B., Rickert, H., Schendler, W.: Electrochim. Acta *21*, 319 (1976)
[11.38] Detry, D. et al.: Z. phys. Chem. *55*, 314 (1967)
[11.39] Birks, N., Rickert, H.: Ber. Bunsenges. phys. Chem. *67*, 501 (1963)
[11.40] Rickert, H., Tostmann, K.-H.: Werkstoffe, Korrosion *21*, 965 (1970)
[11.41] Mrowec, S., Rickert, H.: Z. Elektrochem. *66*, 14 (1962)
[11.42] Kobayashi, H., Wagner, C.: J. Chem. Phys. *26*, 1609 (1957)
[11.43] Roy, P., Schmalzried, H.: Ber. Bunsenges. phys. Chem. *71*, 201 (1967)
[11.44] Bechthold, E., Schwabe, U.: Z. phys. Chem. N.F. *81*, 230 (1972)
[11.45] Knudsen, M.: Ann. Phys. (Leipzig) *29*, 179 (1909)
[11.46] Ratchford, R. J., Rickert, H.: Z. Elektrochem. *66*, 497 (1962)
[11.47] Detry, D. et al.: Z. phys. Chem. N.F. *55*, 314 (1967) Inghram, M. G., Drowart, J.: High-Temperature Technology, p. 219. New York: McGraw-Hill 1960

[11.48] West, L. A., Menzies, A. W.: J. Phys. Chem. *33*, 1880 (1929)
[11.49] Keller, H. et al.: Z. phys. Chem. N.F. *75*, 273 (1971)
[11.50] Keller, H.: Dissertation, University of Dortmund 1970
[11.51] Chu, W. F., Rickert, H., Weppner, W.: Proceedings of the Advanced Study Institute "Fast-ion transport in solids, solid-state batteries and devices" Belgirate September 5—15 1972, p. 181. Amsterdam: North-Holland Publ. Comp. 1973
[11.52] Rickert, H., Weppner, W.: Z. Naturforsch. *29a*, 1849 (1974)
[11.53] Hauffe, K.: Reaktionen in und an festen Stoffen, p. 395. Berlin—Heidelberg—New York: Springer 1966
[11.54] Rizzo, H. F., Gordon, R. S., Cutler, I. B.: Mass Transport in Oxides. Wachtman, J. B., Jr., Franklin, A. D. (eds.), p. 129. NBS Spec. Publ. 296, 1968
Rizzo, H. F., Smith, J. V.: J. Phys. Chem. *72*, 485 (1968)
Sockel, H. G., Schmalzried, H.: Ber. Bunsenges. phys. Chem. *72*, 745 (1968)
[11.55] Crank, J.: The Mathematics of Diffusion, p. 17. Oxford: Univ. Press London 1957
[11.56] Himmel, L., Mehl, R. F., Birchenall, C. E.: Trans. AIME *197*, 827 (1953)
[11.57] Carslaw, H. S., Jaeger, J. C.: Conduction of Heat in Solids, p. 97. Oxford: Clarendon Press 1959
[11.58] Rickert, H.: Z. phys. Chem. N.F. *23*, 355 (1960)

12. Non-Isothermal Systems. Soret Effect, Transport Processes, and Thermopowers

If solid compounds are subjected to a temperature gradient, one or more of the following effects are observed:

a) Changes of concentrations, i.e. changes in stoichiometry or in the activities of the components along the sample. This phenomenon is analogous to that existing in solutions of liquid electrolytes where, due to the presence of a temperature gradient, concentration gradients of ions contained in the solution occur after a certain time interval. We shall therefore refer to the appearance of gradients with respect to stoichiometry or activities caused by temperature differences in solids as a Soret effect, just as in the case of liquid electrolytes.

b) Under certain conditions, transport processes are observed in solid compounds subjected to a temperature gradient, which involve the migration of one or several components of the solid from high to low temperatures or vice versa. For example, the following observation has been made: if a temperature gradient is imposed on solid Ag_2S, the latter being surrounded by gaseous sulfur above 200 °C, Ag_2S disappears at the "hot" end of the sample while an equal amount of Ag_2S is formed at the "cold" end [12.1]. Because of the high mobility of silver ions and electrons in Ag_2S, we may suppose that silver ions and electrons migrate within the silver sulfide while molecular sulfur is transported via the gas phase from higher to lower temperature. This is schematically shown in Fig. 12.0.1.

c) If galvanic cells which are constructed symmetrically, concerning the configuration of the solid phases used as electrolytes and electrodes are subjected to a temperature gradient, electrical potential differences result. These potential differences, the so-called thermoelectric powers, are, however, not only as in the case of metals, determined by the substances and temperature differences but also by further experimental boundary conditions, in particular by fixing the

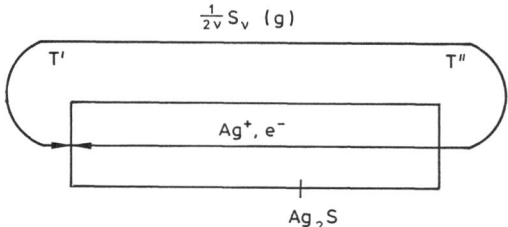

Fig. 12.0.1. Transport processes in Ag_2S which is subjected to a temperature gradient $(T' < T'')$ and surrounded by sulfur vapour

activities of the components at the two electrodes; this will be discussed in more detail in Sect. 12.4.

These three phenomena may be quantitatively treated using the equations of irreversible thermodynamics, the principles of which will first be discussed below.

12.1 Basic Equations of Irreversible Thermodynamics

For the following discussion we shall consider a mixed conducting metal/non-metal compound in which cations of valency z_1 and electrons are the mobile species. A generalization for mobile anions is easily possible. The fluxes of cations and electrons in mol per unit area and unit time denoted by j_1 and j_2 and the energy flux j_3 are allowed to take place only in x-direction, since the temperature gradient imposed is assumed to exist in this direction only. These fluxes are given by the phenomenological equations of irreversible thermodynamics [12.2] as follows:

$$j_1 = L_{11}X_1 + L_{12}X_2 + L_{13}X_3, \tag{12.1.1}$$

$$j_2 = L_{21}X_1 + L_{22}X_2 + L_{23}X_3, \tag{12.1.2}$$

$$j_3 = L_{31}X_1 + L_{32}X_2 + L_{33}X_3. \tag{12.1.3}$$

The quantities X_i ($i = 1, 2, 3$) represent the so-called generalized forces and the L_{ik}'s ($i, k = 1, 2, 3$) the so-called Onsager coefficients correlating forces and fluxes. The forces must be defined appropriately to make the rate of entropy production per unit volume equal to $\sum_{i=1}^{3} j_i X_i$. Therefore, the forces are:

$$X_1 = -\frac{d}{dx}\left(\frac{\eta_1}{T}\right), \tag{12.1.4}$$

$$X_2 = -\frac{d}{dx}\left(\frac{\eta_2}{T}\right), \tag{12.1.5}$$

$$X_3 = \frac{d}{dx}\left(\frac{1}{T}\right) = -\frac{1}{T^2}\frac{dT}{dx}. \tag{12.1.6}$$

Here η_1 and η_2 are the electrochemical potentials of cations and electrons respectively and T the temperature. It is known from experience that in practically all cases we are dealing with the motions of cations and electrons are virtually independent of one another. It then holds:

$$L_{12} = L_{21} = 0. \tag{12.1.7}$$

This was the basic assumption for the derivation of the transport equations in Chap. 6 (see especially Sect. 6.1.9). These equations of Chap. 6 could be used in Chap. 10 to treat diffusion-controlled solid-state reactions, e.g. tarnishing reactions according to Wagner [12.3].

Insertion of Eqs. (12.1.4), (12.1.5) and (12.1.7) into Eqs. (12.1.1) and (12.1.2) leads to the following expressions for the fluxes j_1 and j_2:

$$j_1 = -\frac{L_{11}}{T}\frac{d\eta_1}{dx} + (\eta_1 L_{11} - L_{13})\frac{1}{T^2}\frac{dT}{dx}. \tag{12.1.8}$$

$$j_2 = -\frac{L_{22}}{T}\frac{d\eta_2}{dx} + (\eta_2 L_{22} - L_{23})\frac{1}{T^2}\frac{dT}{dx}. \tag{12.1.9}$$

As shown in Sect. 6.1.9, the Onsager coefficients L_{11} and L_{22} are related to the partial conductivities of cations and electrons σ_1 and σ_2:

$$L_{11} = \sigma_1 T/z_1^2 F^2, \tag{12.1.10}$$

$$L_{22} = \sigma_2 T/F^2. \tag{12.1.11}$$

Taking into account the Onsager reciprocity relations [12.2] which read here

$$\begin{aligned} L_{13} &= L_{31}, \\ L_{23} &= L_{32} \end{aligned} \tag{12.1.12}$$

we can obtain an expression for the energy flux j_3 by inserting Eqs. (12.1.4—6) and (12.1.12) into relation (12.1.3)

$$\begin{aligned} j_3 = &-\frac{L_{13}}{T}\frac{d\eta_1}{dx} + L_{13}\frac{\eta_1}{T^2}\frac{dT}{dx} \\ &-\frac{L_{23}}{T}\frac{d\eta_2}{dx} + L_{23}\frac{\eta_2}{T^2}\frac{dT}{dx} - \frac{L_{33}}{T^2}\frac{dT}{dx}. \end{aligned} \tag{12.1.13}$$

Under isothermal conditions ($T = \text{const.}$) and at constant electrochemical potential η_2 of the electrons, i.e. if only a gradient of the electrochemical potential η_1 of the ions is present, the ratio of energy flux j_3 to ion flux j_1 can be obtained from Eqs. (12.1.8) and (12.1.13):

$$\left(\frac{j_3}{j_1}\right)_{\eta_2, T} = \frac{L_{13}}{L_{11}}. \tag{12.1.14}$$

This quotient of energy flux to particle flux may also be expressed as the sum of the following quantities: the partial molar internal energy $\overline{U}_1$ of particles 1, their electrical energy $z_1 F\varphi$, their heat of transport Q_1^*, also called heat of transfer, and the mechanical work $p \cdot \overline{V}_1$ the system performs affected by the addition of one mole of particles 1, where $\overline{V}_1$ is the corresponding partial molar volume. The quantity $\overline{U}_1 + p\overline{V}_1$ is equal to the partial molar enthalpy $\overline{H}_1$ of the particles considered, in this case the cations. We thus have the following expression:

$$\left(\frac{j_3}{j_1}\right)_{\eta_2, T} = \overline{U}_1 + z_1 F\varphi + Q_1^* + p\overline{V}_1 = \overline{H}_1 + z_1 F\varphi + Q_1^*. \tag{12.1.15}$$

It then follows from Eqs. (12.1.10), (12.1.14) and (12.1.15) that

$$L_{13} = (\overline{H}_1 + z_1 F\varphi + Q_1^*)\,\sigma_1 T/z_1^2 F^2. \tag{12.1.16}$$

Thus, the coupling coefficient L_{13} is expressed in caloric quantities. Similarly, we may obtain the expression for L_{23}. Exchanging index 2 for 1, it follows from Eq. (12.1.16)

$$L_{23} = (\bar{H}_2 - F\varphi + Q_2^*) \, \sigma_2 T/F^2. \tag{12.1.17}$$

Since the electrochemical potentials η_1 and η_2 are given by the relations

$$\eta_1 = \mu_1 + z_1 F\varphi = \bar{H}_1 - T\bar{S}_1 + z_1 F\varphi, \tag{12.1.18}$$

$$\eta_2 = \mu_2 - F\varphi = \bar{H}_2 - T\bar{S}_2 - F\varphi \tag{12.1.19}$$

where $\bar{S}_1$ and $\bar{S}_2$ are the partial molar entropies of the metal ions and electrons respectively, the following expressions result for j_1 and j_2 by inserting Eqs. (12.1.10), (12.1.11) and (12.1.16—19) into Eqs. (12.1.8) and (12.1.9)

$$j_1 = -\frac{\sigma_1}{z_1^2 F^2}\left[\frac{d\eta_1}{dx} + \left(\bar{S}_1 + \frac{Q_1^*}{T}\right)\frac{dT}{dx}\right], \tag{12.1.20}$$

$$j_2 = -\frac{\sigma_2}{F^2}\left[\frac{d\eta_2}{dx} + \left(\bar{S}_2 + \frac{Q_2^*}{T}\right)\frac{dT}{dx}\right]. \tag{12.1.21}$$

Equations (12.1.20) and (12.1.21) yield the particle fluxes of cations and electrons for a solid compound subjected to a temperature gradient. They permit the quantitative treatment of the following three phenomena:

a) the Soret effect,
b) transport processes in solid compounds subjected to a temperature gradient,
c) thermoelectric powers.

12.2 The Soret Effect

The Soret effect, i.e. the stationary gradient of concentrations and thus of deviations from ideal stoichiometry caused by a temperature gradient, can be treated by setting the fluxes j_1 and j_2 given by Eqs. (12.1.20) and (12.1.21) equal to zero, since transport processes have ceased to take place. Thus, on condition that the partial conductivities σ_1 and σ_2 are finite, the terms in the square brackets of Eqs. (12.1.20) and (12.1.21) must be equal to zero. Summation of these two expressions gives the following equation

$$\frac{d(\eta_1 + z_1\eta_2)}{dx}\bigg|_{j_1, j_2 = 0} + \left(\bar{S}_1 + z_1\bar{S}_2 + \frac{Q_1^* + z_1 Q_2^*}{T}\right)\frac{dT}{dx} = 0 \tag{12.2.1}$$

or, since $\eta_1 + z_1\eta_2$ is equal to the chemical potential μ_{Me} of the metal and $\bar{S}_1 + z_1\bar{S}_2$ is equal to the partial molar entropy $\bar{S}_{Me}$ of the metal, and since the term $Q_1^* + z_1 Q_2^*$ may be replaced by Q_{Me}^*, the heat of transport of the neutral metal, it follows from Eq. (12.2.1):

$$\left(\frac{d\mu_{Me}}{dx}\right)_{j_1, j_2 = 0} + \left(\bar{S}_{Me} + \frac{Q_{Me}^*}{T}\right)\frac{dT}{dx} = 0 \tag{12.2.2}$$

or

$$\left(\frac{d\mu_{Me}}{dT}\right)_{j_1, j_2 = 0}\frac{dT}{dx} + \left(\bar{S}_{Me} + \frac{Q_{Me}^*}{T}\right)\frac{dT}{dx} = 0. \tag{12.2.3}$$

Division of Eq. (12.2.3) by dT/dx leads to the expression

$$\left(\frac{d\mu_{Me}}{dT}\right)_{j_1,j_2=0} + \left(\bar{S}_{Me} + \frac{Q_{Me}^*}{T}\right) = 0. \tag{12.2.4}$$

According to Eq. (12.2.4) the change of the chemical potential of the metal with temperature for the case, that no transport processes occur, i.e. j_1, $j_2 = 0$, may be expressed by the partial molar entropy $\bar{S}_{Me}$ and the heat of transport Q_{Me}^* of the metal in the compound considered. Since the relation

$$\frac{d\mu_{Me}^0}{dT} = -S_{Me}^0 \tag{12.2.5}$$

applies to the standard state, i.e. for the pure metal, subtraction of Eq. (12.2.5) from Eq. (12.2.4) affords

$$\left[\frac{d(\mu_{Me} - \mu_{Me}^0)}{dT}\right]_{j_1,j_2=0} = \left[\frac{d(RT \ln a_{Me})}{dT}\right]_{j_1,j_2=0}$$
$$= -\left[(\bar{S}_{Me} - S_{Me}^0) + \frac{Q_{Me}^*}{T}\right]. \tag{12.2.6}$$

The quantities appearing on the left-hand side of Eq. (12.2.6) can be determined experimentally.

Silver sulfide belongs to those solids, for which measurements of the Soret effect have been carried out [12.1]. The experimental arrangement is sketched in Fig. 12.2.1. The solid-state galvanic cell

$$\text{Pt} \mid \text{Ag} \mid \text{AgI} \mid \text{Ag}_2\text{S} \mid \text{Pt} \tag{12.2.I}$$

was used to determine the quantity $RT \ln a_{Ag}$, since the emf E of this cell is, according to Eq. (8.2.2), a measure of $\mu_{Ag} - \mu_{Ag}^0$, i.e. of $RT \ln a_{Ag}$. The gradient of $RT \ln a_{Ag}$ was determined by means of two isothermal cells of the type (12.2.I). Without special assumptions, the gradient of $RT \ln a_{Ag}$ is not only determined by the temperature difference $(T'' - T')$ but also by the value of a_{Ag} at the colder or warmer end of the sample or at an intermediate position. Thus, using the experimental set-up shown in Fig. 12.2.1, it is possible to fix the value of $a_{Ag}(T')$ at the colder end at temperature T'; this may be done either

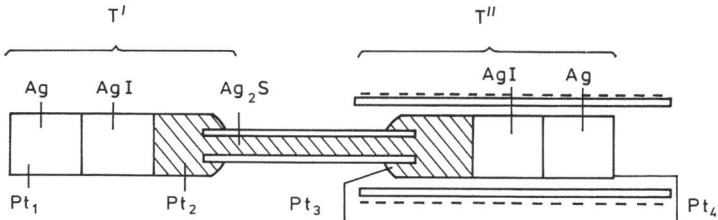

Fig. 12.2.1. Experimental arrangement for the determination of the activities $a_{Ag}(T')$ and $a_{Ag}(T'')$ in Ag_2S subjected to a temperature gradient under stationary conditions ($j_1 = j_2 = 0$), i.e. at negligibly small sulfur partial pressures (with auxiliary heating on the *right-hand side* for adjustment of temperature T'')

Table 12.2.1. Emf measurements using the arrangement shown in Fig. 12.2.1 for the determination of $(d\ RT\ \ln a_{Ag}/dT)_{j_1, j_2=0}$

$E(T' = 250\,°C)/mV$	0	30	60	90	115	140
$E(T'' = 350\,°C)/mV$	28	59	91	124	150	177

by applying a certain potential difference $E(T')$ between contacts Pt_1 and Pt_2 using a potentiometer or by carrying out a coulometric titration [12.4] until a certain value $a_{Ag}(T')$ has been reached. Under stationary conditions, $E(T'')$ is then measured as the potential difference between contacts Pt_4 and Pt_3. The emf measurements for the determination of $(d\ RT\ \ln a_{Ag}/dT)_{j_1, j_2=0}$ are listed in Table 12.2.1. It must be noted that, using the set-up shown, measurements can only be carried out for those values of a_{Ag}, to which negligibly small sulfur pressures above the free surfaces of the Ag_2S sample correspond; if this is not the case, silver migrates within Ag_2S, and sulfur is transported via the gas phase according to the mechanism sketched in Fig. 12.0.1, so that $j_1, j_2 \neq 0$. Values corresponding to low silver activities, i.e. to higher sulfur activities, may be obtained by extrapolation. Using the data contained in Table 12.2.1, it is possible to calculate the quantity $\left(\bar{S}_{Ag} - S^0_{Ag} + \dfrac{Q^*_{Ag}}{T}\right)$ for silver in silver sulfide; values of about 30 J/K mol are obtained.

In view of the small temperature coefficients of the partial conductivities of Ag^+ ions and electrons in Ag_2S, corresponding to the small apparent heats of activation, the heats of transfer are only of the order of RT. Thus, the partial molar entropy of silver in silver sulfide minus that of the standard state, $\bar{S}_{Ag} - S^0_{Ag}$, for the equilibrium with silver is about 21 J/K mol. This value agrees with that obtained by measuring the temperature dependence of the emf of the isothermal cell (12.2.I) [12.5].

12.3 Stationary Transport Processes in Solids Subjected to a Temperature Gradient

Equations (12.1.20) and (12.1.21) may be taken as the starting point in the discussion of transport processes in solids which are subjected to a temperature gradient. The derivation of these equations was based on the assumption that in the solid considered only cations and electrons are mobile, the anions, however, being virtually immobile. Thus, the nonmetal component could — if at all — only be transported via the gas phase, as shown schematically for Ag_2S in Fig. 12.0.1. On the grounds of electroneutrality the fluxes of metal ions and electrons, j_1 and j_2 respectively, are related by the expression

$$j_1 = \frac{j_2}{z_1}. \tag{12.3.1}$$

Taking this equation into account the following expression for j_1 results by multiplying Eq. (12.1.20) by σ_2 and Eq. (12.1.21) by σ_1 and summing the two equations:

$$j_1 = j_{Me} = -\frac{\sigma_1 \sigma_2}{(\sigma_1 + \sigma_2)\, z_1^2 F^2}\left[\frac{d(\eta_1 + z_1\eta_2)}{dx}\right.$$
$$\left. + \left(\bar{S}_1 + z_1\bar{S}_2 + \frac{Q_1^* + z_1 Q_2^*}{T}\right)\frac{dT}{dx}\right]. \tag{12.3.2}$$

We may express the thermodynamic quantities of the cations and electrons in terms of those of the neutral metal, since the following relationships are valid:

$$\eta_1 + z_1\eta_2 = \mu_{Me}, \tag{12.3.3}$$
$$\bar{S}_1 + z_1\bar{S}_2 = \bar{S}_{Me}, \tag{12.3.4}$$
$$Q_1^* + z_1 Q_2^* = Q_{Me}^*. \tag{12.3.5}$$

Hence, the following equation applies to the flux of the metal:

$$j_1 = j_{Me} = -\frac{\sigma_1 \sigma_2}{(\sigma_1 + \sigma_2)\, z_1^2 F^2}\left[\frac{d\mu_{Me}}{dx} + \left(\bar{S}_{Me} + \frac{Q_{Me}^*}{T}\right)\frac{dT}{dx}\right]. \tag{12.3.6}$$

In order to arrive at measurable quantities, the following expression holding for the standard state is added to the term in square brackets of Eq. (12.3.6)

$$-\frac{d\mu_{Me}^0}{dx} - S_{Me}^0 \frac{dT}{dx} = 0. \tag{12.3.7}$$

Thus, the relationship

$$j_1 = j_{Me} = -\frac{\sigma_1 \sigma_2}{(\sigma_1 + \sigma_2)\, z_1^2 F^2}\left[\frac{d(\mu_{Me} - \mu_{Me}^0)}{dx}\right.$$
$$\left. + \left(\bar{S}_{Me} - S_{Me}^0 + \frac{Q_{Me}^*}{T}\right)\frac{dT}{dx}\right] \tag{12.3.8}$$

results or, after replacing the difference $\mu_{Me} - \mu_{Me}^0$ by $RT \ln a_{Me}$,

$$j_1 = j_{Me} = -\frac{\sigma_1 \sigma_2}{(\sigma_1 + \sigma_2)\, z_1^2 F^2}\left[\frac{dRT \ln a_{Me}}{dx}\right.$$
$$\left. + \left(\bar{S}_{Me} - S_{Me}^0 + \frac{Q_{Me}^*}{T}\right)\frac{dT}{dx}\right]. \tag{12.3.9}$$

As can be seen by comparison with Eq. (12.2.6) the second term in square brackets of Eq. (12.3.9), i.e. $\bar{S}_{Me} - S_{Me}^0 + \dfrac{Q_{Me}^*}{T}$, is equivalent to the gradient of $RT \ln a_{Me}$, generated by the Soret effect ($j_1 = j_2 = 0$). Thus, Eq. (12.3.9) may be rewritten to give

$$j_1 = j_{Me} = -\frac{\sigma_1 \sigma_2}{(\sigma_1 + \sigma_2)\, z_1^2 F^2}\left[\frac{dRT \ln a_{Me}}{dx}\right.$$
$$\left. - \left(\frac{dRT \ln a_{Me}}{dT}\right)_{j_1, j_2 = 0} \frac{dT}{dx}\right]. \tag{12.3.10}$$

Equations (12.3.8—10) represent general solutions to the problem dealt with. However, the conductivity of the cations is often much larger than that of the electrons or vice versa. We shall consider the case $\sigma_2 \gg \sigma_1$ which applies, for example, to Ag_2S, on which measurements have been carried out. Then, Eq. (12.3.10) can be simplified to give

$$j_1 = j_{Me} = -\frac{\sigma_1}{z_1^2 F^2}\left[\frac{dRT \ln a_{Me}}{dx} - \left(\frac{dRT \ln a_{Me}}{dT}\right)_{j_1, j_2 = 0}\frac{dT}{dx}\right]. \quad (12.3.11)$$

Thus, the particle flux is determined by the partial conductivity of the less mobile particles, e.g. by that of silver ions in the case of Ag_2S, and further by the difference between the gradient of $RT \ln a_{Me}$ present due to the external conditions and that gradient expected if the condition $j_1 = j_2 = 0$ applies, i.e. if the Soret effect occurs. A special case results, if e.g. Ag_2S, being subjected to a temperature gradient, is brought into equilibrium with metallic silver on both sides of the sample, i.e. at the hot and at the cold end. Hence, throughout the whole sample the activity of silver is equal to unity, i.e. $a_{Ag} = 1$. This is schematically shown in Fig. 12.3.1. Since the gradient of $RT \ln a_{Me}$ is now equal to zero, Eq. (12.3.11) can be simplified. Noting that $Me = Ag$ in this case, the following equation applies:

$$j_1 = j_{Ag} = \frac{\sigma_1}{F^2}\left(\frac{dRT \ln a_{Ag}}{dT}\right)_{j_1, j_2 = 0}\frac{dT}{dx}. \quad (12.3.12)$$

The term $(dRT \ln a_{Ag}/dT)_{j_1, j_2 = 0}$ may approximately be determined using the arrangement shown in Fig. 12.2.1 and applying a finite temperature interval $T'' - T'$; $E(T')$ is set equal to zero whereby $E(T'')$ is measured on condition that the particle fluxes have vanished. Noting the validity of $E = -\frac{RT}{F}\ln a_{Ag}$, we obtain the following expression for the flux of silver j_{Ag} in a sample of length Δx.

$$j_1 = j_{Ag} \cong \frac{\sigma_1^*}{\Delta x F}[-E(T'')]_{j_1, j_2 = 0, E(T') = 0} \quad (12.3.13)$$

where σ_1^* is the mean partial conductivity of silver ions in Ag_2S in the temperature range between T' and T''.

With the aid of this equation the fluxes of silver ions and electrons in Ag_2S at constant activity of silver, i.e. $a_{Ag} = 1$, can be calculated.

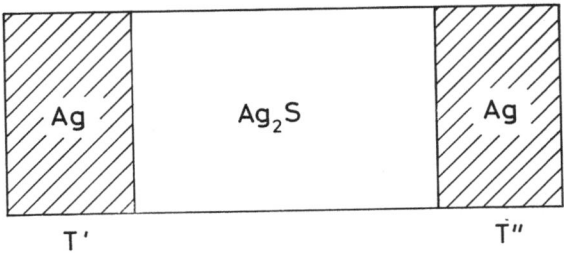

Fig. 12.3.1. Silver sulfide subjected to a temperature gradient in contact with metallic silver at both ends of the sample

The experimental determination of the transport rate can be carried out using the same arrangement as shown in Fig. 12.2.1. For this leads Pt_1 and Pt_2 may be short-circuited, and leads Pt_3 and Pt_4 connected via a shunt of 1 Ohm. The electrical current corresponding to the transport of silver can then be measured with the aid of the voltage drop occurring at this shunt, since the silver transport is equivalent to a flux of Ag^+-ions through AgI and thus also to the corresponding flux of electrons through leads Pt_3, Pt_4 and the shunt. The measured value of the current density was in sufficient agreement with the calculated one [12.1]. As another limiting case the transport rate of silver in silver sulfide subjected to a temperature gradient but surrounded by sulfur vapour was investigated [12.1] and has given good agreement between calculated and measured rates.

12.4 Thermoelectric Powers

In this section thermopowers of solid compounds are discussed, which, from the experimental point of view, can be determined by the use of thermocouples containing as one side leg the compound to be studied. In Fig. 12.4.1 the set-up of such a thermocouple is sketched where, as shown, the other side leg consists of platinum.

The end phases at the lower temperature T are also composed of platinum and denoted as $Pt(\alpha)$ and $Pt(\beta)$ respectively.

This thermocouple represents a galvanic cell which may be written as follows

$$Pt(\alpha) \mid MeX_\nu, MeX_\nu \mid Pt, \quad Pt(\beta) \qquad (12.4.I)$$
$$\underset{T}{\longrightarrow} \quad \underset{T + \Delta T}{\longrightarrow} \quad \underset{T}{\longrightarrow}$$

Thus, thermocouples of this kind are generally referred to as thermo(galvanic)-cells.

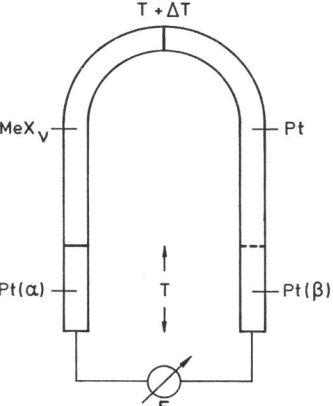

Fig. 12.4.1. Schematic representation of a thermocouple containing a solid compound as one side leg

As can be seen, the configuration of the phases occurring within cell (12.4.I) is symmetrical so that the emf exhibited by the cell, i.e. the potential difference established between Pt(β) and Pt(α) on the condition of open circuit is solely due to the temperature gradient imposed on the cell.

From the theoretical point of view such thermogalvanic cells may be treated in different ways. According to the approach generally adopted in the literature, the potential difference measurable between the terminals of the thermocouple — in the following this quantity will be simply denoted as $\varphi(\beta, T) - \varphi(\alpha, T)$ — is built up by several contributions, some due to heterojunctions (phase boundaries) existing in the cell and others occurring in the homogeneous parts of that phases which are subjected to the temperature gradient. The other way of treating thermocouples which contain solid conductors of either ionic or mixed conductivity, is that one where electrochemical potentials and their gradients in the different phases are used. The latter approach offers some advantages and we will first apply this one. C. Wagner has given an account on this subject [12.7] using electrochemical potentials.

Thermocouples with at least one side leg composed of a solid compound show important differences compared to those consisting only of metals or alloys. The most important difference lies in the fact that the emf of thermocells like (12.4.I) is not only determined by the materials used but also depends on further boundary conditions which are necessarily to be known in order to properly define the state of the compound from the thermodynamic point of view. The following four limiting cases are of particular importance:

a) Throughout the whole length of compound MeX_ν there exists a constant ratio of the concentrations of the components, i.e. the stoichiometry does not vary from point to point. However, since a temperature gradient exists, the thermodynamic activities or the chemical potentials of the components are no longer constant throughout the whole sample but are subject to local variations.

b) There is a stationary gradient of metal excess or metal deficiency corresponding to the Soret effect (compare the discussion of the Soret effect in Sect. 12.2).

c) The activity a_{Me} of the metal Me in compound MeX_ν is constant.

d) There is a constant partial pressure p_{Me} of the metal Me or a constant partial pressure p_{X_2} of the nonmetal in the gas phase surrounding the sample.

Case a) will be observed experimentally, if, for example a thermocell (12.4.I), being at isothermal equilibrium, is suddenly subjected to a temperature gradient; immediately after imposing the temperature difference, the stoichiometry of the compound has not changed, i.e. it is constant throughout the whole sample and equal to that applying under isothermal equilibrium conditions. The new stationary state of compound MeX_ν, which will be attained after a certain time has elapsed, is — since now transport processes have ceased — a stationary gradient of metal excess or metal deficiency corresponding to the Soret effect. This phenomenon has been discussed in Sect. 12.2. Case b) is characterized by this situation.

The condition of constant activity a_{Me} of the metal Me, corresponding to case c), is for example fulfilled, if compound MeX_ν is brought into equilibrium with metal Me at the cold and hot end. This may be achieved by using the following thermocell:

$$Pt(\alpha) \mid Me \mid MeX_\nu, MeX_\nu \mid Me \mid Pt, Pt(\beta) \tag{12.4.II}$$

$$\underset{T}{\xleftarrow{\hspace{2cm}}} \quad \underset{T + \Delta T}{\xleftarrow{\hspace{1.5cm}}} \quad \underset{T}{\xleftrightarrow{\hspace{0.5cm}}}$$

The examples discussed in the following refer to cases b) and c), i.e. we are concerned with the stationary gradient of the activities of the components according to the Soret effect and with the condition of constant activity of the metal; these cases are also of considerable practical importance; most considerations, however, are also generally valid. Just as in previous sections of this chapter we shall assume in the following that only electrons and metal ions are mobile.

The thermoelectric power to be calculated is defined according to the relation

$$\varepsilon = \lim_{\Delta T \to 0} \frac{E(T, \Delta T)}{\Delta T} = \lim_{\Delta T \to 0} \frac{\varphi(\beta, T) - \varphi(\alpha, T)}{\Delta T} = \frac{d[\varphi(\beta, T) - \varphi(\alpha, T)]}{dT} \tag{12.4.1}$$

where E is the emf of cell (12.4.I) or cell (12.4.II), i.e. the value of the potential difference between both end phases under zero-current condition whereby both leads, i.e. α and β, are at the same temperature. The latter has been taken into account by the notation used. Since both leads α and β consist of the same metal, the chemical potential μ_2 of the electrons in lead β is equal to that in lead α, i.e. $\mu_2(\beta) = \mu_2(\alpha)$. Therefore, we have the following identity

$$\varphi(\beta, T) - \varphi(\alpha, T) = -\frac{1}{F} [\mu_2(\beta, T) - F\varphi(\beta, T) - \mu_2(\alpha, T) + F\varphi(\alpha, T)] \tag{12.4.2}$$

which results from multiplying and dividing the left hand side of Eq. (12.4.2) by Faraday's constant F and adding the term $-\frac{1}{F}[\mu_2(\beta, T) - \mu_2(\alpha, T)]$, which is equal to zero. Since the electrochemical potential of the electrons is given by $\eta_2 = \mu_2 - F\varphi$, we can replace the term in square brackets by the difference of the electrochemical potentials η_2 of the electrons in both end phases:

$$\varphi(\beta, T) - \varphi(\alpha, T) = -\frac{1}{F} [\eta_2(\beta, T) - \eta_2(\alpha, T)]. \tag{12.4.3}$$

Under isothermal conditions and for the equilibrium at a phase boundary the electrochemical potential of the electrons is the same on both sides of the phase boundary (see also Sect. 4.10). Therefore, the following holds for the phase boundaries $MeX_\nu/Pt(\alpha)$ at temperature T and $MeX_\nu/Pt(\beta)$ at temperature $T + \Delta T$:

$$\eta_2(MeX_\nu, T) = \eta_2(\alpha, T) \quad \text{and} \quad \eta_2(MeX_\nu, T + \Delta T) = \eta_2(\beta, T + \Delta T). \tag{12.4.4}$$

Hence we can replace $\eta_2(\alpha, T)$ in Eq. (12.4.3) by $\eta_2(MeX_\nu, T)$. By further adding the term $F^{-1}[\eta_2(MeX_\nu, T + \Delta T) - \eta_2(\beta, T + \Delta T)]$ which is equal to zero, we can rewrite Eq. (12.4.3) and obtain

$$\varphi(\beta, T) - \varphi(\alpha, T) = -F^{-1}[\eta_2(\beta, T) - \eta_2(\beta, T + \Delta T)$$
$$+ \eta_2(MeX_\nu, T + \Delta T) - \eta_2(MeX_\nu, T)]. \tag{12.4.5}$$

Thus, it is evident that the difference between the electrochemical potentials of the electrons in Pt(β, T) and Pt(α, T), i.e. the term occurring in square brackets on the right-hand side of Eq. (12.4.3), which is identical with the corresponding term of Eq. (12.4.5), contains variations of the electrochemical potential of the electrons within the platinum, i.e. [$\eta_2(\beta, T) - \eta_2(\beta, T + \Delta T)$] and within the compound, i.e. [$\eta_2(MeX_\nu, T + \Delta T) - \eta_2(MeX_\nu, T)$]. At the phase boundaries no differences of the electrochemical potential of the electrons occur. According to these considerations the following holds for the thermopower if we apply the definition given in Eq. (12.4.1):

$$\varepsilon = \frac{d[\varphi(\beta, T) - \varphi(\alpha, T)]}{dT} = -\frac{1}{F}\left[\frac{d\eta_2(MeX_\nu)}{dT} - \frac{d\eta_2(\beta)}{dT}\right]. \qquad (12.4.6)$$

The right-hand side of this equation contains two terms

$$-\frac{1}{F}\frac{d\eta_2(MeX_\nu)}{dT} = \varepsilon(MeX_\nu) \qquad (12.4.7)$$

and

$$-\frac{1}{F}\frac{d\eta_2(\beta = Pt)}{dT} = \varepsilon(Pt) \qquad (12.4.8)$$

which may be defined as the absolute thermoelectric powers of compound MeX_ν and platinum respectively. Incorporating definitions (12.4.7) and (12.4.8) into Eq. (12.4.6), we obtain:

$$\varepsilon = \varepsilon(MeX_\nu) - \varepsilon(Pt). \qquad (12.4.9)$$

Since in an open circuit no transport processes occur within the platinum, in particular no flux of electrons, it follows from setting j_2 in relation (12.1.21) equal to zero and using Eq. (12.4.8)

$$\varepsilon(Pt) = \frac{1}{F}\left[\bar{S}_2(Pt) + \frac{Q_2^*(Pt)}{T}\right]. \qquad (12.4.10)$$

This equation connects the absolute thermoelectric power of platinum with caloric quantities. The variation of expression (12.4.10) with temperature is proportional to the Thomson coefficient which can experimentally be obtained from the Thomson heat. The Thomson coefficient $\tau(Pt)$ of platinum is related to the absolute thermoelectric power of platinum $\varepsilon(Pt)$ according to the following equation:

$$\tau(Pt) = T\left(\frac{\partial \varepsilon(Pt)}{\partial T}\right). \qquad (12.4.11a)$$

For a derivation of this relation the reader is referred to the literature [12.8].

Insertion of the right-hand side of Eq. (12.4.10) into Eq. (12.4.11a) leads to

$$\tau(Pt) = \frac{T}{F}\frac{\partial\left[\bar{S}_2(Pt) + \dfrac{Q_2^*(Pt)}{T}\right]}{\partial T}. \qquad (12.4.11b)$$

For a determination of the absolute thermoelectric power the Thomson coefficient should be measured as a function of temperature down to 0 K. In practice, however, this has been done for lead in order to get the absolute thermoelectric power of this metal. When using lead it is only necessary to carry out these measurements down to the transition temperature corresponding to the transition to the super-conductive state. The absolute thermoelectric power of platinum may then be obtained from measurements of the thermoelectric power of a lead/platinum thermocouple.

In order to calculate the absolute thermoelectric power of compound MeX_ν, we first consider case b): i.e. a stationary gradient of metal excess or deficiency corresponding to the Soret effect prevails. This, however, means that no transport process takes place under these conditions, in particular no flow of electrons within the compound. Hence, it follows that the flux of electrons j_2 given by Eq. (12.1.21) and thus the term in square brackets of this equation is equal to zero. According to this the change of the electrochemical potential of the electrons in the compound with temperature and thus according to Eq. (12.4.7), the absolute thermoelectric power ε_b of compound MeX_ν is then given by

$$\varepsilon_b(MeX_\nu) = \frac{1}{F}\left[\bar{S}_2(MeX_\nu) + \frac{Q_2^*(MeX_\nu)}{T}\right] \qquad (12.4.12)$$

The index b used emphasizes the fact that this thermoelectric power corresponds to case b).

In case c) the activity a_{Me} of metal Me in the compound, which is related to the chemical potential μ_{Me} of the metal in this phase and that in the standard state μ_{Me}^0 according to the equation

$$RT \ln a_{Me} = \mu_{Me} - \mu_{Me}^0 \qquad (12.4.13)$$

is set constant. Hence, a stationary gradient of the activity, which is required for the Soret effect to occur, cannot be generated, i.e. fluxes of ions and electrons, j_1 and j_2 respectively, take place in the stationary state and have to be accounted for. On the grounds of electroneutrality, these fluxes are connected with one another as follows:

$$z_1 j_1 = j_2. \qquad (12.4.14)$$

Insertion of Eqs. (12.1.20) and (12.1.21) into relation (12.4.14) yields

$$\frac{\sigma_1}{z_1}\left[\frac{d\eta_1}{dx} + \left(\bar{S}_1 + \frac{Q_1^*}{T}\right)\frac{dT}{dx}\right] = \sigma_2\left[\frac{d\eta_2}{dx} + \left(\bar{S}_2 + \frac{Q_2^*}{T}\right)\frac{dT}{dx}\right]. \qquad (12.4.15)$$

Since the chemical potential μ_{Me} of the metal in the compound can be split into the electrochemical potentials η_1 of the ions and η_2 of the electrons in this phase, the gradient of the electrochemical potential of the ions can be replaced by the gradient of the chemical potential of the metal and the that of the electro-chemical potential of the electrons.

$$\frac{d\eta_1}{dx} = \frac{d\eta_1}{dT}\frac{dT}{dx} = \left(\frac{d\mu_{Me}}{dT} - z_1\frac{d\eta_2}{dT}\right)\frac{dT}{dx}. \qquad (12.4.16)$$

Next we solve Eq. (12.4.15) for $\dfrac{d\eta_2}{dT}$, thereby noting the validity of Eq. (12.4.16).
Applying the general definition of $\varepsilon(MeX_\nu)$ given by Eq. (12.4.7), we obtain:

$$\varepsilon_c(MeX_\nu) = \frac{1}{F}\left[\frac{\sigma_1}{(\sigma_1 + \sigma_2)\,z_1}\left(-\frac{d\mu_{Me}}{dT} - \bar{S}_1 - \frac{Q_1^*}{T}\right)\right.$$
$$\left. + \frac{\sigma_2}{\sigma_1 + \sigma_2}\left(\bar{S}_2 + \frac{Q_2^*}{T}\right)\right]. \tag{12.4.17}$$

According to this equation the absolute thermoelectric power of compound MeX_ν for case c) is expressed in terms of caloric quantities and partial conductivities of electrons and ions. Equation (12.4.17) must be further rewritten to obtain a relationship involving measurable quantities. For the quantity $d\mu_{Me}/dT$ we can write

$$\frac{d\mu_{Me}}{dT} = \frac{d\mu_{Me}^0}{dT} + R\ln a_{Me} = -S_{Me}^0 + R\ln a_{Me} \quad \text{for} \quad a_{Me} = \text{const.} \tag{12.4.18}$$

where S_{Me}^0 is the molar entropy of the pure metal, which is appropriately chosen as standard state. Incorporation of Eq. (12.4.18) into Eq. (12.4.17) yields

$$\varepsilon_c(MeX_\nu)_{a_{Me}=\text{const.}} = \frac{1}{F}\left[\frac{1}{z_1}\frac{\sigma_1}{\sigma_1 + \sigma_2}\left(S_{Me}^0 - \bar{S}_1 - R\ln a_{Me} - \frac{Q_1^*}{T}\right)\right.$$
$$\left. + \frac{\sigma_2}{\sigma_1 + \sigma_2}\left(\bar{S}_2 + \frac{Q_2^*}{T}\right)\right]. \tag{12.4.19}$$

This equation is the general expression for the absolute thermoelectric power of compound MeX_ν in case c), i.e. for constant activity of the metal. Equation (12.4.19) becomes less complicated if we refer our considerations to compounds exhibiting predominantly ionic conductivity. In this case the second term in square brackets is vanishingly small and, in addition, the factor containing the partial conductivities before the first term will approximately be equal to unity. For the special case considered we thus obtain the following equation which has often been discussed in the literature [12.9]:

$$\varepsilon_c(MeX_\nu)_{a_{Me}=1,\sigma_1 \gg \sigma_2} = \frac{1}{z_1 F}\left(S_{Me}^0 - \bar{S}_1 - \frac{Q_1^*}{T}\right). \tag{12.4.20}$$

If the compound MeX_ν contained within cell (12.4.II) predominantly exhibits ionic conductivity, i.e. $\sigma_1 \gg \sigma_2$, and if the activity of the metal a_{Me} in the compound equals unity, the following holds for the total thermoelectric power of this cell whereby Eqs. (12.4.9), (12.4.10) and (12.4.20) have been used:

$$\varepsilon_c(\text{cell } 12.4.II)_{a_{Me}=1,\sigma_1 \gg \sigma_2}$$
$$= \frac{1}{z_1 F}\left[S_{Me}^0 - \bar{S}_1(MeX_\nu) - \frac{Q_1^*(MeX_\nu)}{T} - z_1\bar{S}_2(Pt) - \frac{z_1 Q_2^*(Pt)}{T}\right]. \tag{12.4.21}$$

The quantities on the right-hand side of Eq. (12.4.21) refer to various phases. For clarity, this has been taken into account by the notation used, i.e. by writing the phase in parantheses behind the corresponding quantity.

If instead of cell (12.4.II) the following thermogalvanic cell (12.4.III) is used, then, as can be seen, the right-hand side leg consists of metal Me instead of platinum, where the former is attached to both sides of sample MeX_v at T and T $+ \Delta T$ respectively.

$$Me(\alpha) \mid MeX_v, MeX_v \mid Me, Me(\beta) \qquad (12.4.III)$$

$$\overset{\longleftarrow \quad \longrightarrow}{T} \quad \overset{\longleftarrow \quad \longrightarrow}{T + \Delta T} \quad \overset{\longleftarrow \quad \longrightarrow}{T}$$

Thus, we only have to replace on the right-hand side of Eq. (12.4.21) the partial molar entropy of the electrons in platinum $\bar{S}_2(Pt)$ and their heat of transport in this phase $Q_2^*(Pt)$ by the corresponding quantities of the metal, i.e. $\bar{S}_2(Me)$ and $Q_2^*(Me)$, thus arriving at the following expression for the thermopower ε_c of cell (12.4.III):

$$\varepsilon_c(\text{cell } 12.4.III)_{a_{Me}=1, \sigma_1 \gg \sigma_2}$$
$$= \frac{1}{z_1 F} \left[S_{Me}^0 - \bar{S}_1(MeX_v) - \frac{Q_1^*(MeX_v)}{T} - z_1 \bar{S}_2(Me) - \frac{z_1 Q_2^*(Me)}{T} \right].$$
$$(12.4.22)$$

We will now give another derivation of this equation by considering the various electrical potential differences, which either occur at the different phase boundaries or within the phases. We assume that the compound MeX_v in cell (12.4.III) is a pure ionic conductor with cationic conductivity. This special example, which is also relatively simple, is chosen to demonstrate the other way of treating thermogalvanic cells. This approach is usually described in the literature. Returning to the definition of thermopower, defined by Eq. (12.4.1), we can write

$$\varepsilon = \lim_{\Delta T \to 0} \frac{E(T, \Delta T)}{\Delta T} = \lim_{\Delta T \to 0} \frac{\varphi(\beta, T) - \varphi(\alpha, T)}{\Delta T}. \qquad (12.4.23\,a)$$

The electrical potential difference $\varphi(\beta, T) - \varphi(\alpha, T)$ in Eq. (12.4.1) can be built up by several contributions, some of them occurring within homogeneous phases, others at phase boundaries. Therefore, the following is valid

$$\varphi(\beta, T) - \varphi(\alpha, T) = [\varphi(\beta, T) - \varphi(\beta, T + \Delta T)]$$
$$+ [\varphi(\beta, T + \Delta T) - \varphi(MeX_v, T + \Delta T)]$$
$$+ [\varphi(MeX_v, T + \Delta T) - \varphi(MeX_v, T)]$$
$$+ [\varphi(MeX_v, T) - \varphi(\alpha, T)]. \qquad (12.4.23\,b)$$

The potential differences within the square brackets on the right-hand side of Eq. (12.4.23b) may be grouped into two contributions:

a) $\qquad [\varphi(\beta, T + \Delta T) - \varphi(MeX_v, T + \Delta T)] - [\varphi(\alpha, T) - \varphi(MeX_v, T)]$

b) $\qquad [\varphi(MeX_v, T + \Delta T) - \varphi(MeX_v, T)] - [\varphi(\beta, T + \Delta T) - \varphi(\beta, T)]$

The terms within the square brackets of expression a) represent potential differences at the phase boundary Me/MeX_v at higher and lower temperature respectively. The difference between these terms gives the temperature dependence of the contact voltage (Galvani potential difference between two phases) of the heterojunction Me/MeX_v. The contribution to the total thermopower of the cell resulting from term a) is thus called the heterogeneous thermopower denoted as ε_{het}. The terms within the square brackets of expression b) represent potential differences in the homogeneous phases MeX_ν and Me respectively. The contribution to the total thermopower of the cell resulting from expression b) is called the homogeneous thermopower denoted as ε_{hom}. We will now derive an expression for the heterogeneous thermopower ε_{het}. We therefore regard the ionic equilibrium for the metal ions at the phase boundary Me/MeX_ν. According to the equilibrium condition, the electrochemical potential η_1 of the metal ions is equal in both phases:

$$\eta_1(MeX_\nu) = \eta_1(Me). \tag{12.4.24}$$

Using the definition of the electrochemical potential of the metal ions, $\eta_1 = \mu_1 + z_1 F \varphi$, Eq. (12.4.24) can be rewritten as follows:

$$\mu_1(MeX_\nu) + z_1 F \varphi(MeX_\nu) = \mu_1(Me) + z_1 F \varphi(Me). \tag{12.4.25}$$

Rearranging Eq. (12.4.25) and solving for the term $\varphi(Me) - \varphi(MeX_\nu)$, we obtain:

$$\varphi(Me) - \varphi(MeX_\nu) = \frac{1}{z_1 F} [\mu_1(MeX_\nu) - \mu_1(Me)], \tag{12.4.26}$$

i.e. an expression for the electrical potential difference at the phase boundary Me/MeX_ν, which is also called Galvani potential difference. The chemical potential $\mu_1(Me)$ of the metal ions within the metal can be expressed by the chemical potential μ_{Me}^0 of the metal and that of the electrons in the metal $\mu_2(Me)$. This can be done by splitting the chemical potential μ_{Me}^0 of the metal into the contribution of the metal ions and that of the electrons according to

$$\mu_{Me}^0(Me) = \mu_1(Me) + z_1 \mu_2(Me) \tag{12.4.27}$$

and solving for $\mu_1(Me)$. With this we can write instead of Eq. (12.4.26)

$$\varphi(Me) - \varphi(MeX_\nu) = \frac{1}{z_1 F} [\mu_1(MeX_\nu) - \mu_{Me}^0(Me) + z_1 \mu_2(Me)]. \tag{12.4.28}$$

This equation permits to obtain an expression for the heterogeneous thermopower ε_{het}

$$\varepsilon_{\text{het}} = \lim_{\Delta T \to 0} \frac{\varphi(Me) - \varphi(MeX_\nu)}{\Delta T} = \frac{1}{z_1 F} \frac{\partial}{\partial T} [-\mu_e^0(Me) + \mu_1(MeX_\nu) + z_1 \mu_2(Me)]. \tag{12.4.29}$$

The temperature dependences of the various chemical potentials are the molar entropy and the corresponding partial molar entropies respectively. Therefore, we can write instead of Eq. (12.4.29)

$$\varepsilon_{\text{het}} = \frac{1}{z_1 F} [S_{Me}^0 - \bar{S}_1(MeX_\nu) - z_1 \bar{S}_2(Me)]. \tag{12.4.30}$$

According to the conditions prevailing in cell (12.4.III) the partial molar entropy $\bar{S}_1(MeX_\nu)$ of the metal ions in the compound MeX_ν is that one corresponding to equilibrium between the compound and the metal. But it is reasonable to assume that in ionic conductors the quantity $\bar{S}_1(MeX_\nu)$ is independent of the chemical potential of the metal or nonmetal or of deviations from stoichiometry. The total thermopower is the sum of the heterogeneous thermopower ε_{het} and the homogeneous thermopower ε_{hom}. We will not give an independent derivation of the homogeneous thermopower ε_{hom} but we obtain an expression for ε_{hom} by comparing Eq. (12.4.22) for the total thermopower ε_c with Eq. (12.4.30) for the heterogeneous thermopower ε_{het}:

$$\varepsilon_{\text{hom}} = \varepsilon_c - \varepsilon_{\text{het}} = -\frac{Q_1^*(MeX_\nu)}{z_1 FT} - \frac{Q_2^*(Me)}{FT}, \tag{12.4.31}$$

The sum of ε_{het} and ε_{hom} gives ε_c according to Eq. (12.4.22). The term $\dfrac{Q_2^*(Me)}{FT}$ in this equation is often omitted because it is very small compared to the other ones. Therefore, one often finds the equation

$$\varepsilon_c \ (\text{cell } 12.4.\text{III})_{a_{Me}=1,\sigma_1 \gg \sigma_2}$$
$$= \frac{1}{z_1 F} \left[S_{Me}^0 - \bar{S}_1(MeX_\nu) - \frac{Q_1^*(MeX_\nu)}{T} - z_1 \bar{S}_2(Me) \right]. \tag{12.4.32}$$

In the following we will discuss the thermoelectric power of cells of type (12.4.III) containing silver ion conductors and silver metal. For such cells we have $z_1 = 1$ and $Me = Ag$ and Eq. (12.4.32) has to be written in the form

$$\varepsilon_c \ (\text{cell } 12.4.\text{III})_{a_{Ag}=1,\sigma_{Ag^+} \gg \sigma_e}$$
$$= \frac{1}{F} \left[S_{Ag}^0 - \bar{S}_{Ag^+}(AgX) - \frac{Q_{Ag^+}^*(AgX)}{T} - \bar{S}_e(Ag) \right]. \tag{12.4.33}$$

Investigations of thermopowers of cells, which contain pure silver ion conductors and silver metal, have mainly been performed by Shahi et al. [12.10], Takahashi et al. [12.11] and Magistris et al. [12.12].

In Fig. 12.4.2 the absolute thermopowers of several silver ion conductors are plotted versus the reciprocal of absolute temperature.

As can be seen the temperature dependence of the thermopower may conveniently be described by a straight line according to

$$-\varepsilon = \frac{A}{T} + B \tag{12.4.34}$$

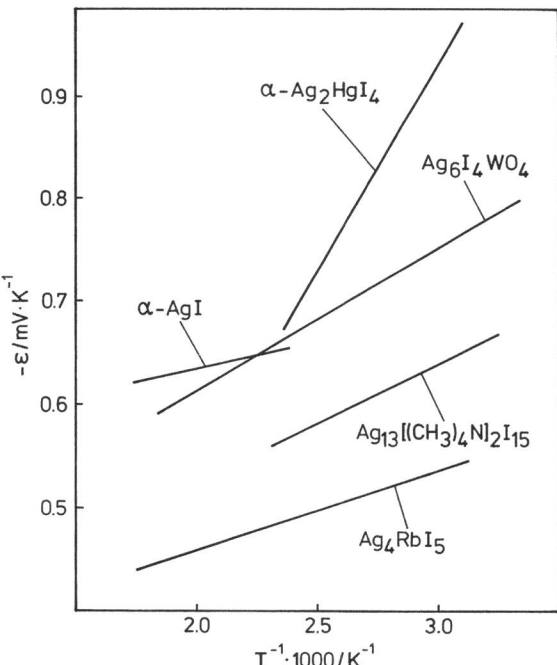

Fig. 12.4.2. Absolute thermoelectric power of several silver ion conductors as a function of the reciprocal of absolute temperature T

Table 12.4.1. Values of A and B (see Eq. (12.4.34)) and of the heat of transport of the Ag^+-ions for several silver-ion conductors obtained by measuring the thermopowers of cells of type (12.4.III) over the temperature range cited

Silver-ion conductor	$A(= Q^*_{Ag^+}/F)/$ mV	$B/mV\ K^{-1}$	$Q^*_{Ag^+}/$ kJ mol^{-1}	Temperature range/°C	Ref.
α-AgI	52	0.531	5.02	146−300	[12.10]
Ag_4RbI_5	78	0.303	7.53	—	[12.11]
α-Ag_2HgI_4	402	−0.278	38.79	50−150	[12.12]
$Ag_6I_4WO_4$	138	0.337	13.32	27−270	[12.10]
$Ag_{13}[(CH_3)_4N]_2I_{15}$	115	0.295	11.10	35−160	[12.10]

where A and B are virtually constant. The reason for this quite simple functional relationship lies in the fact that starting from Eq. (12.4.20) both $Q^*_{Ag^+}$ and the difference $S^0_{Ag} - \bar{S}_{Ag^+}$ are nearly constant since S^0_{Ag} and $\bar{S}_{Ag^+}$ exhibit similar temperature dependences. Both constants A and B may be determined from such plots, the former being obtained as the slope and the latter as the intercept of the line with the ε-axis. Table 12.4.1 contains the results for several solid ionic conductors with structural disorder in the silver-ion partial lattice. The samples used for such measurements were polycrystalline.

The heats of transport of the silver ions listed in Table 12.4.1 are nearly equal to the heats of activation for the transport of silver ions where the latter quantities may be obtained from measurements of the temperature dependence of the conductivity. Values of the heats of activation are not compiled in Table 12.4.1 but can be found in a more extensive paper by Shahi [12.13]. It is very interesting that the heats of transport in the fast ion conductors α-AgI and Ag_4RbI_5 are very low — of the order of magnitude of RT — that is 3.3 kJ mol^{-1} at T = 400 K.

12.5 References

[12.1] Rickert, H., Wagner, C.: Ber. Bunsenges. phys. Chem. *67*, 621 (1963)
[12.2] DeGroot, S. R.: Thermodynamics of Irreversible Processes. Amsterdam: North-Holland Publ. Comp. 1951
DeGroot, S. R., Mazur, P.: Non-equilibrium Thermodynamics. Amsterdam: North-Holland Publ. Comp. 1962
Denbigh, K. G.: The Thermodynamics of the Steady State. London: Methuen 1951
Prigogine, I.: Étude thermodynamique des phenomènes irreversibles. Paris: Dunod 1947
[12.3] Wagner, C.: Z. phys. Chem. *B21*, 25 (1933)
[12.4] Wagner, C.: J. Chem. Phys. *21*, 1819 (1953)
[12.5] Rickert, H., Sattler, V., Wedde, Ch.: Z. phys. Chem. N.F. *98*, 339 (1975)
[12.6] Rickert, H.: Z. phys. Chem. N.F. *23*, 355 (1960)
[12.7] Wagner, C.: Progress in Solid-State Chemistry. Reiss, H., McCaldin, J., (eds.), Vol. 7. Oxford: Pergamon Press 1972
[12.8] Delahay, P., Tobias, C. W.: Advances in Electrochemistry and Electrochemical Engineering, Vol. 3, p. 69. New York: Interscience Publ. 1963
Haase, R.: Thermodynamik der Mischphasen. Berlin−Göttingen−Heidelberg: Springer 1956

[12.9] Agar, J. N.: Advances in Electrochemistry and Electrochemical Engineering,
        Vol. 3, p. 31. New York: Interscience Publ. 1963
        Haase, R.: Thermodynamik der irreversiblen Prozesse. Darmstadt: DR. Dietrich
        Steinkopff-Verlag 1963
        Lidiard, A. B.: Handbuch der Physik, Encyclopedia of Physics, Vol. 20. Berlin—
        Göttingen—Heidelberg: Springer 1957
        Pitzer, K. S.: J. Chem. Phys. *65*, 147 (1961)
[12.10] Shahi, K., Chandra, S.: Z. Naturforsch. *30a*, 1055 (1975); phys. stat. sol. (a) *28*,
        653 (1975); Shahi, K., Sanyal, Nitish, K., Chandra, S.: J. Phys. Chem. Solids *36*,
        1349 (1975)
[12.11] Takahashi, T., Yamamoto, O., Nomura, E.: Denki Kagaku *38*, 360 (1970)
[12.12] Magistris, A., Pezzati, E., Sinistri, C.: Z. Naturforsch. *27a*, 1379 (1972)
[12.13] Shahi, K.: phys. stat. sol. (a) *41*, 11 (1977)

Author Index

Subject Index